Ecological data possess several special properties; the presence or absence of species or semi-quantitative abundance values; non-linear relationships between species and environmental factors; and high inter-correlations among species and among environmental variables. The analysis of such data is important in the interpretation of relationships within plant and animal communities and with their environments.

In this corrected version of *Data analysis in community and landscape ecology*, the contributors demonstrate the methods that have proven most useful, with examples, exercises and case-studies, without using complex mathematics. Chapters explain in an elementary way powerful data analysis techniques such as logit regression, canonical correspondence analysis, and kriging.

DATA ANALYSIS IN COMMUNITY AND LANDSCAPE ECOLOGY

DATA ANALYSIS IN COMMUNITY AND LANDSCAPE ECOLOGY

Edited by

R. H. G. JONGMAN, C. J. F. TER BRAAK &
O. F. R. VAN TONGEREN

CAMBRIDGE
UNIVERSITY PRESS

Published by the Press Syndicate of the University of Cambridge
The Pitt Building, Trumpington Street, Cambridge CB2 1RP
40 West 20th Street, New York, NY 10011-4211, USA
10 Stamford Road, Oakleigh, Melbourne 3166, Australia

First published by Pudoc (Wageningen) 1987
New edition with corrections published by Cambridge University Press 1995

Printed in Great Britain at Biddles

A catalogue record for this book is available from the British Library

Library of Congress cataloguing in publication data

Data analysis in community and landscape ecology / R.H.G. Jongman,
    C.J.F. ter Braak, and O.F.R. van Tongeren, editors. – New ed.
        p.   cm.
    Includes bibliographical references (p.   ) and index.
    ISBN 0 521 47574 0 (pbk.)
        1. Biotic communities–Research–Methodology.   2. Landscape
    ecology–Research–Methodology.   3. Biotic communities–Research–
    Statistical methods.   4. Landscape ecology–Research–Statistical
    methods.   5. Biotic communities–Research–Data processing.
    6. Landscape ecology–Research–Data processing.   I. Jongman, R. H. G.
    II. Braak, C. J. F. ter.   III. Van Tongeren, O. F. R.
    QH541.2.D365   1995
    574.5′0285–dc20   94-31639   CIP

This reprint is authorized by the original publisher and
copyright holder, Centre for Agricultural Publishing and
Documentation (Pudoc), Wageningen, 1987.

ISBN 0 521 47574 0   paperback

Contents

Contributors

Dr A. Barendregt
Department of Environmental Studies
University of Utrecht
Heidelberglaan 2
P.O. Box 80.115
3508 TC Utrecht
the Netherlands

Professor Dr P. Burrough
Geographical Institute
University of Utrecht
Heidelberglaan 2
P.O. Box 80.115
3508 TC Utrecht
the Netherlands

Dr J. C. Jager
Centre for Mathematical Methods
National Institute of Public Health
and Environmental Hygiene
Anthonie van Leeuwenhoeklaan 9
P.O. Box 1
3720 BA Bilthoven
the Netherlands

Dr R. H. G. Jongman
Department of Physical Planning
Wageningen Agricultural University
Generaal Foulkesweg 13
6703 BJ Wageningen
the Netherlands

Ir C. Looman
Department of Public Health
Erasmus University
P.O. Box 1738
3000 DR Rotterdam
the Netherlands

Dr Ir Th. J. van de Nes
Department of Water Management
Province of Gelderland
P.O. Box 9090
6800 GX Arnhem
the Netherlands

Dr J. A. F. Oudhof
Department of Plant Ecology
University of Utrecht
RAET
Software & Computer Services
Eendrachtlaan 10
Utrecht
the Netherlands

Dr C. J. F. ter Braak
Agricultural Mathematics Group
P.O. Box 100
6700 AC Wageningen
the Netherlands

Dr O. F. R. van Tongeren
Data-analyse Ecologie
Wamelaut 27
6931 HS Westervoort
the Netherlands

Preface to first edition

This book has been written for the researcher who uses computers to analyse field data on plant and animal communities and their environment. The book originates from a post-graduate course, held at the Wageningen Agricultural University in 1983, 1984 and 1985, for biologists, geographers, agronomists and landscape architects active in nature management, water management, environmental impact assessment and landscape planning. We have included topics that are useful in a wide range of ecological field studies: regression, to model the relation between a particular species and environmental variables, and to detect whether a species can be an indicator for a particular environmental variable; calibration, to infer about the environment from (indicator) species; ordination and cluster analysis, to summarize data on communities of species; and spatial analysis, to model and display the spatial variation in the environment and the communities studied.

A major aim of the book is to bridge the gap between exploratory methods, e.g. ordination and cluster analysis, and methods that allow statistical hypotheses to be tested, such as regression analysis and analysis of variance. This is important because environmental impact studies, for example, require careful design and analysis that is directed to detecting the effect of the impact variable. The by now traditional methods of ordination and cluster analysis are simply not sufficient; they may fail to detect the effect of the impact variable because of natural sources of variation. Other methods are appropriate here: regression by generalized linear models to detect effects at the species level and canonical ordination to detect effects at the community level. We therefore give an elementary, yet full and up-to-date account of regression, starting from the classical methods of analysis of variance, multiple regression, chi-square tests to logit regression and log-linear regression – all of them regression methods that are appropriate for ecological data. Canonical ordination is introduced in the ordination chapter as a natural combination of multiple regression and regular ordination. One of these methods, canonical correspondence analysis, is shown to avoid most of the problems that have hampered previous applications of canonical correlation analysis to ecological data.

The methods are explained without the use of matrix algebra, but elementary knowledge of algebra, geometry and statistics at a first-year undergraduate level is assumed. To make the book suitable for post-graduate courses and for self-instruction, exercises with solutions have been added to Chapters 2-7. The exercises do not require use of a computer. Some exercises start with computer output and the reader is asked to verify the results by computer, if one is available, but the essence of those exercises is to draw conclusions from the output. It is an advantage if the reader has access to one or more of the computer programs that are mentioned in the text, but certainly not essential. Bibliographic notes are included at the end of Chapter 1-7 to allow readers to put the material into its historical perspective and to introduce them to the literature.

Most chapters are largely self-contained. The order in which they are presented is not necessarily the order in which they should be read. We chose the order of presentation with the following in mind:

- the design of a study and data collection are the first and most important steps in research
- understanding regression is a help to understanding calibration
- understanding regression and calibration is a help to understanding ordination and canonical ordination
- some of the cluster analysis techniques are based on ordination, so a basic understanding of ordination is necessary to follow their explanation
- spatial analysis requires some knowledge of regression; the domain of application is on the landscape level rather than the community and species level.

R. Jongman
C. ter Braak
O. van Tongeren

Wageningen
September 1987

Acknowledgement

We are greatly indebted to all those colleagues who commented on earlier versions of the various chapters, especially I.C. Prentice, R. Hengeveld, N. Gremmen, R. Barel, J. Oude Voshaar, P. Vereijken, L. Barendregt, J. Jansen, M. Jansen, P. Verdonschot, H. van Dobben, H. Siepel, P. Opdam, J. Berdowski, F. Metselaar, G. Hoek, W. Cramer, E. de Hullu, K. van de Kerkhof, P. Laan, A. van de Berg, E. Lieftink-Koeijers, R. Webster, G. Oerlemans, W. Hockey-Wielinga, W. Slob and W. J. van der Steen. Data were provided by M. Batterink, G. Wijffels and P. Ketner of the Department of Vegetation Science, Plant Ecology and Weed Science of the Wageningen Agricultural University and M. Kilic. Technical assistance was given by J. van de Peppel, M. de Vries and J. van Keulen. M. Mijling and W. Laoh-Gieskes did excellent work in typing and correcting several drafts. G. Kleinrensink did the drawings. Last, but not least, we wish to thank Dr J. H. de Ru and Professor Dr Ir R. Rabbinge for stimulating us to write this book and our institutes for giving us the opportunity to do so.

The Editors

List of symbols

ANOVA   analysis of variance.

b_k value of the regression coefficient for the k-th species; used when only one regression coefficient is involved.

$b_0, b_1, b_2, \ldots$ coefficients (parameters) in a regression equation; b_0 is usually the intercept.

c, c_k maximum of a response curve. Can be indexed by a species number (k).

$c_0, c_1, c_2, \ldots$ coefficients (parameters) in a regression equation.

$C(h)$ covariance function, i.e. function giving the covariance between $Z(x_1)$ and $Z(x_2)$, where x_1 and x_2 are two points.

d.f.   degrees of freedom.

$\exp(x)$, e^x the exponential function or antilog of x ($\exp(\log(x))=x$).

Ey or $E(y)$ expected value of a random variable y.

F variance ratio in an ANOVA table.

h lag, distance.

i, j indices numbering sites in the data ($i = 1, 2, \ldots, n; j = 1, 2, \ldots, n$), often used as subscript, e.g. x_i, the value of variable x in the i-th site.

k, l indices numbering the species in the data ($k = 1, 2, \ldots, m; l = 1, 2, \ldots, m$), often used as subscript, e.g. y_k, the value of variable y of the k-th species.

$\log_e(x)$ Naperian or natural logarithm of x (for $x > 0$).

m number of response variables (often equal to number of species).

m.s.   mean square in an ANOVA table.

n number of sites (statistical sampling units, objects, etc.).

p probability of occurrence of a species.

$p(x)$ probability of occurrence of a species as a function of the variable x.

p_{max}	maximum probability of occurrence.
P	P value of a statistical test, e.g. $P < 0.05$.
q	number of explanatory variables (often equal to number of environmental variables).
r	coefficient of correlation (in a sample).
R	multiple correlation coefficient in regression; R^2 is termed the coefficient of determination.
R^2_{adj}	adjusted R^2, also termed percentage variance accounted for. A recommended modification of R^2 to adjust for the number of parameters fitted by regression. For large sample sizes R^2_{adj} is approximately equal to R^2.
s	standard deviation of a sample, or residual standard deviation in regression.
s^2	variance of a sample, or residual variance in regression.
s.e.	standard error.
s.d.	standard deviation of a sample. In Chapter 5, standard deviation of a unimodal response curve.
s.s.	sum of squares in an ANOVA table.
t, t_k	tolerance, a measure of ecological amplitude, is the parameter for curve width in the Gaussian logit response model. It can be indexed by a species number (k).
$t, t (v)$	t or Student's distribution with v degrees of freedom: shows both the random variable and a particular or observed value. $t_\alpha (v)$, or t_α, is the critical value of a t distribution in a two-sided (two-tailed) statistical test with significance level α.
u, u_k	optimum of a response curve, i.e. the value for which the response curve under consideration attains its maximum (when uniquely defined). Can be indexed by a species number (k).
var(y)	variance of a random variable y, also denoted by σ^2 or V.
vr	variance ratio in an ANOVA table.
$\bar{x}$	arithmetic mean of the variable x in a sample.
x_0	value of the variable x at a particular site.
x_i	value of the variable x at the i-th site.
$x_1, x_2, \ldots$	explanatory variables in a regression equation, often observed environmental variables. Also used for latent variables in ordination (i.e. theoretical, environmental variables) or variables indicating spatial position.

y_i	value of a particular response variable y in the i-th site, used when it is clear from the context which variable is being considered.
y_k	value from the k-th response variable (species) in a particular site, used when it is clear from the context which site is considered.
y_{ki}	value of the k-th response variable (species) in the i-th site.
y_{+i}, y_{k+}	sum of y_{ki} over the index $k = 1, \ldots, m$; sum of y_{ki} over the index $i = 1, \ldots, n$.
z_j	j-th environmental variable ($j = 1, 2, \ldots, q$).
z_{ji}	value of j-th environmental variable in the i-th site.
$Z(x)$	a spatial random variable, with x denoting the spatial position. $z(x)$ is the value of $Z(x)$ observed in x
$\gamma(h)$	semivariance. $2\gamma(h) = \mathrm{var}[Z(x_1) - Z(x_2)]$, where x_1 and x_2 are points at distance h apart.
δ_{ij}	dissimilarity between sites i and j.
ε	error term in a regression equation. ε is the random variable $y - E(y)$.
$\hat{\theta}$	estimator of the parameter θ.
λ, λ_s	eigenvalue of a cross-product matrix; λ_s is the eigenvalue of the s-th axis or eigenvector.
$\mu(x)$	expected value of a random variable.
ν	degrees of freedom.
$\rho(h)$	correlation function $\rho(h) = C(h)/C(0)$.
σ	standard deviation of a random variable.
σ^2	variance of a random variable; often used for the error variance in regression.
$\sum_{i=1}^{n} y_i$	$y_1 + y_2 + \ldots + y_n$
$\chi^2(\nu)$	Chi-square distribution with ν degrees of freedom: shows both the random variable and a particular or observed value. $\chi^2_\alpha(\nu)$ is the critical value of a chi-square distribution in a statistical test with significance level α.

Dune Meadow Data

In this book the same set of vegetation data will be used in the chapters on ordination and cluster analysis. This set of data stems from a research project on the Dutch island of Terschelling (Batterink & Wijffels 1983). The objective of this project was to detect a possible relation between vegetation and management in dune meadows. Sampling was done in 1982. Data collection was done by the Braun–Blanquet method; the data are recorded according to the ordinal scale of van der Maarel (1979b). In each parcel usually one site was selected; only in cases of great variability within the parcel were more sites used to describe the parcel. The sites were selected by throwing an object into a parcel. The point where the object landed was fixed as one corner of the site. The sites measure 2x2 m². The sites were considered to be representative of the whole parcel. From the total of 80 sites, 20 have been selected to be used in this book (Table 0.1). This selection expresses the variation in the complete set of data. The names of the species conform with the nomenclature in van der Meijden et al. (1983) and Tutin et al. (1964-1980).

Data on the environment and land-use that were sampled in this project are (Table 0.2):
- thickness of the A1 horizon
- moisture content of the soil
- grassland management type
- agricultural grassland use
- quantity of manure applied.

The thickness of the A1 horizon was measured in centimetres and it can therefore be handled as a quantitative variable. In the dunes, shifting sand is a normal phenomenon. Frequently, young developed soils are dusted over by sand, so that soil development restarts. This may result in soils with several A1 horizons on top of each other. Where this had occurred only the A1 horizon of the top soil layer was measured.

The moisture content of the soil was divided into five ordered classes.It is therefore an ordinal variable.

Four types of grassland management have been distinguished:
- standard farming (SF)
- biological farming (BF)
- hobby-farming (HF)
- nature conservation management (NM).

The grasslands can be used in three ways: as hayfields, as pasture or a combination of these (intermediate). Both variables are nominal but sometimes the use of the grassland is handled as an ordinal variable (Subsection 2.3.1). Therefore a ranking order has been made from hay production (1), through intermediate (2) to grazing (3).

The amount of manuring is expressed in five classes (0-4). It is therefore an ordinal variable.

All ordinal variables are treated as if they are quantitative, which means that the scores of the manure classes, for example, are handled in the same way as the scores of the A1 horizon. The numerical scores of the ordinal variables are given in Table 0.2. There are two values missing in Table 0.2 . Some computer programs cannot handle missing values, so the mean value of the corresponding variable has been inserted. The two data values are indicated by an asterisk.

Table 0.1. Dune Meadow Data. Unordered table that contains 20 relevées (columns) and 30 species (rows). The right-hand column gives the abbreviation of the species names listed in the left-hand column; these abbreviations will be used throughout the book in other tables and figures. The species scores are according to the scale of van der Maarel (1979b).

```
                                  00000000011111111112
                                  12345678901234567890

 1  Achillea millefolium          13..222..4......2...        Ach mil
 2  Agrostis stolonifera          ..48...43..45447...5        Agr sto
 3  Aira praecox                  ...............2.3.         Air pra
 4  Alopecurus geniculatus        .272...53..85..4....        Alo gen
 5  Anthoxanthum odoratum         ....432..4......4.4.        Ant odo
 6  Bellis perennis               .3222....2......2..         Bel per
 7  Bromus hordaceus              .4.32.2..4..........        Bro hor
 8  Chenopodium album             ............1.......        Che alb
 9  Cirsium arvense               ...2................        Cir arv
10  Eleocharis palustris          .......4.....458...4        Ele pal
11  Elymus repens                 44444...6...........        Ely rep
12  Empetrum nigrum               ................2.          Emp nig
13  Hypochaeris radicata          ...........2.....2.5.       Hyp rad
14  Juncus articulatus            .......44.....33...4        Jun art
15  Juncus bufonius               ......2.4..43.......        Jun buf
16  Leontodon autumnalis          .52233332352222.2562        Leo aut
17  Lolium perenne                75652664267......2..        Lol per
18  Plantago lanceolata           ....555..33.....23..        Pla lan
19  Poa pratensis                 44542344444.2...13..        Poa pra
20  Poa trivialis                 2765645454.49..2....        Poa tri
21  Potentilla palustris          ............22.....         Pot pal
22  Ranunculus flammula           .......2....2222...4        Ran fla
23  Rumex acetosa                 ....563.2..2........        Rum ace
24  Sagina procumbens             ...5...22.242.....3.        Sag pro
25  Salix repens                  ................335         Sal rep
26  Trifolium pratense            ....252.............        Tri pra
27  Trifolium repens              .52125223633261..22.        Tri rep
28  Vicia lathyroides             .........12......1..        Vic lat
29  Brachythecium rutabulum       ..2226222244..44.634        Bra rut
30  Calliergonella cuspidata      ............4.3...3         Cal cus
```

Table 0.2. Environmental data (columns) of 20 relevées (rows) from the dune meadows. The scores are explained in the description of the Dune Meadow research project above; asterisk denotes mean value of variable.

Sample number	A1 horizon	Moisture class	Management type	Use	Manure class
1	2.8	1	SF	2	4
2	3.5	1	BF	2	2
3	4.3	2	SF	2	4
4	4.2	2	SF	2	4
5	6.3	1	HF	1	2
6	4.3	1	HF	2	2
7	2.8	1	HF	3	3
8	4.2	5	HF	3	3
9	3.7	4	HF	1	1
10	3.3	2	BF	1	1
11	3.5	1	BF	3	1
12	5.8	4	SF	2	2*
13	6.0	5	SF	2	3
14	9.3	5	NM	3	0
15	11.5	5	NM	2	0
16	5.7	5	SF	3	3
17	4.0	2	NM	1	0
18	4.6*	1	NM	1	0
19	3.7	5	NM	1	0
20	3.5	5	NM	1	0

1 Introduction

R.H.G. Jongman

1.1 Types of ecological research

This book deals with the analysis of ecological data. Before going into that it is wise to define the area in which we are working: Ecology is part of biology and deals with the interrelationships between populations, communities and ecosystems and their environment, but draws on knowledge from many other disciplines, for example climatology, physical geography, agronomy and pedology. Odum (1971) prefers the definition 'Ecology is the study of structure and function of nature'. He stresses the role of ecosystem research in relation to the use of nature by man. Another definition, emphasizing population dynamics, describes ecology as the scientific study of the interactions that determine the distribution and abundance of organisms in nature (Krebs 1978).

Two types of ecological research are autecological and synecological studies. Autecology is the study of one species in relation to its environment, which comprises other organisms and abiotic factors. Synecology, or community ecology, is the study of many species simultaneously in relation to their environment. The number of species or, more generally, taxa concerned can vary from a few to hundreds. For instance, a study on the Black Woodpecker alone belongs to autecology, while a study on forest bird communities including the Black Woodpecker belongs to synecology.

On a larger spatial scale landscape ecology focuses on spatial patterns and the processes related to them; it considers the development of spatial heterogeneity, and spatial and temporal interactions across heterogeneous landscapes. It attempts to answer questions about land-use and land management. Biogeography studies the large scale spatial distributions of species in relation to climate and soil.

Until recently synecological studies have mostly been carried out on terrestrial vegetation. Recently they have been extended to animal communities and aquatic systems. For example, there have been studies on bird communities (Kalkhoven & Opdam 1984, Wiens & Rotenberry 1981), and aquatic ecosystems have been classified on the basis of communities of macrofauna (Section 8.3).These classifications are the basis for landscape analysis. For example, vegetation classifications and soil classifications can be used to classify landscapes. Once we understand the patterns and structures in the landscape, we are able to pay attention to the processes that determine them. The influence of mankind on these processes changes the landscape to a great extent. Especially in landscape ecology several sciences meet: geology, geography, biology and soil science (Risser et al. 1984).

Sometimes it is not easy to define the difference between apparently distinct

1

parts of ecological science as autecology and synecology. Floating vegetation of rather deep waters may consist mostly of one or few species, e.g. the vegetation known in Europe as the community of *Nymphoidetum peltatae* Oberd. et Th. Muell 1960, in which *Nymphoides peltata* is the only species. The study of this vegetation can be defined as either synecology or autecology:

- if the aim is to classify and define the vegetation as a plant community, it can be called synecology
- if the aim is to find the environmental amplitude of *Nymphoides peltata* it might be defined as autecology
- the river system with *Nymphoides peltata* vegetation might be the object of a landscape ecology study

The preceding discussion on the subdivisions of ecology is not purely academic: it indicates the growing complexity in ecological research. With the increasing complexity of the systems studied the restrictions to be met when defining research objectives and methods increase. For instance, the determination of the effect of pollutants on a single species may be a research goal on the short term and causal interactions might be found. However, this gives only sparse information on the changes that might occur in nature. Spatial or long-term ecological research in landscape systems might provide that information, but it will be hard to find causal relations at this higher level of complexity. The integration of autecology and synecology with landscape ecology within one project can provide better insight into the complex relations within the landscape.

The methods appropriate for analysis of ecological data show the same tendency of increasing complexity. This introductory chapter gives an overview for those who are not familiar with ecological research and data-analysis in general, a section on the terminology used in this book and some historical notes. All topics are discussed in detail in Chapters 2 to 7. Chapter 2 describes data collection and related topics: problem recognition, and the formulation of research objectives and related hypotheses. Regression and calibration are explained in Chapters 3 and 4. Multivariate analysis is treated in Chapters 5 (Ordination) and 6 (Cluster analysis). Chapter 7 explains the analysis of spatial data. In Chapter 8 the use of some methods is illustrated by case-studies.

1.2 Steps in research in community and landscape ecology

In general, in every research project several steps can be distinguished. Problem recognition and the formulation of a research objective and related hypotheses are the first steps in any study. Then data must be collected in a correct way. Data analysis follows, to summarize the data or to find causal or descriptive relations.

Experimental research in ecology is difficult to carry out, especially at the more complex levels of communities, ecosystems and landscapes. Most studies at these levels are descriptive. The sets of data being analysed are usually large; they are commonly gathered during field surveys. The analytical techniques used are determined by the objectives of the project; the results are influenced by what is sampled and the way sampling is carried out.

Many data in ecological research originate from field surveys. A field survey starts – after its objectives are clearly stated – by planning how, where and when to take samples. Sampling strategy is important, not only to retrench expenses, but also to get interpretable data.

In many countries mapping projects are carried out for planning and nature-conservation management. Mapping can be described as the classification and generalization of field data so that they can be depicted on a map. Data for a map are collected by a field survey, often in combination with a satellite or aerial survey. The more detailed the research done, the more information can be given about the area that was investigated, by smaller, clearly defined mapping units. The objective of the study determines the scale of the mapping project.

Monitoring can be defined as the process of repetitive observation of one or more environmental variables (Harvey 1981; Meijers 1986). The analysis of data stemming from a monitoring project may indicate, for example, changes in levels of pollution or results of nature-conservation management. The variables can be biological, chemical or physical: for example, lichens, carbon dioxide, or the water-table, respectively. In a monitoring project, data collection is repeated over a certain period of time at well-defined, permanent locations and at specified time intervals.

Data used in community and landscape ecology are mostly multivariate, i.e. each statistical sampling unit is characterized by many attributes. This means that (Gauch 1982):
 – data are complex, showing noise, redundancy, internal relations and outliers
 – data are bulky
 – some information in the data is only indirectly interpretable.
Multivariate methods make such data easier to handle.

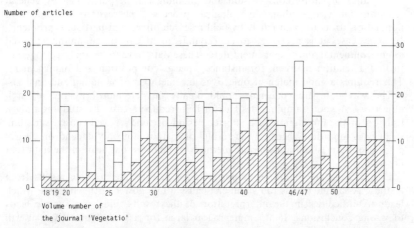

Figure 1.1 Growth of articles on multivariate analysis in the journal Vegetatio over the last 20 years. □ total number of articles. ▨ number of articles on multivariate analysis.

3

These problems were recognized by vegetation scientists and as a result they began looking for adequate data analysis techniques. This led to the founding, in 1969, of the Working Group for Data Processing in Phytosociology of the International Society for Vegetation Science, which has played an important part in stimulating and co-ordinating the development of data analysis techniques (van der Maarel et al. 1976).

Multivariate methods in ecology can be divided into three groups: direct gradient analysis, or regression analysis; indirect gradient analysis, or ordination; and classification, or cluster analysis. The development of multivariate methods can be illustrated by the number of articles in which these methods are used in the journal Vegetatio; these have increased from one or two per volume in 1968 to about 60% of all articles per volume in 1984 (Figure 1.1). In these articles all approaches of vegetation science are present: direct and indirect gradient analysis, as well as cluster analysis.

1.3 Implementing ecological research

Ecological research is used to support and evaluate nature conservation at international, national and local levels; it is applied in planning. In the Netherlands, for example, landscape ecological research provides essential information for physical planning at national, regional and local levels. Most of the Dutch provinces have developed programmes of mapping and monitoring. These programmes may entail research on the ground-water table, vegetation, birds, macrofauna, and the like. The data from these mapping and monitoring projects are used for many purposes, for example: planning locations for human activities (e.g. outdoor recreation or transportation); to predict the effect of ground-water extraction; to check water for pollutants.

The aims of nature conservation and planning can be partly met by general scientific knowledge about hydrological processes, succession, isolation and population structure, and partly by field research directed at particular problems. In many cases it is necessary to collect data on the biotic community as well as environmental and management data. These data have to be analysed together to find a relation between, for instance, species composition and management. This requires a good match of objectives and methods of sampling and analysis.

The way data are collected should ideally be determined by the research objectives. A good sampling design can help reducing costs by studying only a fraction of the population in a shorter period of time. A well-chosen sampling strategy is important, because the results of a study not only depend on clearly defined objectives and appropriate methods of analysis, but also on the data that are used.

Correct data are of utmost importance for getting interpretable results from a study. A sample that is too small can cause a low power of the analysis, which leads to difficulties in the interpretation of the results. Bias in a sample leads to wrong conclusions. If a sample is too large for a study's objectives, it will cause a waste of effort, time and money. Clearly the sampling phase is a crucial one in any research project. For this reason, data collection and its relation with

4

objectives and methods of analysis is the first topic of this book (Chapter 2).

There are several ways to use multivariate analysis in research. Multivariate methods are scientific tools to process data collected during field surveys. To apply these methods detailed knowledge of the mathematical principles upon which they are based is not required, but at least some background is necessary to know when a method can be used. This book is not a mathematics textbook; our purpose is to make the ecologist aware of what he or she is doing. When you understand a numerical method and what it can be used for, it is possible to choose the best method to solve your problem; it also helps ensure that you use the method correctly.

To provide a common theme for the explanation and application of techniques presented, a common set of data will be used in Chapters 3 to 6. This is part of the data of a research project to study management–vegetation relations of dune meadows on the Dutch island of Terschelling (Batterink & Wijffels 1983), in the Wadden Sea. This set of data, referred to as Dune Meadow Data, is described in a preliminary section of the book. A summary of the data is printed on a loose card that is placed inside the book.

In practical terms, then, what techniques can provide answers to what questions about the dune-meadow management carried out on the island of Terschelling? One may want to explain the variation in abundance of a particular species, for example the abundance of *Agrostis stolonifera* in Table 0.1, which varies between 0 and 7. The abundance of *A. stolonifera* is plotted against moisture in Figure 1.2. Moisture is seen to explain some of the variation as *A. stolonifera* is absent in the driest parcels (moisture=0) and common in the wetter parcels. The relation displayed in Figure 1.2 can be analysed quantitatively by regression analysis (Chapter 3). More generally, regression analysis can be used for assessing which environmental variables best explain the species' abundance and which environmental variables appear to be unimportant. Once the relations between species and environmental variables have been quantified, they can be used to predict the species abundance from observed values of one or more observed environmental variables (prediction, Chapter 3) or, conversely, to predict values of environmental variables from observed abundance values of species (calibration, Chapter 4). Instead of abundance, one may have recorded merely the species presence. Chapters 3 and 4 also describe how to analyse presence–absence data with regression and calibration techniques.

Alternatively, one may want to explain the variation in the abundance of a number of species of an ecological community. The first step to achieve this aim can be to summarize the species data by searching for the dominant patterns of variation in community composition. The abundances of species usually covary in a more or less systematic way, because they react to the same environmental variables. Without knowing the environmental variables, one may therefore still attempt to reconstruct such variables from the species data alone. This is called ordination (Chapter 5). In ordination, sites and species are arranged along axes that represent theoretical variables in such a way that these arrangements optimally summarize the species data. Figure 5.7 shows such an arrangement for the 20 sites and the 30 species in the Dune Meadow Data. The second step in the analysis

Abundance

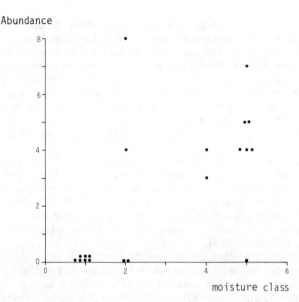

Figure 1.2 Relation of observed numbers of *Agrostis stolonifera* present and moisture class; from Dune Meadow Data.

is to relate the arrangements obtained to environmental variables or to known characteristics of the species.This is a powerful, indirect way of exploring possible relations between community composition and environment. Chapter 5 also introduces the so-called canonical ordination techniques, which search in a more direct way for relations between community data and environmental variables. The techniques of Chapter 3, 4 and 5 are also known in ecology under the name of gradient-analysis techniques.

Another way of summarizing species data is to use cluster analysis (Chapter 6). Cluster analysis is based on the idea that community types exist, and that each can be characterized by characteristic species combinations. Cluster analysis for the purpose of describing community types thus attempts to form groups of sites in such a way that the community–composition of sites varies most between groups and varies least within groups. This is illustrated with the Dune Meadow Data in Table 6.7. Subsequently, the groups may be interpreted in terms of environmental variables. This is another indirect way to explore species–environment relations. Cluster analysis is especially useful for defining mapping units. For example, on the basis of floristic data recorded at a limited number of sites, a vegetation classification is made. During subsequent field work, the distinguished vegetation types are mapped to obtain a vegetation map of a region.

It is unrealistic to believe that units delineated on a map are really homogeneous. An environmental variable, such as precipitation, will vary gradually over a region.

One may want to interpolate its values between sampling points, or make a contour map. This can be achieved by using trend-surface analysis (Chapter 7). A special aspect of spatial data is that deviations from the trend over short distances often are highly correlated, but essentially random and therefore unpredictable. In other words, there is no easy direct relation between site location and the value of an attribute; the value of an attribute at a certain site might be better predicted by the values of nearby sites than by the general spatial trend. Chapter 7 also describes methods to quantify this variability. This may lead to more insight in the variable studied. It can also be useful to obtain better interpolations by methods such as kriging, and to choose a better sampling strategy.

Ordination and cluster analysis are often used in the early exploratory phase of an ecological investigation. The results may suggest relations to be studied in more detail in subsequent research. Regression may help in the study of more specific questions in the later phases of research. This order – ordination and cluster analysis first, regression analysis later – can also be used to analyse an existing set of data. Ordination and cluster analysis are useful to explore the data for possible patterns of community variation. Subsequently, one may attempt to explain the abundances of some species of particular interest in more detail by regression analysis, using measured environmental variables as explanatory variables. Nevertheless, in this book regression and calibration are introduced before ordination, because most ordination techniques require a basic understanding of regression and calibration. Ordination is covered before cluster analysis because some cluster analysis techniques are based on ordination.

1.4 Terminology

Terminology used to describe various concepts of sampling is a source of confusion in ecology because ecologists often use terms that differ from those common to mathematical statistics. This is particularly so for the term sample, used by many ecologists (e.g. Gauch 1982) as a synonym for 'statistical sampling unit'. Also, ecologists from various branches of ecology use their own, different, terms for the 'statistical sampling unit'. For clarity, it is necessary to define several terms as they will be used in this book. We have adopted the following conventions:
- The word sample is used in its common statistical meaning, i.e. a collection of statistical sampling units.
- The statistical sampling unit is termed in this book as site (except in Chapter 2, where the exact statistical terminology is used).

Unfortunately, this convention introduces another source of confusion, because the word site is also often used to indicate a land unit, for example in geomorphology. To avoid confusion with this other meaning of the word site, we use words like location, land unit, stand or parcel when we want to indicate where the site comes from. So in one stand we can collect several sites, distributed in space and time to obtain a sample of this stand. We can use these sites to describe the stand, its spatial heterogeneity, and its temporal changes.

Examples of what we call sites are:

- the quadrat of a vegetation as used in vegetation studies (the so-called relevé is the description of a quadrat)
- a slice from a sediment core used by a palynologist to count pollen
- the catch in a net used by limnologists and entomologists to estimate species abundances
- the plot used by a forester to estimate production
- a soil core used by a soil scientist to determine chemical and physical soil properties
- a volume of water collected to measure chemical composition or estimate species abundances
- the catcn in a trap over a certain period of time used by animal ecologists.
Some other terminological points must be made:
- In contrast to Pielou (1984), we treat the word clustering as being more or less synonymous with classification. The word cluster is used to indicate a class of points, no matter how arbitrarily it's boundaries are chosen.
- Quantities of species can be measured (or estimated) in different ways, depending on the kind of organism concerned and the objectives of the study. We use the word occurrence to indicate presence–absence data, abundance to indicate numbers of individuals, cover to indicate the estimated proportion of soil shaded from a vertical light source by any species
- In a few cases we could not avoid using the term sample instead of site, but there it is clear from the context what is meant.

1.5 Historical notes

Ecology is an old science. It goes back as far as ancient Greek and Roman times, when philosophers such as Plato, Hippocrates, Plinius, Strabo and Posidonius wrote treatises on the relation between man and nature. These studies were partly philosophical in approach, but they were also in part a record of their empirical observations of changes in nature caused by man (Glacken 1967). Despite this long history, the development of ecology as a modern science began modestly. It has it's origins in the scientific discoveries of the Renaissance. For example, it was in the wake of that period that Anthoni van Leeuwenhoek (1632-1723) made his calculations on the reproductivity of insects (Egerton 1968), Linnaeus (1707-1778) developed his system of plant classification, and Mentzel (1622-1701) coined the term plant geography. Later, in the nineteenth century, the development of modern autecology and population dynamics benefited greatly from the impulse they received from the theories and research of Haeckel, Malthus and Darwin.

Synecology developed from plant geography, from efforts to classify vegetation. Humboldt (1807) made the first vegetation classifications on a physiognomic base. At the end of the nineteenth century an approach in vegetation science was developed in Northern Europe that is based on research of vegetations with a small number of species – the 'Nordic' approach. The relatively small number of species in the Scandinavian vegetation resulted in emphasis of vegetation layers (strata) and their major or dominant species. At about the same time a floristic–sociological approach was developing in Southern Europe. The major exponent of this 'French–Swiss' school was J.J. Braun–Blanquet. This school emphasizes the classification of vegetation based on its floristic composition. Quantitative methods fit in well with this approach and as a result it has developed into one of the major quantitative approaches in community ecology.

Beside these two, schools of vegetation science have developed in other parts of the world: Russia, Great Britain, Australia and the United States (Whittaker 1973). The local vegetation structures and the vegetation patterns, together with differences in scientific traditions, have greatly influenced the schools that have developed there. The Anglo-American school had great influence on the development of gradient analysis as one of the major quantitative approaches in ecology.

Its geographical aspect brought vegetation science in contact with geography. A convergent development in physical geography and vegetation science in Germany and Great Britain led in the thirties to the development of landscape ecology (Neef 1982). After a long period of silence, growth started anew between 1960 and 1970, at a time of rediscovery of nature and nature conservation.

Parallel to the development of the ecological sciences there has been a development of statistical methods: the principle of maximum likelihood (19th century), most univariate statistical methods (1930s and 1940s), principal components analysis (1930s), most other multivariate methods (1950s-1980s), and generalized linear models (1970s and 1980s).

The introduction of the computer, which made methods requiring voluminous computations practicable, has contributed largely to the development of new methods for data analysis.

2 Data collection

J.C. Jager and C.W.N. Looman

2.1 Data collection, not an isolated or undirected activity

In this chapter several statistical methodological aspects of community and landscape ecology will be discussed. The advanced methods of analysis dealt with in this book are of great use to the researcher of ecological communities. One would expect that technical knowledge of these methods offers sufficient assurance for an objective and efficient application of these methods. However this is not the case. The advanced character of the methods and the complexity of ecological research can cause the researcher to become caught up in unexpected methodological pitfalls.

Whole filing-cabinets full of information about the composition of the flora and fauna in a certain area do not necessarily offer a useful starting point for research. Samples taken from a vegetation in a certain area to establish the natural values of that area must be differently placed (in time and space) from samples taken to detect the disturbing influences of nearby industries. In the first case, one could select the sampling units randomly from the area to be studied. In the second case, it could be far better to sample at systematically determined distances from the industrial plant. The purpose of a research project puts demands on the way in which data are collected. A set of data suitable for one purpose may be totally unsuitable for another. The source of the observation data on which one bases one's research determines which questions can be answered and the nature of the ensuing answers to those questions.

The execution of research within the framework of an empirical science can be taken to be an iterative process of adjustment (Box 1976; Box et al. 1978) in which two activities, theoretical speculation and data collection, must be closely connected. Indeed, theory and practice are continually linked by this process. Initial hypotheses (conjectures, models, theories) may have logical consequences that can be compared against data (facts, phenomena). Discrepancies may lead to adjustment of the hypotheses, which become the starting point for a new cycle of adjustment. The collection of the data is not an isolated, undirected activity. In the interaction between theory and practice statistical methodology plays an important part. In statistics, methods have been developed for the description and classification of data of observations, for data collection (sampling theory), for the building of mathematical models (modelling), for comparing models with observations and for inferring (generalized) conclusions (hypothesis testing, parameter estimation). Statistics offer us (cognitive) methods and techniques that make it possible to allow the above mentioned iterative process to run smoothly.

Statistical textbooks do inform researchers in detail about the many methods available to them. There are several textbooks specially intended for the biologist (e.g. Sokal & Rohlf 1981). Illustrated with details from real biological research, these books give an inventory of the well-stocked tool-kit of statistics. Such books are valuable to the researcher; they show how statistical methods can be applied. But the researcher is also in need of guidelines – to apply these tools in good coherence: the tuning of the structure of a study to the analyses on the research questions and the objectives of the research. It is very difficult to treat this aspect of statistical methodology in a textbook.

In a well-thought-out project the following activities are essential:
- the stating of the problem and the objectives of the project (Section 2.2)
- the planning of data collection (Section 2.3)
- the analysis of the observations and the interpretation of the results (Sections 2.4 and 2.5).

These are the theoretical activities that support the *execution* of the research project – the data collection. The lack of textbooks in which strategies for research are explicitly discussed, and the existence of a number of publications (Section 2.6) that clearly show serious deficiencies in the application of statistics (for example in ecology) justify including a chapter on data collection in this book. In our attempt to localize the most dangerous pitfalls we will make comments on the above-mentioned activities.

2.2 Objectives and data collection, start by defining the finish

Statisticians have pointed out repeatedly, and with increasing stress, that a research project must be based on an explicit statement of the problem and the objectives. 'Without this, it is easy in a complex survey to forget the objectives when engrossed in the details of planning and to make decisions that are at variance with the objectives' (Cochran 1963). During the planning of a research project the statement of objectives is the guideline for the decisions one has to make about the definitions of possible treatments, explanatory variables and response variables, by which expensive, irrelevant or unobtainable aims can be avoided (Cochran 1983). Box et al. (1978) write: 'In any investigation it is of utmost importance (1) to *define clearly the objectives* of the study to be undertaken, (2) to be sure that all interested parties agree on these objectives, (3) to agree on what criteria will determine that the objectives have been reached, and (4) to arrange that, if the objectives change, all interested parties will be made aware of this fact and will agree on the new objectives and criteria. It is surprising how often these steps are either ignored or not given the careful attention they deserve, a circumstance that often leads to difficulties and sometimes to disaster' (Box et al. 1978, p. 15).

If the advice above applies to researchers in general, it definitely applies to the landscape ecologist in particular. Several characteristics of research in landscape ecology are responsible for this. First, the landscape ecologist is usually involved in multidisciplinary research (incorporating, for example, ecology, geography and agriculture; cf. Risser et al. 1984) in which usually several researchers and institutions have to co-operate. Moreover, landscape ecology is frequently aimed at policy-making. Then, the various parties in a research project are often guided by diverging interests: scientific ones and policy-making ones. The ecologist may be interested in, for example, the causal background of the structure of a vegetation, while a researcher on behalf of the policy-maker may prefer to work as quickly as possible towards formulating a policy from the research. In the absence of the same objectives for all concerned, it all too easily happens that during the research the objectives change and diverge. Suppose, for example, researchers started a project to describe in some ecological sense an area. Then it is easy for a conflict to occur when they are asked to use the data to predict the effects of, say, building a motorway or establishing new recreation facilities. The ecologist is then forced to analyse a set of data, that was not, or not optimally, designed for the new objectives.

Second, landscape ecology commonly requires long-term field research. The time-span between the start and finish of the relevant research is so great that changes in the original problem and associated aims can easily occur. For example, research into the relation between landscape structure and the density of breeding bird populations requires observations over a long period. In contrast to field research, laboratory research often easily yields results with experiments that take a short time. Since research in landscape ecology usually involves an expensive observation process, the researcher is frequently forced to answer questions from a set of data that was not put together expressly to answer these questions. The desire to use multi-purpose surveys is understandable here.

Third, research into natural communities is methodologically complex. The background to this complexity will be dealt with in Section 2.7. The multivariate analysis required in this context also demands sets of data to be specifically designed for the objectives. This conflicts with the desire to use multi-purpose surveys.

2.3 Planning of data collection

2.3.1 Study design

Once the problem and objectives of a research project have been stated, it is necessary to develop a schema that summarizes accurately the procedure to be executed from start to finish. The development of this schema – often called the 'experimental design' or the 'study design' – implies detailed attention to a whole series of activities, which are summarized below in the conventional terminology of statistics (see Cox (1958) and Cochran (1977)):
  - The statement in statistical terms of the problem to be solved. To what (statistical) population(s) does one direct the research? Which (statistical) hypotheses are to be considered for testing? Which parameters does one want to estimate?
  - The description of the objects for study (the experimental unit or the sampling unit).
  - The specification of response (dependent) variables, explanatory (independent) variables and treatments (in terms of factors and factor levels).

12

- The design of a (random) procedure for the allotment of treatments to experimental units or a (random) procedure for the selection of sampling units (e.g. completely random sampling, stratified sampling).
- The specification of the variables to be observed, the relevant observation methods and the measurement scales to be used.
- The determination of the desired accuracy and precision of the observation methods to be used. In this activity one should not restrict oneself to the specification of the accuracy and precision of the observations to be done with a measuring instrument (e.g. with how many digits after the decimal point does one have to register a pH), but also one has to determine the taxonomic level (species, subspecies) of the organisms in the ecological survey.
- The establishment of the desired duration and size (number of study objects) of the research. The statistical power analysis (see below) can be useful with this.
- The specification of the statistical methods to be used for analysis, the setting up of an scheme of analysis and a framework for the representation of the results (tables, graphs, diagrams).
- The evaluation of the objectives and relevant criteria in the light of the above-mentioned activities.

This list of activities is not complete. Besides, it is not always possible to realize all these activities. For example, in non-experimental research the assignment of treatments to study objects (experimental units) does not occur. In addition, as already mentioned, the landscape ecologist is frequently forced to make use of observational material that he did not collect himself as a starting point for his research. It is then impossible for him to tune the observation process to the problem under consideration.

Whenever possible, a good researcher will – almost intuitively – pay attention to the features listed in the schema. Researchers do, however, according to their nature, stress different aspects. For example, one researcher may try to improve his research by increasing the number of observations, while another will concentrate on the technical refinement of the instruments to be used. However the best approach would be for the researcher to take all the activities listed into consideration and to consciously weigh up where he will direct his attention in the setting up and conduction of the research.

A well-thought-out design is the foundation for meaningful application of statistical techniques, especially statistical tests of hypotheses. The hypothesis being tested – the so-called null hypothesis – is a statement (assumption) about the collection of all elements under investigation, i.e. the statistical population. This statement is mostly given as a zero difference. For example: 'there is no difference between the mean values of the height of oaks in Forest A and B'. It can be formulated more generally as

$$\mu_1 - \mu_2 = 0.$$

Another example: 'the means of two populations differ by 4 units', i.e.

$$\mu_1 - \mu_2 - 4 = 0.$$

These two examples refer to population means, but null hypotheses can be formulated for many other parameters, such as variances, regression coefficients and correlation coefficients.

A statistical test is meant to enable us to make a statement about populations based on the study of samples from those populations. Two types of error can occur in a statistical test: Type I error – the rejection of a true null hypothesis; or Type II error – the acceptance of the null hypothesis when it is false. The probabilities of making false decisions are the probability α of making a Type I error and the probability β of making a Type II error, respectively. An important concept in testing is the power of a statistical test, i.e. the probability of rejecting the null hypothesis when it is false, $1-\beta$. The probabilities α, β or $1-\beta$ and the sample size (the number of sampling units involved in the study) are related. An increase in sample size reduces β and thus increases $1-\beta$. The balanced assessment of α, β, $1-\beta$ and the (required) sample size is the subject matter of so-called statistical power analysis. A short survey of statistical power analysis is provided by Cohen (1973). A statistical power analysis is an essential part of a study design.

2.3.2 Study type

When the study design is completed, it is then possible to determine what type of study it is. Studies can be classified according to their purpose. Cox & Snell (1981) advise to typify any study using the dichotomy 'explanatory' vs. 'pragmatic', i.e. research directed to obtaining insight into one or more phenomena vs. research with a practical/technological aim. Another important dichotomy concerns exploratory objectives vs. confirmatory objectives of a study. The first sort are aimed at the detection of relations that represent a starting point for later research devoted to the testing of hypotheses; the second sort are aimed at obtaining conclusive statements (statistical proof), made possible by the application of hypothesis testing and parameter estimation.

Apart from this classification of studies by intended purpose, a typology based on a characterization of the study design is important. It is important because the study type indicates to the researcher what type of conclusions he may expect to be able to make from the application of statistical techniques on the data concerned, once he has reached the interpretation phase of the study. Cox & Snell (1981) mention the following main types:
- *Experiments.* The investigator sets up and controls the system under study, and influences the procedures to be conducted. Especially if objective randomization is involved in the assignment of treatments to the experimental units, it will be possible to conclude that significant differences in response between treatments are a consequence of the treatments.
- *Pure observational studies.* The investigator has no control over the collection of the data. It may be possible to detect differences and relations between measured variables, but interpretation of the differences requires caution. The

14

real explanation for the differences may not become apparent; it may not have been even measured.

- *Sample surveys.* Under the investigator's control, a (random) sample is drawn from well-defined statistical populations. Research results in a good description of the populations and the differences between those populations, but similar to observational studies, the explanation of relations between observed variables remains problematic.
- *Controlled prospective studies.* The researcher selects units in which various variables considered to be explanatory are measured. The units are then monitored to see whether some particular event occurs. If all potentially important explanatory variables can be measured, which is never totally possible, prospective studies lead to conclusive statements on the explanatory character of the measured variables (cf. experiments).
- *Controlled retrospective studies.* A characteristic response has been registered and subsequently the history of the units is traced to detect relevant explanatory variables.

Note that this classification of main types does not meet methodological criteria of an adequate classification: a classification must be exclusive and extensive (van der Steen 1982). This classification is not exclusive: research of the fourth or fifth type, controlled prospective or retrospective studies, for example, can fall under research of the third type, sample surveys. Hurlbert (1984) proposes a different terminology to distinguish between experiments, which he calls 'manipulative experiments', and surveys (including prospective studies), which he calls 'mensurative experiments'. A classification of study types that meets the formal demands of classification should be connected to the range of possible conclusions that may result from each study type. Problems that will be discussed in Section 2.6 are then less likely to occur. Unfortunately such a classification has yet to be developed.

2.3.3 Sampling strategy

As stated in Subsection 2.3.1 development of a study design implies development of a procedure for the allotment of treatments to experimental units or a procedure for the selection of sampling units. The sampling strategy represents the skeleton of a study design. Randomized procedures, i.e. random allotment of treatments to experimental units or random selection of sampling units, are almost always preferable to non-randomized procedures. Randomization provides modern research a particularly powerful tool to increase the range of validity of the conclusions emerging from a research project (cf. Fisher 1954): randomization eliminates systematic errors (bias) and provides the basis for warranted application of inferential statistics (testing, estimation).

It is not necessary for the sampling procedure to be completely randomized. In the study of natural populations it can be of profit (i.e. more precise estimation) to make use of so-called stratified sampling (e.g. Snedecor & Cochran 1980). For the reader of this book it is interesting to take note of what Bunce et al. (1983) wrote on this matter. They developed a stratification system for ecological sampling in Britain based on environmental strata. Other, sometimes useful, sampling strategies that deviate from completely random sampling contain a systematic element, e.g. the line transect method, where sites are located at regular distances along a randomly (!) selected transect. Systematic sampling strategies require, however, justification by specification of assumptions about the type of variation in the sampled population to arrive at statistically sound conclusions.

In certain cases, for instance in a research project involving treatments or observations that could cause irreparable damage, it is not desirable to sample randomly. Anderson et al. (1980) discuss the possible reasons for non-randomized studies in the field of medicine. Their discussion – translated into ecological terms – is also of interest to ecologists.

For several methodological aspects of the sampling strategy used in ecological field studies the reader is referred to Hurlbert (1984a), Pielou (1976) and Southwood (1978).

2.4 Analysis of data and data collection

2.4.1 Introduction

In this section we will discuss some subjects that are related to the analysis of data. Analysis involves many activities:
- The inspection of the data to determine their quality. Do outliers occur? Are there any gross errors? Are there any missing values? If so, what can be done about it?
- Investigation (by graphing and testing) whether the data meet the requirements and assumptions of the analytic methods used. If necessary, values of observations may be transformed.
- The preparation of the observations for computer processing and further analysis.
- The application of the analytic methods chosen in the study design and the presentation of the results (tables, diagrams, graphs).

We shall not discuss the content of the analytic methods that are dealt with in later chapters. But we do want to bring to the reader's attention some general aspects of the analysis that precede the application of these methods, namely the measurement scale, the transformation of observed variables and the typification of the frequency distributions of the variables to be analysed.

2.4.2 Measurement scales

To prevent unnecessary work and disappointment it is important to decide upon the appropriate measurement scale in the planning phase of a project. Measurement scale will therefore – of all the aspects of study design (Subsection 2.3.1) – receive special attention.

Measuring the abundance of a species or the status of an environmental factor means assigning a value to each sampling unit. These values contain the information about the samples we have collected. Values can be related to each other in different

ways, which are called scales of measurement. In this book we use four kinds of scales:
- nominal scales
- ordinal scales
- interval scales
- ratio scales.

The first scale is referred to as being 'qualitative'; the last two are referred to as 'quantitative'.

The number of restrictions, or constraints, of each scale increase in this order, i.e. nominal scales have fewer restrictions than ordinal scales, ordinal scales fewer than interval scales and interval scales fewer than ratio scales. A scale with fewer restrictions is 'weaker' than a scale with more restrictions, which is 'stronger'.

A nominal scale is the least restrictive. The values have no relation to each other and are often referred to as classes. An example is the factor soil type, which may have the classes 'clay', 'peat' and 'sand'. Values assigned to these classes, for example 10, 11 and 17, respectively, do not imply any order and differences between them do have no function. The respective values only represent, in effect, the names of the classes.

An ordinal scale implies more than a nominal scale, namely a rank order between the values or classes. A well-known ordinal scale is the Braun–Blanquet scale, for measuring abundances of plants. This scale was extended by Barkman et al. (1964) and recoded to numeric values by van der Maarel (1979b) to provide a scale for use in numerical analyses. Table 2.1 presents these scales. The possible values or symbols in this scale can be ranked from low to high abundance, but the differences between the values are not fixed. It can't be said that the difference between the values '1' and '2' is less or greater than the difference between '4' and '5'. Because of this feature the calculation of the mean or standard deviation can give misleading results (see Exercise 2.1). Another example of an ordinal scale is when we have nothing but an ordering of objects, for instance when a person has ranked objects like rivulets or hedges according to their value for nature conservation.

The third type of scale, the interval scale, possesses a constant unit of measurement. The differences between values can be compared with each other. An example is temperature measured in degrees centigrade. A difference of say 5 degrees means the same all over the scale. Yet with an interval scale the position of the zero value is arbitrary. This leads to the fact that we can say that a certain difference is twice as large as an other, but not that a certain object is twice as large as an other when corresponding values are in the ratio 2 to 1. Means and standard deviations can be calculated as usual.

A ratio scale is like an interval scale, but with a fixed zero point. So it is meaningful to calculate ratios. Abundance measures for species are often ratio scales, for instance number of individuals or proportions (usually expressed as a percentage).

When choosing a measurement scale one should be guided by the analytic method to be used. Some analytic methods demand a strong scale and cannot be used for the analysis of data measured on a weak scale. Though reduction to a weaker

Table 2.1 Original cover scale of Braun–Blanquet, extended to a combined cover/abundance scale by Barkman et al. (1964) and then recoded by van der Maarel (1979b).

Braun–Blanquet		Barkman		van der Maarel's scale
Symbol	Cover (%)	Symbol	Cover (%) or Abundance	
		r	rare	1
		+	few	2
1	< 5%	1	many	3
		2m	abundant	4
2	5-25 %	2a	5-12.5 %	5
		2b	12.5-25%	6
3	25-50%	3		7
4	50-75%	4		8
5	> 75 %	5		9

scale can be carried out later on, after measuring, it is preferable to use the weakest scale possible, since obtaining measurements using a weak scale requires considerably less effort than measurements using a strong scale. Presence–absence data (nominal scale), for example, are far more easy to obtain than data on proportions (ratio scale).

2.4.3 Frequency distributions

For the statistical analysis of the data it is important to know how the data are distributed. We distinguish between distributions of discrete variables and distributions of continuous variables. When the smallest possible difference between two values is essentially greater than zero, the distribution is called discrete; when every real value in an interval is possible the variable is called continuous. When we count the number of animals in a sampling unit, the variable resulting will be discrete; when we measure a pH, the variable will be continuous.

The normal distribution (Figure 2.1a) is a distribution for continuous variables. Its major feature is that the relative frequency of observing a particular value is symmetric around the mean. The normal distribution is described by two parameters, the mean (μ) and the standard deviation (σ). The standard deviation is a measure for the spread of the values around the mean. About two-third of the values from a normal distribution fall between μ-σ and μ+σ and about 95% of the values fall between $\mu-2\sigma$ and $\mu+2\sigma$.

When we make a histogram of the values of, for example, the proportions of a species in a particular set of data, the histogram will often suggest that the underlying theoretical distribution is not symmetric but skewed. A particular kind of skewed distribution is the log-normal distribution (Figure 2.1b). A variable

18

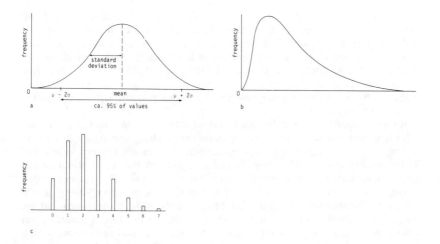

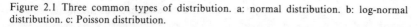

Figure 2.1 Three common types of distribution. a: normal distribution. b: log-normal distribution. c: Poisson distribution.

has a log-normal distribution if its logarithm has a normal distribution. Because normal distributions are more convenient to deal with, 'skewed' variables that have a log-normal distribution are usually transformed by taking their logarithm. Throughout this book the Naperian, or natural, logarithm is used. This transformation does not hinder interpretation of results for the following reasons. The mean of a log-transformed variable can be back-transformed by taking its antilog, i.e. exp $\bar{x}$. This gives the geometric mean of the variable. The geometric mean is an estimate of the *median* of the log-normal distribution. Further, the standard deviation of a log-transformed variable (s.d.) is about equal to the *coefficient of variation* of the original variable, at least if s.d. $\leqslant 0.5$. For greater values of the standard deviation, the coefficient of variation is estimated by $\sqrt{\exp(s^2)-1}$ (Aitchison & Brown 1969). (The coefficient of variation is, by definition, equal to the standard deviation divided by the mean).

There are several rules of thumb for deciding whether a continuous variable, which has positive values only, can better be described by a log-normal than by a normal distribution. For large sets of data one can look at histograms. For small sets of data, assuming a log-normal distribution is recommended if the standard deviation is larger than the mean or if the maximum value of the variable is more than twenty times larger than the smallest value.

We shall mention two distributions for discrete variables. The first is the binomial distribution. A binomial distribution occurs when, for example, a population consists of males and females in a ratio of 1 : 3 and we randomly select n individuals. The number of females in the sample of n individuals then follows a binomial distribution. The parameters of this distribution are the total (n) and the probability (p) that a randomly selected individual is a female; in the example $p = 0.75$.

19

The variance of a variable that is binomially distributed is equal to $np(1-p)$. When we randomly select n sites, then the number of occurrences of a species will also follow a binomial distribution, with total n and probability p (the probability of occurrence of the species). When we select only a single site ($n = 1$), then the number of occurrences of the species is either 0 or 1, depending on whether the species is absent or present in the site selected. As $n = 1$, this particular binomial distribution is described by a single parameter, p, the probability of occurrence of the species.

A second distribution for discrete variables is the Poisson distribution (Figure 2.1c). This distribution occurs when we count the number of organisms in a region or the number of occurrences of a particular event in a given period of time. The count can result in the values 0, 1, 2, 3, ..., etc. A Poisson distribution can be described with one parameter: the mean. The variance of a Poisson distribution is simply equal to its mean. The assumption that counts of organisms follow a Poisson distribution is often unwarranted. Counts made of organisms that are located completely at random in a region do follow a Poisson distribution, but often organisms are aggregated (clumped) so that zero counts and extremely high counts are encountered more frequently than in a Poisson distribution with the same mean. For an aggregated population, the variance exceeds the mean.

2.4.4 Transformations

The application of the techniques that are being dealt with in this book frequently requires transformation of the data of basic variables. A transformation consists of a replacement of measured values by other values. Depending on the motivation for the transformation, two groups can be distinguished. The first group serves to make several variables comparable (e.g. abundances of species among samples, values of different environmental factors). The second group ensures a better fit of the values in a regression model, which can then be used for further statistical analysis.

Frequently, we want to express in a number whether an environmental variable takes on a high or low value in a certain sample. Obviously it is easier to inspect the difference between the average over all samples and the measured value from the sample of interest than to inspect all measured values. We replace the value y_{ki} by

$$y_{ki}^* = y_{ki} - y_{k+}/n \qquad\qquad \text{Equation} \quad 2.1$$

for all k. This is called 'centring'. Subsequently one can relate the deviation to the total variation of the variables. This is especially desirable where it concerns variables with different units of measurement. For example, a deviation of one pH unit is obviously quite different from a deviation of one microgram of phosphate. To compare such variables the value y_{ki} is replaced by

$$y_{ki}^* = (y_{ki} - y_{k+}/n)/s_k, \qquad\qquad \text{Equation} \quad 2.2$$

20

where s_k represents the standard deviation of the series. This is called 'standardizing to zero mean and unit variance'. An alternative is to replace the measured values by their rank numbers. Several other methods are available to make variables comparable. Subsection 6.2.4 gives more details.

The second group of transformations ensures a better fit of the data values to a certain distribution or to a certain model. We know for instance that the responses of many organisms to concentrations of cations is logarithmic. For that reason the concentrations of most cations are usually transformed by taking the logarithm or the negative logarithm (e.g. H^+ ions to pH). Other variables also frequently display a logarithmic relation instead of a linear one. Another reason for transformation can be that least-squares regression (Chapter 3) requires that the dependent variable has normally distributed random errors. If the error is proportional to the expected value of the dependent variable (constant coefficient of variation) logarithmic transformation of the dependent variable is required. The log transformation is:

$$y_{ki}^* = \log_e y_{ki}.$$
Equation 2.3a

If zero values occur, one can use instead:

$$y_{ki}^* = \log_e (y_{ki}+1).$$
Equation 2.3b

The cover/abundance scale of van der Maarel (Table 2.1) is comparable to log-transformed cover (Figure 2.2) for the cover values (5-9); the abundance values (1-4) are unrealistically taken to represent the same proportion of cover.

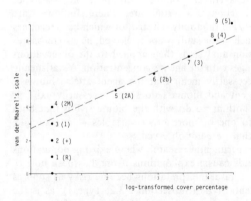

Figure 2.2 Log-transformed cover percentages of Table 2.1 plotted against the scale of van der Maarel.

2.5 Interpretation of results

Ideally a study should meet a number of strict requirements. It should:
– have a clearly stated aim
– have well-defined populations, observation processes, hypotheses to be tested and parameters to be estimated
– use random sampling/random assignment of treatments
– have previously specified probabilities of making false decisions
– use a study design based on the requirements above
– use analytical methods appropriate to the requirements above
– have conclusions that concern only the questions stated in the study.

A study design meeting these requirements will give clear, statistically sound conclusions about the population(s) under study and provide a basis to formulate causal explanations about observed differences (effects) and relations among variables. Evidence of causality is more convincing when treatments can be applied to the objects under study than when they cannot be manipulated. When one is studying existing populations, as in surveys, then the interpretation of the analyses becomes more complicated, because it is possible to come up with more than one explanation.

An example may illustrate this. Assume a researcher is interested in the question as to whether a toxic agent in the ground influences the biomass of growing sunflowers. Sunflower cuttings are (randomly) assigned to pots in which the researcher has established desired levels of the agent, varying over a relevant range. After some time the biomass of each plant is assessed. The study design is a 'completely randomized design with replications' and the results can be analysed by regression analysis. If the regression analysis shows a statistically significant decrease in the biomass for an increase of the agent levels in the ground, then it is safe to conclude the agent affects the biomass – there is a (possibly indirect) causal relation between the toxic agent and biomass. Suppose the researcher conducts his study as follows. He chooses a certain area where sunflowers are abundant. A number of sunflowers are randomly selected, of which he determines the biomass. At the place where each of the sunflowers is selected, he also measures the agent concentration. Regression analysis is then applied to the observations to check whether the biomass changes with the agent's concentration. A statistically significant outcome does not necessarily mean that the agent affects biomass, i.e. a causal relation between agent and biomass does not necessarily exist. It may well be that the agent has nothing to do with the biomass, but that both variables fluctuate in response to one or more other variables – for example, amount of human disturbance – that were not observed.

The example shows that non-experimental research where existing populations are (randomly) sampled only yields causal explanations if one has been able to exclude alternative explanations. Causal interpretations become risky as soon as the study is not of the experimental type (Subsection 2.3.2, Type 1), as is the case in environmental research that is based on field observations of biological communities and environmental variables. The example also shows that the scope of a study is not determined by the analytic technique used – here in both cases a simple regression analysis – but rather by the study type. In both cases, however, one arrives at conclusive statements (statistical proof) on the populations under study (a hypothetical Population-case 1: sunflowers assigned to specified levels of the toxic agent; and a real Population-case 2: all sunflowers present in a certain area) that depend for their validity on the randomization involved. Studies not meeting the randomization requirement, or the other requirements listed at the beginning of this section, do not yield conclusive, i.e. statistically sound, statements about the relevant population(s). Such studies should be given the label 'detection'.

The methodological status of statistical classification methods continues to be a subject of discussion. Hartigan (1975) remarked on cluster techniques that 'they are not yet an accepted inhabitant of the statistical world'. It is usual to label them under the heading of 'exploratory data analysis'. That implies one correctly assumes the methods concerned to be of detective value and not contributive to the testing of hypotheses. One considers the methods to be 'a tool for suggestion and discovery' (Anderberg 1973). The conscious use of the distinction proof-detection is therefore continuously in order when one wants to interpret the results of the application of these techniques. The relevant classification techniques are relatively young. Integration with classical statistical methodology, typified by models for observations (for example analysis of variance and regression analysis), the testing of hypotheses and the estimation of parameters is as yet incomplete. Gordon (1981, p.141) remarks : 'The integration of classification into the main stream of statistics should prove beneficial to both'. The detective character of many multivariate analyses in ecology is attributable, on the one hand, to the nature – the stage of development – of these methods and, on the other hand, to the nature of the study type – non-experimental field research – in which they are applied.

2.6 Sources of misinterpretation

In the previous sections we have tried to indicate how an investigator might plan his research and the conditions he has to meet to make certain types of conclusive statements. However there are various sources of misinterpretation of ecological data, most of which affect the status and range of validity of the conclusions. Our aim in this section is to indicate the implications of not meeting the requirements of a perfect study design.

2.6.1 Incorrect use of statistical analysis

Statistical methodological evaluations of the use of statistics in research offer a sombre picture, which undoubtedly is caused by the relatively meagre attention paid to research strategies in statistical textbooks. Several authors have brought to light a number of the abuses that occur as a result, for example:
- poorly designed and incorrectly analysed field experiments (Hurlbert 1984)
- incorrect application of Student's t test (Innis 1979)
- incorrect use of multiple comparison procedures (Dawkins, 1983; Johnson & Berger 1982; Mead & Pike 1975)

- incorrect or insufficient application of basic statistical principles (Buckland 1982; Tacha et al. 1982)
- lack of attention paid to the methodology of the design and analysis of biological monitoring systems (Meyers et al. 1982).

These authors offer numerous recommendations to remedy this situation; their publications are compulsory reading for the ecologist.

2.6.2 Taking detection to be proof

One of the most frequent misinterpretations is taking detection for proof. Sources of this kind of misinterpretation are manifold. Mertz & McCauley (1980) noticed that in the past field workers were wary of laboratory experiments and that remnants of this attitude still persist today. They also noticed that these field workers made no distinction between correlational data and results of research upon which a researcher was able to impose treatments.

Hurlbert (1984) evaluated the application of statistics in field experiments and concluded that ecological literature is riddled with 'poorly designed and incorrectly analysed experimental work'. He introduced the concept of 'pseudoreplication' for those cases where either treatments are not replicated or replicates are not statistically independent. Often in such experiments a treatment is applied to a plot in which several sites are located, or sites in subsequent years are treated as being independent observations. Tacha et al. (1982) observed the lack of explicit description of the population in ecological literature, which makes it difficult to distinguish between statistical proof (conclusive statements on the sampled population) and detection (statements on the target population). This is very important in designing landscape ecological directives, when one is interested in an area greater than the area studied. The distinction between proof and detection is also impossible to make when sampling units or experimental units are not properly defined, or statistical hypotheses and objectives are not clearly stated.

Deviation from the planned analysis may change the character of the conclusions emerging from a study from proof to detection. When, for instance, one decides to analyse a difference or a relation that is perceived in the data without having planned to do so beforehand, the result has detective value only (the analysis predicts what has already happened: 'prophecy ex eventu').

2.6.3 Misinterpretation caused by biased sampling

By not using random sampling, one can obtain a 'distorted picture' (Snedecor & Cochran 1980) of a population. Biased sampling, such as choosing sites from a vegetation that are considered to be typical (key sites) or that have as many species as possible, introduces an observer-tied bias. Other types of bias may be caused because certain measurements have been performed with a frequency that depends on the season or the location (e.g. near the laboratory vs. far from the laboratory). Then the results of any statistical analysis may be different from the results that would have been obtained if random sampling had been applied. It is worthwhile to go to quite some effort to avoid this kind of bias.

24

2.6.4 Other sources of misinterpretation

Strong (1980) considered the strategy of ecological research. He established that explicitly formulated null hypotheses in a broad sense – 'null hypotheses entertain the possibility that nothing has happened, that a process has not occurred, or that change has not been produced by a cause of interest' (Strong 1980, p. 271) – occur remarkably seldom in ecology. He writes: 'Instead, most research is either phenomenological on one hand or corroborative on the other. Phenomenological ecology collects facts and measurements about populations, communities or ecosystems without ostensible hypotheses. ... Corroborative ecological research is motivated by a particular alternative hypothesis and ignores the null hypothesis. It usually either develops deductive theory in which an alternative is treated as true or collects circumstantial evidence that corroborates an alternative' (Strong 1980, p. 273). Strong analysed examples of research of the latter type. He concluded that ignoring and not stating explicit null hypotheses (null models) hampers ecological research. We notice that reactions on Strong's view (Quinn & Dunham 1983; Simberloff 1983) bring to light controversial points of view with respect to the methodology of ecological research and theory formation; the value of inferential statistics involving specified null hypotheses is not being denied here, however.

Several years after the publication of Strong (1980), Toft & Shea (1983) (see also Cohen 1977) pointed out that the assessment of the power is one of the most important statistical tools for investigating ecological patterns. They found, however, that 'Power statistics are drastically underused in basic and applied ecological research when they could provide objective measures of the sensitivity of null hypothesis tests and thereby strengthen some statistical inferences. Null models are being used increasingly to investigate the causes of community-wide patterns, yet researchers tend to ignore the risk involved of committing the Type II error associated with these models'.

The insufficient use of statistical power analysis – even though the underlying principles have long been known – goes hand in hand with frequent misinterpretation of the concepts 'statistically significant' and 'statistically non-significant'. Often researchers take a statistically significant outcome automatically for a biologically relevant result, while a statistically non-significant outcome is taken for granted as absence of an effect or relation. The execution of a statistical test in research represents a powerful tool, but testing is not the end of the research. Test outcomes need to be evaluated. Absence of significance can be the result of too small statistical power, when one has arrived at significant outcomes the application of estimation techniques (confidence intervals) is in order for evaluating the size of an effect and its reliability.

2.7 Complexity of ecological research

From a methodological point of view ecological investigation into natural communities represents a complex area of research. The ecologist often encounters all possible difficulties of empirical science at the same moment:

– a wide variability in the variables studied
– a complex interaction between explanatory variables and response variables
– uncertainty about the causes of the observed correlations.

These difficulties are closely bound up with the fact that ecological research is mostly field research. The great variability in the variables observed in the field (quantities concerning species in field samples) is caused by the existence of many influencing, but also changing, abiotic and biotic factors. For instance, abundances of species in the vegetation are influenced by the chemical composition of the soil, pH, water-table, manuring, grazing and treading. Perhaps explanatory variables influence each other, as response variables do. For example, the water-table affects the chemical composition of the soil and, in turn, the chemical components affect the pH. This results in correlations between variables, so that causal explanations are difficult to assess.

Due to the many variables involved, the analysis of ecological data requires the application of multivariate statistical techniques. These methods are of relatively recent date and do not yet link up with all aspects of classical statistics, with its elaborate methods of hypothesis testing and parameter estimation.

Both the nature of the research – observational field studies – and the stage of development of the analytic techniques required generate the methodological complexity associated with landscape ecological research. This situation justifies and necessitates a carefully thought-out study design. All the activities summarized in Subsection 2.3.1 explicitly need attention to prevent disappointments during the analysis and interpretation phase of the research project.

2.8 Bibliographical notes

In general statistical textbooks are not easily to read for biologists. The aim of this section is to mention some textbooks that are of importance for ecologists. A selection is made from both statistical handbooks aimed at researchers in general and statistical handbooks aimed at the researcher in ecology.

Well known is the textbook of Sokal & Rohlf (1981), which is specially aimed at the biologist. Clarke (1980) gives an introduction to statistics and experimental design in a series of student texts on contemporary biology. Easily accessible general statistical handbooks have been written by Snedecor & Cochran (1980), Box et al. (1978) and Cox & Snell (1981, especially the first part of the book).

Study design is explained in a number of classic texts on statistics, such as those describing experiments written by Cochran & Cox (1957) and Cox (1958), and a book on the design of sampling surveys written by Cochran (1977).

The application of statistical methods in ecology is treated in textbooks by Pielou (1977), Green (1979) and Southwood (1978). Slob (1987) discusses strategical aspects of applying statistics in ecological research.

2.9 Exercise

Exercise 2.1   Scales and averages

Two researchers go into the field to try to find out where the problem of heathlands changing into grasslands is more serious, in the Netherlands or in Scotland. Each visits the same locations in both countries, but they use different scales: Dr A. estimates proportion of cover (%) and Dr B. uses the combined cover/abundance scale of Braun–Blanquet, coded numerically according to van der Maarel (1979) (see Table 2.1). The results of both investigations are presented below.

Location	Country	Dr A.'s estimates		Dr B.'s estimates	
		Calluna	*Deschampsia*	*Calluna*	*Deschampsia*
a	N	70	4	8	1
b	N	10	70	5	9
c	N	7	80	5	8
d	N	90	2	9	1
e	S	70	4	8	4
f	S	80	10	7	5
g	S	80	20	9	7
h	S	85	5	9	5

Compute averages and standard deviations obtained by both investigators for each country and each species. Look at the differences and state your conclusions.

2.10 Solution to exercise

Exercise 2.1   Scales and averages

The results are worked out completely for *Deschampsia* as % estimated by Dr A. For the Netherlands he finds a mean proportion of cover of: $(4+70+80+2)/4 = 39\%$. The standard deviation is: $[(4-39)^2 + (70-39)^2 + (80-39)^2 + (2-39)^2/3]^{1/2} = 41.8\%$. For Scotland he finds a mean proportion of cover of: $(4+10+20+5)/4 = 9.8\%$. The standard deviation is : $[(4-9.8)^2 + (10-9.8)^2 + (20-9.8)^2 + (5-9.8)^2/3]^{1/2} = 7.3\%$. His conclusion is that (although not significant) the proportion of cover of *Deschampsia* is lower in Scotland than in the Netherlands. The rest of the results are shown below.

Country		Dr A.'s results		Dr B.'s results	
		Calluna	*Deschampsia*	*Calluna*	*Deschampsia*
Netherlands	mean	44.3	39.0	6.8	4.8
	s.d.	42.1	41.8	2.1	4.4
Scotland	mean	78.8	9.8	8.2	5.3
	s.d.	6.3	7.3	1.0	1.3

Both investigators agree about *Calluna*: it is more dominant in Scotland than in the Netherlands. However they disagree about *Deschampsia*, since Dr B. concludes that *Deschampsia* is more important in Scotland than in the Netherlands. Note also the very high standard deviations in the Dutch samples, which might be caused by a skewed or bimodal distribution.

3 Regression

C.J.F. ter Braak and C.W.N. Looman

3.1 Introduction

3.1.1 Aim and use

Regression analysis is a statistical method that can be used to explore relations between species and environment, on the basis of observations on species and environmental variables at a series of sites. Species may be recorded in the form of abundances, or merely as being present. In contrast with ordination and cluster analysis, we cannot analyse data on all species simultaneously; in regression analysis, we must analyse data on each species separately. Each regression focuses on a particular species and on how this particular species is related to environmental variables. In the terminology of regression analysis, the species abundance or presence is the response variable and the environmental variables are explanatory variables. The term 'response variable' stems from the idea that the species react or respond to the environmental variables in a causal way; however, causality cannot be inferred from a regression analysis. The goal of regression analysis is more modest, namely to describe the response variable as a function of one or more explanatory variables. This function, termed the response function, usually cannot be chosen such that the function predicts responses without errors. By using regression analysis, we attempt to make the errors small and to average them to zero. The value predicted by the response function is then the expected response: the response with the error averaged out.

Regression analysis is well suited for what Whittaker (1967) termed 'direct gradient analysis'. In ecology, regression analysis has been used mainly for the following:
- estimating parameters of ecological interest, for example the optimum and ecological amplitude of a species
- assessing which environmental variables contribute most to the species' response and which environmental variables appear to be unimportant. Such assessment proceeds through tests of statistical significance
- predicting the species' responses (abundance or presence–absence) at sites from the observed values of one or more environmental variables
- predicting the values of environmental variables at sites from observed values of one or more species. Such prediction is termed calibration and is treated separately in Chapter 4.

3.1.2 Response model and types of response variables

Regression analysis is based on a response model that consists of two parts: a systematic part that describes the way in which the expected response depends on the explanatory variables; and an error part that describes the way in which the observed response deviates from the expected response.

The systematic part is specified by a regression equation. The error part can be described by the statistical distribution of the error. For example, when fitting a straight line to data, the response model (Figure 3.1) is

$$y = b_0 + b_1 x + \varepsilon \qquad\qquad \text{Equation 3.1}$$

with
y the response variable
x the explanatory variable
ε the error
b_0 and b_1 fixed but unknown coefficients; they are the intercept and slope parameter, respectively.

The expected response, denoted by Ey, is equal to $b_0 + b_1 x$. The systematic part of the model is thus a straight line and is specified by the regression equation

$Ey = b_0 + b_1 x.$

The error part is the distribution of ε, i.e. the random variation of the observed response around the expected response. The aim of regression analysis can now be specified more precisely. The aim is to estimate the systematic part from data while taking account of the error part of the model. In fitting a straight line, the systematic part is simply estimated by estimating the parameters b_0 and b_1.

In the most common type of regression analysis, least-squares regression, the

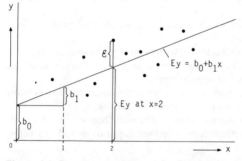

Figure 3.1 Response model used in fitting a straight line to data points (•) by least squares regression. For explanation see Subsection 3.1.2.

30

distribution of the error is assumed to be the normal distribution (Subsection 2.4.3). Abundance values of a species commonly show a skew distribution that looks like a log-normal distribution (Subsection 2.4.3), with many small to moderate values and a few extremely large values. Abundance values often show this type of distribution even among sites whose environmental conditions are apparently identical. By transforming the abundance values to logarithms, their distribution becomes more like a normal distribution (Williamson 1972). To analyse abundance values by least-squares regression, it is therefore often more appropriate to use log-abundance values. A problem then arises when the species is absent, because the abundance is then zero and the logarithm of zero is undefined.

A regression technique appropriate for presence–absence data is logit regression. Logit regression attempts to express the probability that a species is present as a function of the explanatory variables.

3.1.3 Types of explanatory variables and types of response curves

The explanatory variables can be nominal, ordinal or quantitative (Subsection 2.4.2). Regression techniques can easily cope with nominal and quantitative environmental variables, but not with ordinal ones. We suggest treating an ordinal variable as nominal when the number of possible values is small, and as quantitative when the number of possible values is large.

Regression with a single quantitative explanatory variable consists of fitting a curve through the data. The user must choose in advance how complicated the fitted curve is allowed to be. The choice may be guided by looking at a scatter

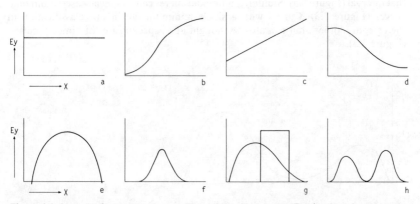

Figure 3.2 Shapes of response curves. The expected response (Ey) is plotted against the environmental variable (x). The curves can be constant (a: horizontal line), monotonic increasing (b: sigmoid curve. c: straight line), monotonic decreasing (d: sigmoid curve), unimodal (e: parabola. f: symmetric, Gaussian curve. g: asymmetric curve and a block function) or bimodal (h).

31

plot of the response variable against the explanatory variable or can be guided by available knowledge and theory about the relation. We denote the environmental variable by the letter x and the expected response by Ey, the expected value of the response y. We distinguish the following types of curves, often referred to as response curves (Figure 3.2):

- constant: Ey is equal to a constant; the expected response does not depend on x (Figure 3.2a).
- monotonically increasing (or decreasing): Ey increases (or decreases) with increasing values of x. Examples are straight lines and sigmoid curves (Figure 3.2b,c,d).
- unimodal (single-peaked): Ey first increases with x, reaches a maximum and after that decreases. Examples are the parabola with a maximum (Figure 3.2e), and a bell-shaped curve like the Gaussian curve (Figure 3.2f). The value of x where Ey reaches its maximum is termed the mode or optimum. The optimum does not need to be unique when the curve has a 'plateau' (Figure 3.2g). A unimodal curve can be symmetric (with the optimum as point of symmetry) or asymmetric (Figure 3.2g).
- bimodal: Ey first increases with x, reaches a maximum, then decreases to a minimum, after which Ey increases to a new maximum, from which Ey finally decreases again (Figure 3.2h).
- other: Ey has another shape.

The types of curves are listed in order of their complexity. Only in the simplest case of a constant response curve does the environmental variable have no effect on the response of the species. Monotonic curves can be thought of as special cases of unimodal curves; when the optimum lies outside the interval that is actually sampled, then the unimodal curve is monotonically increasing or decreasing within that interval (Figure 3.3). Similarly, unimodal curves can be special cases of bimodal curves (Figure 3.3). Curves with a single minimum fall in our classification in the category 'other', but can also be thought of as special cases of bimodal curves (Figure 3.3).

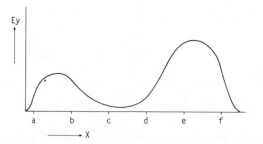

Figure 3.3 Response curves derived from a bimodal curve by restricting the sampling interval. The curve is bimodal in the interval a-f, unimodal in a-c and in d-f, monotonic in b-c and c-e and almost constant in c-d. In the interval b-e, the curve has a single minimum. (Ey, expected response; x, environmental variable).

32

3.1.4 Outline of the chapter

In this chapter, we introduce regression techniques for analysing quantitative abundance data (least-squares regression, Section 3.2) and presence–absence data (logit regression, Section 3.3). In both sections, we first present a model in which the explanatory variable is nominal, and then models in which the explanatory variable is quantitative, in particular models that are based on straight lines and parabolas.

From the straight line and the parabola, we derive curves that are more useful in ecological data analysis. For abundance data, we derive from them the exponential curve and the Gaussian curve, respectively, and for the analysis of presence–absence data, the sigmoid curve and the Gaussian logit curve. The curves based on a parabola allow estimation of the indicator value (optimum) and the ecological amplitude (tolerance) of the species. Problems involved in analysing quantitative data containing many zero values are dealt with in Section 3.4. In Section 3.5, both least-squares regression and logit regression are extended to multiple regression. Multiple regression can be used to study the effect of many environmental variables on the response by the species, be it quantitative or of presence–absence type. The topic of Section 3.6 is model choice and regression diagnostics. Finally, we leave regression and introduce the method of weighted averaging, which is a simple method for estimating indicator values of species. This method has a long tradition in ecology; it has been used by Gause (1930). We compare the weighted averaging method with the regression method to estimate indicator values.

3.2 Regression for quantitative abundance data: least-squares regression

3.2.1 Nominal explanatory variables: analysis of variance

The principles of regression are explained here using a fictitious example, in which we investigate whether the cover proportion of a particular plant species at sites systematically depends on the soil type of the sites. We distinguish three soil types, namely clay, peat and sand. The observed cover proportions showed a skew distribution within each soil type and therefore we decided to transform them by taking logarithms. Before taking logarithms, we added the value 1 to the cover expressed in percentages, to avoid problems with the two zero values in the data. Figure 3.4 displays the resulting response values for each soil type.

Our response model for this kind of data is as follows. The systematic part simply consists of three expected responses, one for each soil type, and the error part is the way in which the observed responses within each soil type vary around the expected responses in each soil type. On the basis of Figure 3.4, it appears not unrealistic to assume for the error part of the response model that the transformed relative covers within each soil type follow a normal distribution and that the variance of this distribution is the same for each of the three soil types. We further assume that the responses are independent. These assumptions constitute the response model of the analysis of variance (ANOVA), which is

33

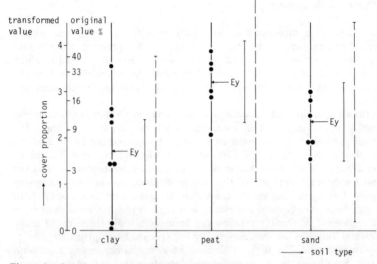

Figure 3.4 Relative cover (log-transformed) of a plant species (•) in relation to the soil types clay, peat and sand. The horizontal arrows indicate the mean value in each type (Table 3.1). The solid vertical bars show the 95% confidence interval for the expected values in each type and the dashed vertical bars the 95% prediction interval for the log-transformed cover in each type. (fictitious data).

one particular form of least-squares regression.

The first step in regression analysis is to estimate the parameters of the model. The parameters are here the expected responses in the three soil types. We estimate them by using the least-squares principle. We choose values for the parameters such that the sum (over all sites) of the squared differences between observed and expected responses is minimal. The parameter values that minimize this sum of squares, are simply the mean values of the transformed relative covers for each soil type. The expected response as fitted (estimated) by regression is, therefore just the mean of the response in each soil type. The fitted values are indicated by arrows in Figure 3.4. The difference between an observed response (a dot in Figure 3.4) and the fitted value is termed a residual, which in Figure 3.4 is a vertical distance. Least-squares thus minimizes a sum of squared vertical distances; the minimum obtained is called the residual sum of squares.

Regression analysis by computer normally gives not only parameter estimates but also an analysis-of-variance table (ANOVA table). From the ANOVA table (Table 3.1), we can derive how well the regression equation explains the response variable. In the example, the fraction of variance accounted for (R^2_{adj}) is 0.25, which means that only a quarter of the variance in the responses is explained by the differences between soil types. The ANOVA table can further be used for testing statistically whether the expected responses differ among soil types; that is, whether the mean values for each soil type differ more than could be

34

Table 3.1 Means and ANOVA table of the transformed
relative cover of Figure 3.4.

Term	mean	s.e.	95% confidence interval
Clay	1.70	0.33	(1.00, 2.40)
Peat	3.17	0.38	(2.37, 3.97)
Sand	2.33	0.38	(1.53, 3.13)

Overall mean 2.33

ANOVA table

	d.f.	s.s	m.s.	F
Regression	2	7.409	3.704	4.248
Residual	17	14.826	0.872	
Total	19	22.235	1.170	

$R^2_{adj} = 0.25$

expected by chance if soil type did not affect the relative cover. For this test,
the variance ratio F (Table 3.1) must be compared with the critical value of an
F distribution with 2 and 17 degrees of freedom in the numerator and denominator,
respectively (2 and 17 are the degrees of freedom for the regression and residual
in the ANOVA table). The critical value (at the 5% significance level) is 3.59.
(Consult for this a table of the F distribution, for instance in Snedecor & Cochran
1980.) In the example, the variance ratio (4.248) is larger than 3.59. Under the
null hypothesis of equal expected responses, this happens in 5% of the cases only.
So it is unlikely that the expected responses are equal. From the ANOVA table,
we can thus conclude that the expected responses do differ and we say that the
cover proportions differ significantly between soil types at the 5% level ($P <
0.05$, F test).

How precisely have we estimated the expected responses? An indication for
this is the standard error of the estimates (Table 3.1). The standard error can
be used to construct a confidence interval for the expected response. The end-
points of a 95% confidence interval for a parameter (and the expected response
is a parameter in this example) are given by

$$(\text{estimate}) \pm t_{0.05}(v) \times (\text{standard error of estimate}) \qquad \text{Equation 3.2}$$

The symbol $\pm$ is used to indicate addition or subtraction in order to obtain upper
and lower limits. The symbol $t_{0.05}(v)$ denotes the 5% critical value of a two-tailed
t test. The value of $t_{0.05}(v)$ depends on the number of degrees of freedom (v)
of the residual and can be obtained from a t table (e.g. Snedecor & Cochran
1980). In our example, $v = 17$ and $t_{0.05}(17) = 2.11$, which gives the intervals
shown in Figure 3.4 and Table 3.1.

We may also want to predict what responses are likely to occur at new sites of a particular soil type. A prediction interval for new responses is much wider than the confidence interval for the expected response. To construct a prediction interval, we need to know the residual standard deviation. This is the standard deviation of the residuals and is obtained from the ANOVA table by taking the square root of the 'mean square of the residual'. We obtain from Table 3.1 the residual standard deviation of $\sqrt{(0.872)} = 0.93$. In the example, the residual standard deviation is simply an estimate of the standard deviation within soil types. The prediction interval within which 95% of the new responses fall is now given by

$$\text{(estimated response)} \pm t_{0.05}(\nu) \sqrt{(\text{s.d.}^2 + \text{s.e.}^2)} \qquad\qquad \text{Equation 3.3}$$

where s.d. is the residual standard deviation and s.e. the standard error of the estimated response. Equation 3.3 yields for clay the interval

$$1.70 \pm 2.11 \sqrt{(0.93^2 + 0.33^2)} = (-0.38, 3.78).$$

If we had done many observations, the estimated response would be precisely the expected response. Then s.e. $= 0$ and $t_{0.05}(\infty) = 1.96$, so that Equation 3.3 reduces to: expected response $\pm 1.96 \times$ s.d. Figure 3.4 also displays the prediction intervals for the three soil types.

That procedure is sufficient for ANOVA by computer and for interpretation of the results. But for a better understanding of the ANOVA table, we now show how it is calculated. After we have estimated the parameters of the model, we can write each response as

$$\text{observed value} = \text{fitted value} + \text{residual}. \qquad\qquad \text{Equation 3.4}$$

For example, one of the observed responses on peat is 3.89 (corresponding to a cover of 48%). Its fitted value is 3.17, the mean response on peat, and the residual is thus $3.89 - 3.17 = 0.72$. We therefore write this response as $3.89 = 3.17 + 0.72$.

Each term in Equation 3.4 leads to a sum of squares; we first subtract the overall mean (2.33) from the observed and fitted values and then calculate (over all sites) sums of squares of the observed values and of the fitted values so corrected and of the residuals. These sums of squares are the total sum of squares, the regression sum of squares and the residual sum of squares, respectively, and are given in Table 3.1 in the column labelled with s.s. (sum of squares). The total sum of squares is always equal to the regression sum of squares and the residual sum of squares added together. Each sum of squares is associated with several degrees of freedom (d.f. in Table 3.1). The number of degrees of freedom equals $n - 1$ for the total sum of squares (n being the number of sites), $q - 1$ for the regression sum of squares (q being the number of estimated parameters, the value 1 is subtracted because of the correction for the overall mean) and $n - q$ for the residual sum of squares.

In the example, $n = 20$ and $q = 3$. The column labelled m.s. (mean square) is obtained by dividing the sum of squares by its number of degrees of freedom. The mean square of the residual is a measure of the difference between the observed and the fitted values. It is the variance of the residuals; hence its usual name residual variance. Similarly, the total variance is obtained; this is just the sample variance of the responses, ignoring soil type. The fraction of variance accounted for by the explanatory variable can now be defined as

$$R^2_{adj} = 1 - (\text{residual variance}/\text{total variance}),$$

which is also termed the adjusted coefficient of determination. In the example, $R^2_{adj} = 1 - (0.872/1.170) = 0.25$. The original, unadjusted coefficient of determination (R^2) does not take into account how many parameters are fitted as compared to the number of observations, its definition being

$$R^2 = 1 - (\text{residual sum of squares}/\text{total sum of squares}).$$

When a large number of parameters is fitted, R^2 may yield a value close to 1, even when the expected response does not depend on the explanatory variables. The multiple correlation coefficient, which is the product–moment correlation between the observed values and the fitted values, is just the square root of the coefficient of determination. Finally, the ratio of the mean squares of the regression and the residual is the variance ratio (F). If the expected responses are all equal, the variance ratio randomly fluctuates around the value 1, whereas it is systematically greater than 1, if the expected values differ; hence its use in statistical testing.

3.2.2 Straight lines

In Figure 3.5a, the explanatory variable is mean water-table, a quantitative variable that enables us to fit a curve through the data. A simple model for these data is a straight line with some scatter around the line. The systematic part of the response model is then

$$Ey = b_0 + b_1 x \qquad\qquad \text{Equation 3.5}$$

in which
Ey denotes the expected value of the response y
x denotes the explanatory variable, the mean water-table
b_0 and b_1 are the parameters that must be estimated
b_0 is the intercept (the value at which the line crosses the vertical axis)
b_1 is the slope parameter or the regression coefficient of the straight line
   (Figure 3.1)
b_1 is the expected change in y divided by the change in x.

The error part of the model is the same as for ANOVA (Subsection 3.2.1), i.e.

37

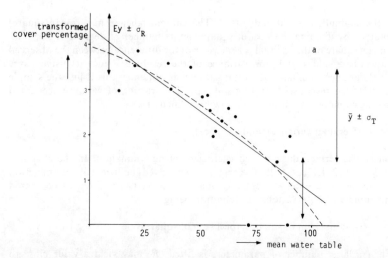

Figure 3.5a Straight line fitted by least-squares regression of log-transformed relative cover on mean water-table. The vertical bar on the far right has a length equal to twice the sample standard deviation σ_T, the other two smaller vertical bars are twice the length of the residual standard deviation (σ_R). The dashed line is a parabola fitted to the same data ($\bullet$).

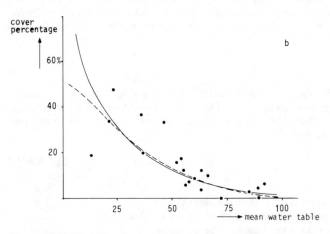

Figure 3.5b Relative cover in relation to water-table with curves obtained by back transformation of the straight line and parabola of Figure 3.5a.

38

the responses are taken to be mutually independent and are normally distributed around their expected values (Ey) as specified by the straight line (Equation 3.5). The errors are thus taken to follow a normal distribution and the variance of the errors to be independent of the value of x.

We again use the least-squares principle to estimate the parameters. That is, we choose arbitrary values for b_0 and b_1, calculate with these values the expected responses at the sites by Equation 3.5, calculate the sum of squared differences between observed and expected responses, and stop when we cannot find values for b_0 and b_1 that give a smaller sum of squared differences. In Figure 3.5a, this procedure means that we choose the line such that the sum of squares of the vertical distances between the data points and the line is least. (Clearly, any line with a positive slope is inadequate!) For many regression models, the estimates can be obtained by more direct methods than by the trial and error method just described. But, the estimates are usually obtained by using a computer program for regression analysis so that we do not need to bother about the numerical methods used to obtain the least-squares estimates. For the straight-line model, we need, for later reference, the equations for estimating b_0 and b_1

$$b_0 = \bar{y} - b_1 \bar{x} \qquad\qquad\qquad \text{Equation 3.6a}$$

$$b_1 = \Sigma_{i=1}^{n}(y_i - \bar{y})(x_i - \bar{x})/ \Sigma_{i=1}^{n} (x_i - \bar{x})^2 \qquad\qquad\qquad \text{Equation 3.6b}$$

where
y_i and x_i are the values of y and x at the i-th site
$\bar{y}$ and $\bar{x}$ are the mean values of y and x, respectively.

Table 3.2 shows standard output of a computer program for regression analysis, in which b_0 is estimated at 4.411 and b_1 at −0.0370. The ANOVA table can in

Table 3.2 Straight line fitted by least-squares: parameter estimates and ANOVA table for the transformed relative cover of Figure 3.5.

Term	Parameter	estimate	s.e.	t
Constant	b_0	4.411	0.426	10.35
Water-table	b_1	−0.0370	0.00705	−5.25

ANOVA table

	d.f.	s.s	m.s.	F
Regression	1	13.45	13.45	27.56
Residual	18	8.78	0.488	
Total	19	22.23	1.170	

$R^2_{adj} = 0.58$

principle be obtained by the rules of Subsection 3.2.1 below Equation 3.4, by using fitted values as calculated from Equation 3.5. The following statistics are derived from the ANOVA table as in Subsection 3.2.1. The residual standard deviation is $\sqrt{0.488} = 0.70$, which is much smaller than the standard deviation of the observed response, $\sqrt{1.170} = 1.08$.

The fraction of variance accounted for by the straight line is 0.58. The multiple correlation coefficient reduces in straight line regression to the absolute value of the product-moment correlation between x and y (0.78 in Figure 3.5a). The variance ratio can again be used for statistical testing, here for testing whether the expected responses depend on the mean water-table. The critical F at the 5% significance level is now 4.41, because there is only 1 degree of freedom for the regression (Snedecor & Cochran 1980). Because the variance ratio (27.56) exceeds this F, the expected response does depend on the mean water-table. An alternative for this F test is to use a two-tailed t test of whether b_1 equals 0; if b_1 were 0, the straight line would be horizontal, so that the expected response would not depend on x. This t test uses the t of b_1, which is the estimate of b_1 divided by its standard error (Table 3.2). This value (–5.25) is greater (in absolute value) than the critical value of a two-tailed t test at the 5% level obtained from a t table: $t_{0.05}(18) = 2.10$, and so b_1 is not equal to zero; thus the relative cover of our species does significantly depend on the mean water-table. Yet another way of testing whether $b_1 = 0$ is by constructing a 95% confidence interval for b_1 with Equation 3.2. The result is the interval $-0.037 \pm 2.10 \times 0.00705 = (-0.052, -0.022)$.

The value 0 does not lie in this interval and 0 is therefore an unlikely value for b_1. Which of the three tests to use (F test, t test or test through the confidence interval) is a matter of convenience; they are equivalent in straight-line regression.

After regression analysis, we should make sure that the assumptions of the response model have not been grossly violated. In particular, it is useful to check whether the variance of the errors depends on x or not, either by inspecting Figure 3.5a or by plotting the residuals themselves against x. Figure 3.5a does not give much reason to suspect such a dependence.

In the analysis, we used transformed relative covers. The data and the fitted straight line of Figure 3.5a are back-transformed to relative covers in Figure 3.5b. The fitted line is curved on the original scale: it is an exponential curve. Note that in Figure 3.5b, the assumption that the error variance is independent of x does not hold; this could have been a reason for using a transformation in the first place. It may be instructive now to do Exercise 3.1.

3.2.3   Parabolas and Gaussian curves

In Subsection 3.2.2, we fitted a straight line to the responses in Figure 3.5a. But wouldn't a concave curve have been better? We therefore extend Equation 3.5 with a quadratic term in x and obtain the parabola (Figure 3.2e)

$$Ey = b_0 + b_1\,x + b_2\,x^2 \qquad\qquad \text{Equation 3.7}$$

Table 3.3 Parabola fitted by least-squares regression: parameter estimates and ANOVA table for the transformed relative cover of Figure 3.5.

Term	Parameter	estimate	s.e.	t
Constant	b_0	3.988	0.819	4.88
Water-table	b_1	-0.0187	0.0317	-0.59
(Water-table)^2	b_2	-0.000169	0.000284	-0.59

ANOVA table

	d.f.	s.s	m.s.	F
Regression	2	13.63	6.815	13.97
Residual	17	8.61	0.506	
Total	19	22.23	1.170	

$R^2_{adj} = 0.57$

We again use the least-squares principle to obtain estimates. The estimates are given in Table 3.3. The estimates for b_0 and b_1 change somewhat from Table 3.2; the estimate for b_2 is slightly negative. The parabola fitted (Figure 3.5, dashed line) gives a slightly smaller residual sum of squares than the straight line. But, with the change in the number of degrees of freedom of the residual (from 18 to 17), the residual variance is greater and the fraction of variance accounted for is lower. In the example, the advantage of the parabola over the straight line is therefore doubtful. A formal way to decide whether the parabola significantly improves the fit over the straight line is by testing whether the extra parameter b_2 is equal to 0. Here we use the t test (Subsection 3.2.2). The t of b_2 (Table 3.3) is much smaller in absolute value than the critical value, 2.11; hence the data provide no evidence against b_2 being equal to 0. We conclude that a straight line is sufficient to describe the relation between the transformed cover proportions and mean water-table; a parabola is not needed.

Generally the values of t for b_0 and b_1 in Table 3.3 are not used, because they do not test any useful hypothesis. For example, the t test of whether b_1 is equal to 0 in Equation 3.7 would test a particular kind of parabola against the general parabola of Equation 3.7, quite different from the meaning of the t test of the slope parameter b_1 in Table 3.2.

In principle, we can extend the response function of Equation 3.7 to higher-order polynomials in x by adding terms in x^3, x^4, ... There is no advantage in doing so for the data in Figure 3.5a. Polynomial regression of species data has limited use except for one special case. *When we fit a parabola to log-transformed abundances, we actually fit a Gaussian response curve to the original abundance data.* The Gaussian response curve has the formula

$$z = c \exp [-0.5(x - u)^2 / t^2]$$
Equation 3.8

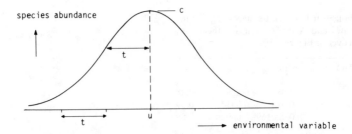

Figure 3.6 Gaussian response curve with its three ecologically important parameters: maximum (c), optimum (u) and tolerance (t). Vertical axis: species abundance.
Horizontal axis: environmental variable. The range of occurrence of the species is seen to be about $4\,t$.

where
z is the original abundance value
c is the species' maximum abundance
u is its optimum (the value of x that gives maximum abundance)
t is its tolerance (a measure of ecological amplitude).

Note that in this chapter the symbol t is used in two ways: the t of a regression coefficient (Subsection 3.2.2) and the t of a Gaussian curve. Which t is intended should be clear from the context of the passages concerned.

Figure 3.6 displays the Gaussian curve and its parameters. The curve is seen to rise and fall over a length of about $4t$. If we take the logarithm on both sides of Equation 3.8, we obtain

$$\log_e z = \log_e (c) - 0.5\,(x - u)^2 / t^2 = b_0 + b_1\,x + b_2\,x^2 \qquad \text{Equation 3.9}$$

where the third form follows by expanding

$$(x - u)^2 = x^2 - 2\,u\,x + u^2$$

and by setting:

$$b_0 = \log_e (c) - u^2 / (2t^2);\; b_1 = u / t^2;\; b_2 = -1 / (2t^2). \qquad \text{Equation 3.10}$$

By fitting a parabola to log-abundances, we obtain least-squares estimates for b_0, b_1 and b_2, from which we can obtain estimates of

the optimum, $u = -b_1 / (2b_2)$ Equation 3.11a

the tolerance, $t = 1 / \sqrt{(-2b_2)}$ Equation 3.11b

the maximum, $c = \exp (b_0 + b_1\,u + b_2\,u^2)$.   Equation 3.11c

42

These equations are derived from Equation 3.10 where $b_2 < 0$. If the estimate of b_2 is positive, the fitted curve has a minimum instead of a maximum. Approximate standard errors of the estimated optimum and tolerance can be derived from the variances and covariances of b_1 and b_2 that are provided as options by statistical packages. A confidence interval for the optimum can also be calculated. Details of these calculations are given in Section 3.9.

It may be instructive now to do Exercise 3.2 (except Part 3.2.8).

3.3 Regression for presence–absence data: logit regression

3.3.1 Nominal explanatory variables: chi-square test

Table 3.4 shows the numbers of dune meadow fields in which the plant species *Achillea ptarmica* was present and in which it was absent. The fields are divided into four classes depending on agricultural use. The relevant question for these data is whether the frequency of occurrence of *Achillea ptarmica* depends systematically on agricultural use. This question is analogous to the question that was studied in Subsection 3.2.1, although here the response of the species is not relative cover, but merely presence or absence. The usual thing to do is to calculate the relative frequency in each class, i.e. the number of fields of a given class in which the species is present divided by the total number of fields of that class (Table 3.4). But relative frequency of occurrence is simply the mean value when we score presence as 1 and absence as 0. Calculating means was what we did in Subsection 3.2.1. The response is thus $y = 1$ or $y = 0$, and the expected response, Ey, is the expected frequency, i.e. the probability of occurrence of the species in a field randomly drawn from all fields that belong to the class. Relative frequency is therefore an estimate of probability of occurrence.

Table 3.4 Numbers of fields in which *Achillea ptarmica* is present and absent in meadows with different types of agricultural use and frequency of occurrence of each type (unpublished data from Kruijne et al. 1967). The types are pure hayfield (ph), hay pastures (hp), alternate pasture (ap) and pure pasture (pp).

Achillea ptarmica	Agricultural use				
	ph	hp	ap	pp	total
present	37	40	27	9	113
absent	109	356	402	558	1425
total	146	396	429	567	1538
frequency	0.254	0.101	0.063	0.016	0.073

If the probabilities of occurrence of *Achillea ptarmica* were the same for all four classes, then we could say that its occurrence did not depend on agricultural use. We shall test this null hypothesis by the chi-square test. This test proceeds as follows. Overall, the relative frequency of occurrence is $113/1538 = 0.073$ (Table 3.4). Under the null hypothesis, the expected number of fields with *Achillea ptarmica* is in pure hayfield $0.073 \times 146 = 10.7$ and in hay pasture $0.073 \times 396 = 29.1$ and so on for the remaining types. The expected number of fields in which *Achillea ptarmica* is absent is therefore in pure hayfield $146 - 10.7 = 135.3$ and in hay pasture $396 - 29.1 = 366.9$.

We now measure the deviation of the observed values (o) and the expected values (e) by the chi-square statistic, that is the sum of $(o - e)^2/e$ over all cells of Table 3.4. We get $(37 - 10.7)^2/10.7 + (109 - 135.3)^2/135.3 + (40 - 29.1)^2/29.1 + \ldots = 102.1$. This value must be compared with the critical value, $\chi_\alpha^2(v)$, of a chi-square distribution with v degrees of freedom, where $v = (r - 1)(c - 1)$, r is the number of rows and c the number of columns in the table. In the example, $v = 3$ and the critical value at the 5% level is $\chi_{0.05}^2(3) = 7.81$. Consult a χ^2 table, for instance Snedecor & Cochran (1980). The chi-square calculated, 102.1, is much greater than 7.81, and we conclude therefore that the probability of occurrence of *Achillea ptarmica* strongly depends on agricultural use. Notice that the chi-square statistic is a variant of the residual sum of squares: it is a weighted sum of squares with weights $1/e$.

The chi-square test is an approximate test, valid only for large collections of data. The test should not be used when the expected values in the table are small. A rule of thumb is that the test is accurate enough when the smallest expected value is at least 1. A remedy when some expected numbers are too small is to aggregate classes of the explanatory variable.

3.3.2 Sigmoid curves

We now look at the situation in which we have a presence–absence response variable (y) and a quantitative explanatory variable (x). Data of this kind are shown in Figure 3.7. Just as in Subsection 3.3.1, the expected response is the probability of occurrence of the species in a site with a particular value of the environmental variable. This probability will be described by a curve. Probabilities always have values between 0 and 1. So a straight-line equation

$$Ey = b_0 + b_1 x \qquad\qquad \text{Equation 3.12}$$

is not acceptable, because $b_0 + b_1 x$ can also be negative. This difficulty could be solved by taking the exponential curve

$$Ey = \exp(b_0 + b_1 x) \qquad\qquad \text{Equation 3.13}$$

However the right side of Equation 3.13 can be greater than 1, so we adapt the curve once more to

$$Ey = p = [\exp(b_0 + b_1 x)]/[1 + \exp(b_0 + b_1 x)] \qquad\qquad \text{Equation 3.14}$$

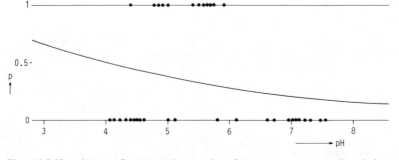

Figure 3.7 Sigmoid curve fitted by logit regression of the presences • at $p = 1$) and absences (• at $p = 0$) of a species on acidity (pH). In the display, the sigmoid curve looks like a straight line but it is not. The curve expresses the probability (p) of occurrence of the species in relation to pH.

This curve satisfies the requirement that its values are all between 0 and 1. The only further reason to take this curve, and not another one, is mathematical convenience. The curves representing Equations 3.12-3.14 are shown in Figure 3.8; Equation 3.14 represents a sigmoid curve. All three curves are monotonic and have two parameters, namely b_0 and b_1. The part $b_0 + b_1 \, x$ is termed the linear predictor. For probabilities, we use the symbol p instead of Ey (Equation 3.14).

The systematic part of the response model is now defined. Next, we deal with the error part. The response can only have two values, hence, the error distribution is the Binomial distribution with total 1 (Subsection 2.4.3). So the variance of y is $p(1 - p)$. We have now completed the description of the model.

To estimate the parameters from data, we cannot use ordinary least-squares regression because the errors are not normally distributed and have no constant variance. Instead we use logit regression. This is a special case of the generalized linear model (GLM, McCullagh & Nelder 1983). The term logit stems from logit transformation, that is the transformation of p

$$\log_e [p/(1 - p)] = \text{linear predictor} \qquad \text{Equation 3.15}$$

which is just another way of writing

$$p = [\exp (\text{linear predictor})]/[1 + \exp (\text{linear predictor})] \qquad \text{Equation 3.16}$$

The solution to Exercise 3.3 shows that Equations 3.15 and 3.16 are equivalent. The left side of Equation 3.15 is termed the link function of the GLM. Logit regression is sometimes called logistic regression.

In GLM, the parameters are estimated by the maximum likelihood principle. The likelihood of a set of parameter values is defined as the probability of the responses actually observed when that set of values were the true set of parameter

45

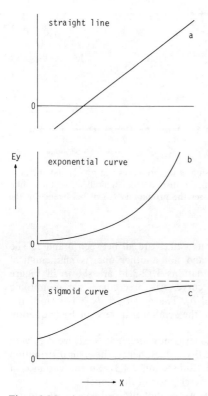

Figure 3.8 Straight line (a), exponental curve (b) and sigmoid curve (c) representing Equations 3.12, 3.13 and 3.14, respectively.

values. The maximum likelihood principle says that we must choose that set of parameter values for which the likelihood is maximum. A measure for the deviation of the observed responses from the fitted responses is the residual deviance, which is $-2 \log_e L$, where L is the maximized likelihood. The residual deviance takes the place of the residual sum of squares in least-squares regression. The least-square principle (Subsection 3.2.1) is equivalent to the maximum likelihood principle, if the errors are independent and follow a normal distribution. Least-squares regression is thus also a special case of GLM. In general, the parameters of a GLM must be calculated in an iterative fashion; provisional estimates of parameters are updated several times by applying repeatedly a weighted least-squares regression, in which responses with a small variance receive a larger weight in the residual sum of squares than responses with a large variance. In logit regression, the variance of the response was $p(1 - p)$. So the weight depends

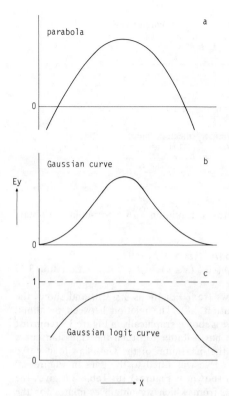

Figure 3.9 Parabola (a), Gaussian curve (b) and Gaussian logit curve (c) representing Equations 3.7, 3.8 and 3.17, respectively.

on the fitted value of p and hence on the parameter estimates; calculations must therefore be iterative. Computer programs for logit regression are available in statistical packages including GLIM (Baker & Nelder 1978), GENSTAT (Alvey et al. 1977), BMDP (Dixon 1981, subprogram PLR) and SAS (Harrell 1980). Ter Braak & Looman (1986) give an example of a program in GLIM.

We fitted the sigmoid curve of Equation 3.14 to the data of Figure 3.7 by logit regression. Table 3.5 shows the estimated parameter and the residual deviance; its number of degrees of freedom is $n - q$, where q is the number of estimated parameters (Subsection 3.2.1). The resulting curve (Figure 3.7) does not differ significantly ($P > 0.05$) from a horizontal line, as judged by a t test of whether b_1 equals 0. All tests in logit regression are approximate, because the error distribution is not normal (cf. the chi-square test of Subsection 3.3.1). Apart from this, there is no difference from the t test described in Subsection 3.2.2.

Table 3.5 Sigmoid curve fitted by logit regression: parameter estimates and deviance table for the presence–absense data of Figure 3.7.

Term	Parameter	estimate	s.e.	t
Constant	b_0	2.03	1.98	1.03
pH	b_1	–0.484	0.357	–1.36

		d.f.	deviance	mean deviance
Residual		33	43.02	1.304

3.3.3 Gaussian logit curves

When we take for the linear predictor in Equation 3.16 a parabola, we obtain the Gaussian logit curve

$$p = [\exp (b_0 + b_1 x + b_2 x^2)]/[1 + \exp (b_0 + b_1 x + b_2 x^2)]$$
$$= c \exp [-0.5 (x - u)^2/t^2]/ [1 + c \exp (-0.5 (x - u)^2/t^2)] \qquad \text{Equation 3.17}$$

The third form of the equation follows from Equations 3.8-3.10 and shows the relation to the Gaussian curve (Equation 3.8). The relation between parabola, Gaussian curve and Gaussian logit curve is shown graphically in Figure 3.9 (contrast Figure 3.8). The Gaussian logit curve has a flatter top than the Gaussian curve but the difference is negligible when the maximum of the Gaussian logit curve is small (< 0.5). The Gaussian logit curve was fitted to the data in Figure 3.7 by using GENSTAT and the result is shown in Figure 3.10. Table 3.6 gives the parameter estimates of b_0, b_1 and b_2, from which we obtain estimates for the optimum and the tolerance by using Equations 3.11a,b. The result is $u = 5.28$ and $t = 0.327$. The maximum of the fitted curve in Figure 3.10 is the (estimated) maximum probability of occurrence of the species (p_{max}) and can be calculated

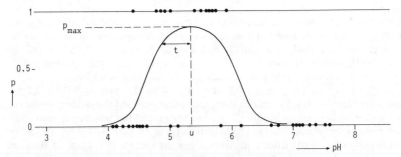

Figure 3.10 Gaussian logit curve fitted by logit regression of the presences (● at $p = 1$) and absences (● at $p = 0$) of a species on acidity (pH). Same data as in Figure 3.7. u = optimum; t = tolerance; p_{max} = maximum probability of occurrence.

Table 3.6 Gaussian logit curve fitted by logit regression: parameter estimates and deviance table for the presence–absence data of Figure 3.10. The data are the same as in Figure 3.7.

Term		Estimate	s.e.	t
Constant	b_0	−128.8	51.1	−2.52
pH	b_1	49.4	19.8	2.50
pH^2	b_2	4.68	1.90	−2.47
		d.f.	deviance	mean deviance
Residual		32	23.17	0.724

from the second form of Equation 3.17 by inserting the value of u (5.28) for x and the values of b_0, b_1 and b_2 from Table 3.6; we obtain $p_{max} = 0.858$.

We can decide whether the Gaussian logit curve significantly improves the fit over the sigmoid curve by testing whether b_2 equals 0. Here we use the t test again (Subsection 3.2.3). The t of b_2 is −2.47 (Table 3.6) and we conclude that the fitted curve differs significantly from a sigmoid curve. It is justified to use a one-tailed t test here, if we only want to detect unimodal curves, i.e. curves with $b_2 < 0$ (Snedecor & Cochran 1980, Section 5.5). If b_2 is significantly smaller than 0, then the optimum is said to be significant. An approximate 95% confidence interval for u is (5.0, 5.8), obtained from Section 3.9.

A more general method of statistical testing in GLM is by the deviance test, in which the residual deviance of a model is compared with that of an extended model. The additional parameters in the latter model are significant when the drop in residual deviance is larger than the critical value of a chi-square distribution with k degrees of freedom, k being the number of additional parameters. As an example, the drop in deviance going from the sigmoid curve to the Gaussian logit curve (Tables 3.5 and 3.6) is $43.02 - 23.17 = 19.85$. This drop is larger than $\chi^2_{0.05}(1) = 3.84$. Hence the single additional parameter b_2 is significant. The deviance test replaces the F test of least-squares regression.

An example of analysing presence–absence data is provided in Exercises 3.4 and 3.5.

3.4 Regression for abundance data with many zero values

Abundance data with many zero values (i.e. absence) always show a skew distribution. So one should transform them before analysing them by least-squares regression. But the logarithmic transformation does not work, because the logarithm of zero is undefined. The value 0 might be caused by rounding error, but even then one often does not know whether the original value was 0.1, 0.01 or even smaller. On a log scale the difference between these values is large, and one does not know which value to choose. A common practice is to add a small value to the abundance data before logs are taken, as was done in Subsection 3.2.1, but this is somewhat arbitrary; different values may lead to different results of analysis if there are many zeros among the data. An additional problem is that in the model abundance values may be negative, which does not make sense (e.g. the prediction interval for clay in Figure 3.4). Other transformations do not work either.

In least-squares regression after logarithmic transformation, the implicit assumption is that the abundance data follow a log-normal distribution. The probability of observing the value 0 from a log-normal distribution is, however, zero. A distribution that allows zero values is the Poisson distribution (Subsection 2.4.3). Observations arising from a Poisson distribution can take the integer values 0, 1, 2, 3, ... and have a variance that is equal to the mean. Counts of the number of animals in a region, for example, take integer values only. We assume for a moment that the data follow a Poisson distribution and seek appropriate response curves. The curves must not be negative, but may rise above the value 1. The exponential transformation used in Equation 3.13 is therefore sufficient. The exponential curve can be fitted to data by log-linear regression, which is again a special case of GLM (Subsection 3.3.2). The regression is termed log-linear because another way of writing Equation 3.13 is

$$\log_e \mathrm{E}y = \text{linear predictor} \hspace{4cm} \text{Equation 3.18}$$

By using $b_0 + b_1 x + b_2 x^2$ in the linear predictor, we again obtain the Gaussian curve provided $b_2 < 0$ (Equations 3.8-3.11). The Gaussian curve can thus be fitted to abundance data with zero values by carrying out a log-linear regression. In this way we circumvent the problem of having to take logarithms of zeros. The optimum, tolerance and maximum are derived from the estimates of b_0, b_1 and b_2 as in Subsection 3.2.3.

The assumption that abundance data follow a Poisson distribution is often false (Subsection 2.4.3). Fortunately, the assumptions of log-linear regression can be relaxed. It is sufficient that the variance in the data is proportional to the mean (McCullagh & Nelder 1983). When this weaker assumption is also inappropriate, a possible ad-hoc method is to transform the species data to presence–absence. This method sacrifices all the quantitative information. The quantitative information can be retained partly by also analysing 'pseudo-species' (Hill et al. 1975). A pseudo-species is a presence–absence variable that is defined, for instance, by a cut-level value y_c. The pseudo-species at cut-level value y_c is present if the abundance of the species exceeds the cut-level value y_c, and is absent if the abundance is less. By choosing a set of cut levels, we get a set of pseudo-species, each of which can be analysed separately by logit regression. An attractive property of the method of pseudo-species is that the response curve of each pseudo-species is unimodal whenever the response curve for the original abundances is unimodal. Then, the tolerances of the response curves of the pseudo-species decrease with increasing value of the cut level; their optima may shift when the response curve for abundance is asymmetric. A disadvantage of the method is that the choice of cut levels is arbitrary and that the results of the separate analyses cannot be combined easily into a simple description of the relation between the abundance of the species and the environmental variable under consideration.

50

In some ecological applications the quantitative information on abundance is of the type 'absent, a few, many'. We suggest transforming such data to 'Is the species present?' and 'Is the species abundant?', and to analyse each variable separately by logit regression. The second variable is a pseudo-species.

3.5 Multiple regression

3.5.1 Introduction

In the previous sections, the response variable was expressed in various ways as a function of a single environmental variable. A species may, however, respond to more than one environmental variable. To investigate such a situation, we need multiple regression. In multiple regression, the response variable is expressed as a function of two or more explanatory variables (response-surface analysis). Separate analyses of the response for each of the environmental variables cannot replace multiple regression if the environmental variables show some correlation with one another and if there are interaction effects, i.e. if the effect of one variable depends on the value of another variable.

We will show how least-squares regression and logit regression can be extended to study the effect of two environmental variables. The extension to more than two variables will then be obvious. Typical cases of multiple regression will be illustrated in the section on multiple logit regression, although they occur equally in multiple least-squares regression. In separate subsections, we will discuss the analysis of interaction effects and the inclusion of nominal explanatory variables in multiple regression.

3.5.2 Multiple least-squares regression: planes and other surfaces

An extension of the straight line to two explanatory variables is a plane (Figure 3.11). A plane has the formula

$$Ey = b_0 + b_1 x_1 + b_2 x_2 \qquad\qquad \text{Equation 3.19}$$

where
x_1 and x_2 are two explanatory variables
b_0, b_1 and b_2 are parameters or regression coefficients.

b_0 is the expected response when $x_1 = 0$ and $x_2 = 0$. b_1 and b_2 are the rates of change in the expected response along the x_1 and x_2 axes, respectively. b_1 thus measures the change in Ey with x_1 for a fixed value of x_2, and b_2 the change in Ey with x_2 for a fixed value of x_1.

The parameters are, again, estimated by the least-squares method, i.e. by minimizing the sum of squares of the differences between the observed and expected response. This means in geometric terms (Figure 3.11) that the regression plane is chosen in such a way that the sum of squares of the vertical distances between the observed responses and the plane is minimum.

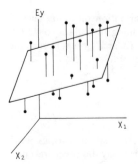

Figure 3.11 Three-dimensional view of a plane fitted by least-squares regression of responses (●) on two explanatory variables x_1 and x_2. The residuals, i.e. the vertical distances between the responses and the fitted plane are shown. Least-squares regression determines the plane by minimization of the sum of these squared vertical distances.

A multiple regression analysis carried out by computer not only gives estimates for b_0, b_1 and b_2, but also standard errors of the estimates and associated values of t (Table 3.3). Fitting a parabola is a special case of multiple regression analysis where $x_1 = x$ and $x_2 = x^2$. The values of t can be used to test whether a coefficient is zero (Subsection 3.2.1), i.e. whether the corresponding variable contributes to the fit of the model in addition to the fit already provided by the other explanatory variable(s).

By extending the parabola we obtain the quadratic surface

$$Ey = b_0 + b_1 x_1 + b_2 x_1^2 + b_3 x_2 + b_4 x_2^2 \qquad \text{Equation 3.20}$$

which has five parameters. When y in this model is the logarithm of abundance, we are fitting through multiple regression a bivariate Gaussian response surface to the observed abundances, provided b_2 and b_4 are both negative. With t tests, we can see whether one of the parameters is equal to zero. In particular, to detect whether the surface is unimodal in the direction of x_1, we test the null hypothesis ($b_2 \geqslant 0$) against the alternative hypothesis ($b_2 < 0$) through the t corresponding to the coefficient b_2, as in Subsection 3.3.3. Similarly, we use the t corresponding to b_4 to test whether the surface is unimodal in x_2.

The optimum and tolerance of the species with respect to x_1 are calculated as in Subsection 3.2.3 by inserting in Equation 3.11a,b the values of b_1 and b_2 obtained from fitting Equation 3.20. Standard errors and the confidence interval for the optimum can still be obtained by using the equations given in Section 3.9. The optimum and tolerance with respect to x_2 are obtained analogously by replacing b_1 by b_3 and b_2 by b_4.

To investigate whether x_2 in this model influences the abundance of a species in addition to x_1, we need to test whether both b_3 and b_4 equal 0. This test requires simultaneous testing of two parameters, which cannot be done with two

52

separate t tests. For this, we need the F test. For an F test, we must fit two regression equations, a simple one with only x_1 and x_1^2 and an extended one, in which x_2 and x_2^2 are added, and compare the residual sum of squares, RSS_1 and RSS_2, respectively, by calculating

$$F = [(RSS_1 - RSS_2)/(df_1 - df_2)]/ (RSS_2/df_2) \qquad \text{Equation 3.21}$$

where df_1 and df_2 are the degrees of freedom of RSS_1 and RSS_2, respectively.

Under the null hypothesis that the additional parameters b_3 and b_4 equal 0, F follows an F distribution with $df_1 - df_2$ and df_2 degrees of freedom (Subsection 3.2.1). The null hypothesis is rejected if the calculated F exceeds the critical value of this distribution. This test can be used whenever simple and extended models are to be compared in multiple least-squares regression. Our previous applications of the F test were special cases of Equation 3.21, in which the simple model was the no-effect model 'Ey is constant'.

3.5.3 Multiple logit regression: logit planes and Gaussian logit surfaces

In Subsection 3.3.2, logit regression was obtained from least-squares regression by replacing Ey by $\log_e [p/(1 - p)]$ and there is no reason not to do so in multiple regression. This replacement transforms the plane of Equation 3.19 into a logit plane defined by the equation

$$\log_e [p/(1 - p)] = b_0 + b_1 x_1 + b_2 x_2 \qquad \text{Equation 3.22}$$

We will now show what multiple regression can add to the information provided by separate regressions with one explanatory variable. Figure 3.12 displays the values of x_1 and x_2 in a sample of 35 sites and also shows which sites an imaginary species is present at. Fitting Equation 3.22 to the data by using GLM (Subsection 3.3.2) gives the results shown in the first line of Table 3.7. Judged by t tests, both b_1 and b_2 differ significantly from 0, and we conclude that the presence of the species depends both on x_1 and x_2. By fitting a model with x_1 only (Equation 3.14), we obtain the second line of Table 3.7. The estimated probability of occurrence increases somewhat with x_1 ($b_1 = 0.16$), but not significantly (the t of b_1 is 1.33). We would thus have concluded wrongly that the presence of the species did not depend on x_1. From fitting a model with x_2 only, we would also have concluded wrongly that x_2 was irrelevant for predicting presence of the species. By comparing the residual deviances of the models fitted (Table 3.7), we see that x_1 and x_2 are good explanatory variables only when taken together. Such variables are said to be complementary in explanatory power (Whittaker 1984).

The values of b_1 and b_2 in the multiple regression clearly describe the pattern of species occurrence in Figure 3.12. In words, for any given value of x_2, the probability of occurrence strongly increases with x_1, and for any given value of x_1, it strongly decreases with x_2. A line drawn at about 45° in Figure 3.12 actually separates most of the species presences from the absences.

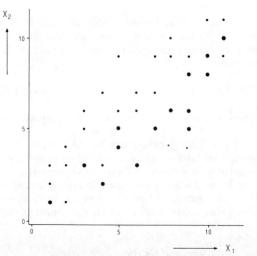

Figure 3.12 Data illustrating that explanatory variables can be complementary in explanatory power. The scatter diagram of x_1 and x_2 shows the sites where a particular species is present (•) and absent (·).

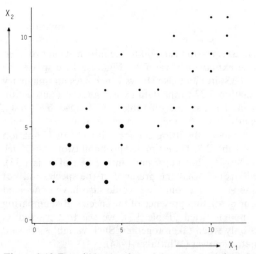

Figure 3.13 Data illustrate that explanatatory variables can replace each other in multiple regression equations. The scatter diagram of x_1 and x_2 shows the sites where a particular species is present (•) and absent (·).

Table. 3.7 Multiple logit regressions of the data of Figure 3.12 to illustrate that explanatory variables can be complementary in explanatory power (d(res) = residual deviance).

Terms in model	b_1	b_2	t value of b_1	t value of b_2	d(res)	d.f.
x_1, x_2	1.53	–1.66	2.98	–2.96	23.99	32
x_1	0.16	–	1.33	–	45.25	33
x_2	–	–0.15	–	–1.17	45.69	33
none	–	–	–	–	47.11	34

Table 3.8 Multiple logit regressions of the data of Figure 3.13 to illustrate that explanatory variables can substitute each other in a model (d(res) = residual deviance).

Terms in model	b_1	b_2	t value of b_1	t value of b_2	d(res)	d.f.
x_1, x_2	–0.61	–0.625	–1.63	–1.59	17.47	32
x_1	–0.94	–	–2.85	–	20.57	33
x_2	–	–1.016	–	–2.88	20.82	33
none	–	–	–	–	41.88	34

Figure 3.13 shows the occurrence of another species. When x_1 and x_2 are used to explain this species' occurrence, the t 's (first line of Table 3.8) show that neither b_1 nor b_2 differs significantly from 0. It should not be concluded now that neither x_1 nor x_2 has an effect on the species' presence. These t tests only say that we do not need both x_1 and x_2 in the model. The fits with x_1 only and with x_2 only show that, taken singly x_1 and x_2 have both an effect. Moreover, these fits give about the same deviance; hence, x_1 can substitute x_2 in the model (Whittaker 1984). We observe in Figure 3.13 that the species occurs at low values of x_1 and x_2, but cannot say which variable this is caused by, because there were too few sites where x_1 was low and x_2 high or vice versa. We cannot distinguish their effects. This problem often arises when explanatory variables are highly correlated in the sample. This problem is known as the multicollinearity problem. For example, we may wish to know whether the probability of occurrence of a certain rare meadow plant decreases with potassium or with phosphate. But, in a survey potassium and phosphate will be strongly correlated, because they are usually applied simultaneously; so the question cannot be answered by a survey. Multicollinearity also arises when the number of explanatory variables is only slightly less than the number of sites.

Figures 3.12 and 3.13 illustrate the cases that create the most surprise at first. Less surprising are the cases in which neither multiple regression nor separate regressions show up any effects, or in which the techniques demonstrate the same

55

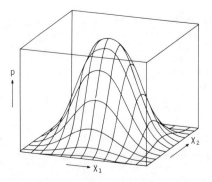

Figure 3.14a Three-dimensional view of a bivariate Gaussian logit surface with the probability of occurrence (p) plotted vertically and the two explanatory variables x_1 and x_2 plotted in the horizontal plane.

effects. Finally, it may also happen that both x_1 and x_2 show an effect on the species in the separate regressions, whereas in multiple regression only one of them shows an effect. This happens, for example, when x_1 is the only effective variable, and x_2 is correlated with x_1. The possible effect of x_2 in the regression with x_2 only is then due to its correlation with x_1, as multiple regression may show.

In multiple regression with more than two explanatory variables, all the previous cases may occur together in one analysis. Further, instead of pairs of variables that are substitutable or complementary, we may have triplets, quadruplets, etc. (Whittaker 1984). These concepts are important when one wants to select the best set of explanatory variables in a regression equation (Montgomery & Peck 1982; Whittaker 1984).

We now proceed to quadratic models. By inserting the quadratic surface of Equation 3.20 in Equation 3.15, we obtain a bivariate Gaussian logit surface, provided both b_2 and b_4 are negative (Figure 3.14a). This surface has ellipses as contour lines (lines of equal probability) with main axes parallel to the x_1 and x_2 axis (Figure 3.14b). The parameters of this models can again be estimated by GLM. Further analysis proceeds as from Equation 3.20, except that the F test must be replaced by the deviance test (Subsection 3.3.3).

3.5.4 Interaction between explanatory variables

Two explanatory variables show interaction of effects if the effect of the one variable depends on the value of the other. We can test for interaction by extending regression equations with product terms, like $x_1 x_2$.

By extending Equation 3.19 in this way, we obtain

$$Ey = b_0 + b_1 x_1 + b_2 x_2 + b_3 x_1 x_2 = (b_0 + b_2 x_2) + (b_1 + b_3 x_2)x_1 \qquad \text{Equation 3.23}$$

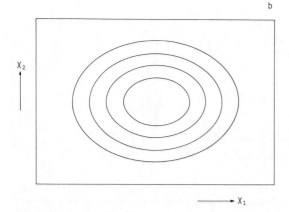

X_2

X_1

Figure 3.14b Elliptical contours of the probability of occurrence p plotted in the plane of the explanatory variables x_1 and x_2. One main axis of the ellipses is parallel to the x_1 axis and the other to the x_2 axis.

The final expression in Equation 3.23, obtained by simple algebra, shows that the relation between Ey and x_1 in this model is still a straight line, but that the intercept and slope and hence the effect of x_1 depend on the value of x_2. Conversely, the effect of x_2 depends on the value of x_1. The parameters b_1, b_2 and b_3 in Equation 3.23 can be estimated by using any multiple regression program and calculating the new variable $x_3 = x_1 x_2$ and specifying x_1, x_2 and x_3 as the explanatory variables. The interaction can be tested by a t test whether b_3 equals 0.

By extending the Gaussian model of Equation 3.20 with a product term, we obtain in the logit case

$$\log_e [p/(1 - p)] = b_0 + b_1 x_1 + b_2 x_1^2 + b_3 x_2 + b_4 x_2^2 + b_5 x_1 x_2 \qquad \text{Equation 3.24}$$

If $b_2 + b_4 < 0$ and $4 b_2 b_4 - b_5^2 > 0$, Equation 3.24 describes a unimodal surface with ellipsoidal contours as in Figure 3.14b, but without the restriction that the main axes are horizontal or vertical. If one of these conditions is not satisfied, it describes a surface with a single minimum or one with a saddle point (e.g. Carroll 1972). When the surface is unimodal, the overall optimum (u_1, u_2) can be calculated from the coefficients in Equation 3.24 by

$$u_1 = (b_5 b_3 - 2 b_1 b_4)/d \qquad \text{Equation 3.25a}$$

$$u_2 = (b_5 b_1 - 2 b_3 b_2)/d \qquad \text{Equation 3.25b}$$

where $d = 4 b_2 b_4 - b_5^2$.

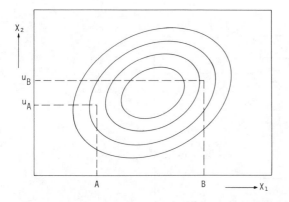

Figure 3.15 Interaction in the Gaussian logit model. The elliptical contours of the probability of occurrence p with respect to the explanatory variables x_1 and x_2 are shown. The main axes of the ellipses are not parallel to either the x_1 axis or the x_2 axis. u_A and u_B are the optima with respect to x_2 that correspond to levels A and B of x_1.

The optimum with respect to x_1 for a given value of x_2 is $-(b_1 + b_5 x_2)/(2 b_2)$ and thus depends on the value of x_2 if $b_5 \neq 0$. The expression is obtained by rearranging Equation 3.25 in the form of a parabola and using Equations 3.10 and 3.11. Figure 3.15 clearly shows this interaction. We can test this interaction by using a t test whether b_5 equals 0.

3.5.5 Nominal explanatory variables

Multiple regression can also be used to study the simultaneous effect of nominal environmental variables or of both quantitative and nominal environmental variables. To show how nominal variables may enter the multiple regression equation, we express the ANOVA model of Subsection 3.2.1 as a regression equation. In the example of Subsection 3.2.1, the nominal variable soil type had three classes: clay; peat; sand. We take clay as the reference class and define for peat and sand two dummy variables, x_2 and x_3, the values of which are either 0 or 1. The dummy variable for peat x_2 takes the value 1 when the site is on peat and the value 0 when the site is on clay or sand. The dummy variable for sand x_3 takes the value 1 when the site is on sand and the value 0 when the site is on clay or peat. A site on clay thus scores the value 0 for both dummy variables, a site on peat scores the value 1 for x_2 and 0 for x_3, etc. The systematic part of the model of Subsection 3.2.1 can be written as

$$Ey = b_1 + b_2 x_2 + b_3 x_3 \qquad \text{Equation 3.26}$$

The coefficient b_1 gives the expected response on the reference class clay, the coefficient b_2 the difference in expected response between peat and clay, and coefficient b_3 the difference between sand and clay. The coefficients b_1, b_2 and b_3 can be estimated by multiple least-squares regression. For the data of Figure 3.4, we obtain $b_1 = 1.70$, $b_2 = 1.47$, $b_3 = 0.63$. The mean is then on clay $b_1 = 1.70$, on peat $b_1 + b_2 = 3.17$ and on sand $b_1 + b_3 = 2.33$, as can be checked with Table 3.1. The ANOVA table of this multiple regression analysis is precisely that of Table 3.1. When a nominal variable has k classes, we simply specify $k - 1$ dummy variables (Montgomery & Peck 1982, Chapter 6).

The next example concerns the presence–absence of the plant species *Equisetum fluviatile* in fresh water ditches in the Netherlands. We will investigate the effect of electrical conductivity (mS m^{-1}) and of soil type (clay, peat, sand) on the species by logit regression, using the model

$$\log_e [p/(1 - p)] = b_0 + b_1 x_1 + b_2 x_1^2 + b_3 x_2 + b_4 x_3 \qquad \text{Equation 3.27}$$

where x_1 is the logarithm of electrical conductivity and x_2 and x_3 are the dummy variables defined in the previous example. Here b_3 and b_4 represent the effect of the nominal variable soil type. Figure 3.16 shows that this model consists of three curves with different maxima but with identical optima and tolerances. The coefficient b_3 is the difference between the logits of the maxima of the curves for peat and the reference class clay; the coefficient b_4 is the analogous difference between the curves for sand and clay. We can test whether the maxima of these curves are different by comparing the residual deviance of the model with x_1 and x_1^2 with the residual deviance of Equation 3.27. The difference is a chi-square with two degrees of freedom if soil type has no effect. This is another example of the deviance test.

To calculate the optimum and tolerance in Equation 3.27, we simply use Equation 3.11; to calculate standard errors and a confidence interval for the optimum, we can use the equations of Section 3.9. Exercise 3.6 may serve as an example.

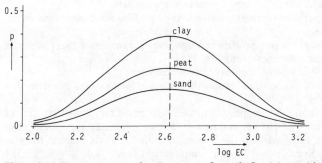

Figure 3.16 Response curves for *Equisetum fluviatile* fitted by multiple logit regression of the occurrence of *E. fluviatile* in freshwater ditches on the logarithm of electrical conductivity (EC) and soil type surrounding the ditch (clay, peat, sand). Data from de Lange (1972).

3.6 Model choice and regression diagnostics

Many things can go wrong in regression analysis. The type of response curve or the error distribution may have been chosen incorrectly and there may be outliers that unduly influence the regression. To detect such faults is the purpose of regression diagnostics (Belsley et al. 1980; Cook & Weisberg 1982; Hocking & Pendleton 1983). What we can do, for instance, is to plot the residuals of a regression against the fitted values or against each of the explanatory variables and look for outliers and systematic patterns in these plots. The references just given deal mainly with regression diagnostics for quantitative response variables. Here we focus on presence–absence data and response curves of species.

One would like to base the shape of a response curve of a species on physiological and ecological theory. But there is no generally accepted theory (Austin 1980) and therefore no ubiquitously applicable response curve. In the absence of theory, one can still proceed by empirical methods and decide upon an applicable curve on the basis of many empirical results. Early studies by Gause (1930), Curtis & Mcintosh (1951) and Whittaker (1956) showed that monotonic response curves are too simple as an ecological response model and that a unimodal model is more appropriate. Simple ecological reasoning shows that also bimodal curves are a realistic option: a species can be outcompeted near its physiological optimum by more competitive species whereas the species may be able to cope with less favourable environmental conditions when competition is less. The response curve applicable to field conditions is then the result of the physiological response curve and competition between species (Fresco 1982). Hill (1977) suggested, however, that a good ecological variable, minimizes the occurrence of bimodal species distributions.

When there are no ideas a priori of the shape of the response curve, one can best divide the quantitative environmental variable into classes and calculate the frequency of occurrence for each class as in Subsection 3.3.1 (Gounot 1969; Guillerm 1971). By inspection of the profiles of the frequencies for several species, one may get an idea which type of response curve is appropriate.

Curves have several advantages over frequency profiles for quantitative environmental variables:

- curves when kept simple, provide through their parameters a more compact description than frequency profiles
- there is no need to choose arbitrary class boundaries
- there is no loss of information because the environmental variable is not divided into classes
- when the Gaussian model applies, statistical tests based on curves have greater power to detect that the environmental variable influences the species than the chi-square test based on the frequency profile. This is because the chi-square test of Subsection 3.3.1 is an omnibus test that is able to detect many types of deviations from the null hypothesis, whereas the t tests and deviance tests of Subsections 3.3.2, 3.3.3 and Section 3.5 test the null hypothesis against a specified alternative hypothesis.

A clear disadvantage of curves is that one is forced to choose a model which

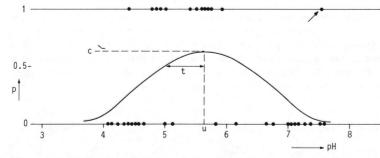

Figure 3.17 The change in a fitted Gaussian logit curve by adding an influential point. Adding a single presence at pH = 7.6 (indicated by an arrow) to Figure 3.10 considerably decreases the estimated maximum and increases the estimated tolerance and optimum.

may be wrong for the data at hand. For example, is the true response curve symmetric? When asymmetry is suspected, one can transform the explanatory variable, for example by taking logarithms, and one can compare the residual deviances before and after transformation. The detection of a deviation from a supposed response curve may aid our understanding of the relation of the species with the environment and in general triggers off a new cycle in the process of model building.

Data points that unduly influence the regression require special attention with presence–absence data. For example, adding a presence to Figure 3.10 at pH 7.6 drastically changes the fitted response curve (Figure 3.17). When there are two or more explanatory variables, we suggest you plot the variables in pairs as in Figures 3.12 and 3.13 and inspect the plots for outlying presences. When such an outlier is spotted, you must attempt to find out whether it is a recording error or whether the site was atypical for the conditions you intended to sample, and decide after such attempts whether or not to retain the outlier in the data. We also suggest that you always try to remove the lowest or highest x where the species is present to check that the fitted response stays roughly the same (cf. the jackknife technique, Efron 1982).

3.7 The method of weighted averaging

This section is devoted to estimation of species indicator values (Ellenberg 1982). In terms of response curves, there are two possible definitions of species indicator value: it is either the optimum or the centroid of the species response curve. These definitions coincide only if the response curve is symmetric. In Subsections 3.2.3 and 3.5.2, we have shown how an optimum can be estimated by fitting a curve or a surface to the species data by regression. In the regression method, we have to assume a particular response curve. Ecologists have long used a simpler method for estimating indicator values (Ellenberg 1948; 1979). This is the method of weighted averaging, which circumvents the problem of having to fit a response

61

curve. When a species shows a unimodal curve against a particular environmental variable, the presences of the species will most frequently occur near the optimum of the curve. An intuitively reasonable estimate of the indicator value is therefore obtained by taking the average of the values of the environmental variable over those sites where the species is present. For abundance data, a weighted average may be taken in which values are weighted proportional to the species' abundance, i.e.

$$u^* = (y_1 x_1 + y_2 x_2 + ... + y_n x_n)/(y_1 + y_2 + ... + y_n)$$ Equation 3.28

where
u^* is the weighted average
$y_1, y_2, ..., y_n$ are the abundances of the species
$x_1, x_2, ..., x_n$ the values of the environmental variable at the Sites 1, 2 ... n.

The weighted average disregards species absences. An unpleasant consequence of this is that the weighted average depends on the distribution of the environmental variable in the sample (Figure 3.18). Highly uneven distributions can even scramble the order of the weighted averages for different species (Figure 3.18).

Ter Braak & Looman (1986) compared the performance of the methods of weighted averaging and of Gaussian logit regression to estimate the optimum of a Gaussian logit curve from presence–absence data. Through simulation and practical examples, they showed that the weighted average is about as efficient as the regression method for estimating the optimum:
- when a species is rare and has a narrow ecological amplitude
- when the distribution of the environmental variable among the sites is reasonably homogeneous over the whole range of occurrence of the species along the environmental variable.

In other situations, weighted averaging may give misleading results (Exercise 3.2.8). Similar conclusions also hold for quantitative abundance data; for quantitative abundance data, the weighted average efficiently estimates the optimum of the Gaussian response curve, if the abundances are Poisson-distributed and the sites are homogeneously distributed over the whole range of the species.

Despite its deficiencies, the method of weighted averaging is a simple and useful method to show up structure in a data table such as Table 0.1 by rearranging species and sites on the basis of an explanatory variable. As an example, we shall demonstrate this by rearranging the Dune Meadow Data in Table 0.1 on the basis of the moisture value of the sites (relevés). For each species, we calculate its weighted average for moisture, e.g. for *Aira praecox*

$$u^* = (2 \times 2 + 3 \times 5)/(2 + 3) = 3.8$$

and arrange the species in order of the values so obtained and the sites in order of their moisture value (sites with equal moisture are arranged in arbitrary order). The result is shown in Table 3.9. *Plantago lanceolata* is clearly restricted to the driest sites, *Ranunculus flammula* to the wettest sites, and *Alopecurus geniculatus*

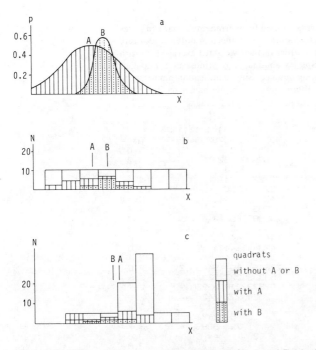

Figure 3.18 The response curves of imaginary species A and B (a); the occurrence of these species in two samples of 80 sites, in which the environmental variable is distributed evenly (b) or unevenly (c). The weighted averages are indicated with lines. The two sampling designs yield weighted averages that are in reverse order. p = probability of occurence; N = number of sites; x = environmental variable.

to sites with intermediate moisture. In Table 3.9, most of the abundance values (>0) are arranged in a band along the 'diagonal'. The method of weighted averaging tends to show up such a diagonal structure in a table, when species show unimodal curves for the environmental variable. This idea is extended in Section 5.2.

3.8 Bibliographic notes

The least-squares technique dates back to the early nineteenth century with the work of K.F. Gauss. The principle of maximum likelihood was founded by R.A. Fisher in the 1920s. The generalized linear model (GLM) was introduced by Nelder & Wedderburn (1972) and made it easy to fit a major class of non-linear models to data. Among the many statistical textbooks on least-squares regression are Draper & Smith (1981), Seber (1977), Montgomery & Peck (1982) and Mosteller & Tukey (1977). Useful more general statistical texts for biologists are Parker (1979), Sokal & Rohlf (1981) and Snedecor & Cochran (1980). Dobson

63

Table 3.9 Weighted averaging used for rearranging species and sites in Table 0.1. The sites (columns) are arranged in order of moisture and the species (rows) in order of their weighted average ($u*$) with respect to moisture. Species abundance is printed as a one-digit number, a blank denoting absence. Site identification numbers are printed vertically. For abbrevations of species names see Table 0.1.

```
      species              sites
                              11   11 1 111112
                           1256718340792834569O     u*
26  Tri  pra               252                      1.0
18  Pla  lan               55533    32             1.2
28  Vic  lat                 21 1                  1.3
 1  Ach  mil               13222    42             1.4
 6  Bel  per               32    2222              1.5
 7  Bro  hor               42 2   34               1.5
23  Rum  ace                563        22          1.7
17  Lol  per               7526672656 2 4          1.7
 9  Cir  arv                       2               2.0
11  Ely  rep               444    44 6             2.0
19  Poa  pra               442344354414 42         2.0
 5  Ant  odo                432     44       4     2.1
20  Poa  tri               27645  654 5449 2       2.6
16  Leo  aut               5333552232223222 62     2.6
27  Tri  rep               525232216 332261 2      2.7
29  Bra  rut                26246222 242  4434     2.9
13  Hyp  rad                  2   2        5       3.4
24  Sag  pro                  2 5  2422    3       3.5
 4  Alo  gen               2     72  3855 4        3.7
15  Jun  buf               2       44 3            3.8
 3  Air  pra                      2         3      3.8
25  Sal  rep                3             35       3.9
 2  Agr  sto                  48  3445447 5        4.1
14  Jun  art                     4 4   33 4        4.8
 8  Che  alb                      1                5.0
10  Ele  pal                   4 458 4            5.0
12  Emp  nig                        2             5.0
21  Pot  pal                     22               5.0
22  Ran  fla                     22222 4          5.0
30  Cal  cus                      4 3 3           5.0

      MOISTURE             11111112222445555555
```

(1983) and McCullagh & Nelder (1983) provide an introduction to GLM.

A major contribution to the analysis of species–environment relations was made by Whittaker (1956; 1967). His direct gradient analysis focused on response curves and surfaces of species with respect to a complex of environmental variables, that changed gradually in geographic space. The term 'gradient' therefore then had a geographical meaning, but in recent use the term is equivalent to 'environmental variable'. Whittaker used simple smoothing methods to fit the curves and surfaces. Following Gleason (1926), Ramensky (1930) and Gause (1930), he stressed that species react 'individualistically' to environmental variables and that response surfaces of species are often unimodal. Whittaker's view opposed the 'integrated-community hypothesis' of Clements (1928), which viewed communities of species as organisms of a higher scale. The integrated-community hypothesis stimulated much work on succession and on the interrelations between species, disregarding environmental variables. Conversely, the individualistic concept (in its most extreme form, at least) disregards direct relations between species. McIntosh (1981) discussed these apparently contrasting views. Fresco (1982) attempted to incorporate species–environment and inter-species relations into a single regression equation.

Whittaker (1956; 1967) dealt with gradients, i.e. ordinal or quantitative environmental variables. Gounot (1969) and Guillerm (1971) proposed methods similar to that of Subsection 3.3.1, which are applicable for presence–absence species data and nominal environmental variables. They divided environmental variables into classes when the variables were quantitative. Our approach of using logit regression makes it possible to deal with quantitative and nominal variables in a single analysis.

An early ecological example of fitting sigmoid curves to presence–absence data is found in Jowett & Scurfield (1949). They applied probit analysis (Finney 1964), an alternative for logit regression that usually gives similar results. Polynomial least-squares regression was advocated by Yarranton (1969; 1970). He noticed the problem of absences of species (zero abundance values). Austin (1971) stressed the power of regression analysis and gave several examples from plant ecology where abundance data were first transformed logarithmically and then analysed by least-squares regression using parabolas and second-order response surfaces. Alderdice (1972) explained and applied second-order response surfaces in marine ecology. Gauch & Chase (1974) provided a computer program to fit the Gaussian response curve by least squares to ecological data that might include zero abundances. Their approach has become outdated with the advent of GLM. Austin et al. (1984) showed the usefulness of GLM in direct gradient analysis, using log-linear regression and logit regression, with second-order polynomials as linear predictors. We believe that GLM (Section 3.5) should become a standard tool in applied ecology. Response surfaces fitted by GLM are particularly useful in models simulating the impact of various options in environmental management.

3.9 Standard errors of estimated optimum and tolerance; confidence interval for the optimum

We denote the variance of the estimates of b_1 and b_2 in Equations 3.9, 3.17, 3.20 or 3.24 by v_{11} and v_{22} and their covariance by v_{12}. Using Taylor expansion, we calculate the approximate variance of the estimated optimum and tolerance:

$$\text{var}\,(\hat{u}) = (v_{11} + 4\,u\,v_{12} + 4\,u^2\,v_{22})/(4\,b_2{}^2) \qquad \text{Equation 3.28}$$

$$\text{var}\,(\hat{t}) = v_{22}/(-8\,b_2{}^3) \qquad \text{Equation 3.29}$$

An approximate $100(1 - \alpha)\%$ confidence interval for the optimum is derived from Fiellers theorem (Finney 1964, p.27-29). Let t_α be the critical value of a two-sided t test at chosen probability level α with $n - 3$ degrees of freedom, where n is the number of sites. For example, $t = 2.00$ for a 95% confidence interval and 63 sites. Calculate

$$g = t_\alpha^2 v_{22}/b_2^2 \qquad\qquad\qquad \text{Equation 3.30a}$$

and

$$D = 4 b_2^2 \text{ var } (\hat{u}) - g(v_{11} - v_{12}^2/v_{22}). \qquad\qquad \text{Equation 3.30b}$$

$$u_{\text{lower}}, u_{\text{upper}} = [\hat{u} + 0.5 \, g \, v_{12}/v_{22} \pm 0.5 \, t_\alpha \, (\sqrt{D})/b_2]/(1 - g) \qquad \text{Equation 3.31}$$

where the symbol $\pm$ indicates addition and subtraction in order to obtain the lower and upper limits of the confidence interval, respectively. If b_2 is not significantly different from zero ($g > 1$), then the confidence interval is of infinite length and, taken alone, the data must be regarded as valueless for estimating the optimum.

3.10 Exercises

Exercise 3.1 Straight line regression

In a study of the impact of acid rain on diatoms, van Dam et al. (1981) collected data on diatom composition and water chemistry in Dutch moorland pools. For each sample, a total of 400 diatomaceous frustules were identified under a microscope. The numbers of frustules of the species *Frustulia rhomboides* var. *saxonica* and the relative sulphate concentrations $S_{\text{rel}} = [SO_4^{2-}]/([Cl^-] + [SO_4^{2-}] + [HCO_3^-])$ in the 16 samples taken in 1977 and 1978 were as follows (van Dam et al. 1981, Tables 2 and 5):

pool	V2	B6	B3	B4	V1	B5B	B8	B1	D6	B7	B2	D3	D2	D1	D5	D6
Frustulia count	0	0	14	3	0	5	6	21	62	26	14	48	97	99	28	202
S_{rel}	0.78	0.64	0.69	0.70	0.64	0.77	0.73	0.77	0.58	0.44	0.44	0.37	0.23	0.19	0.31	0.23

Exercise 3.1.1 Construct a graph of the data, plotting $\log_e [(Frustulia$ count) + 1] on the vertical axis. Note that the relation looks linear.

Exercise 3.1.2 Fit a straight line to the data taking $\log_e$ (*Frustulia* count $+$ 1) as the response variable and the relative sulphate concentration as the explanatory variable. Use a pocket calculator or a computer for least-squares regression to verify the following results.

		estimate	s.e.	t
constant	b_0	5.848	0.806	7.26
S_{rel}	b_1	–5.96	1.41	–4.22

ANOVA table

	d.f.	s.s.	m.s.
regression	1	24.34	24.340
residual	14	19.11	1.365
total	15	43.45	2.897

Exercise 3.1.3 What are the systematic part and the error part of the response model fitted in Exercise 3.1.2? What are the fitted value and the residual for Pool B2?

Exercise 3.1.4 What are the residual sum of squares, the residual variance, the residual standard deviation and the fraction of variance accounted for? How many degrees of freedom are there for the residual sum of squares?

Exercise 3.1.5 Calculate a 95% confidence interval for the regression coefficient b_1. Is the estimate of b_1 significantly ($P < 0.05$) different from 0?

Exercise 3.1.6 Estimate the expected responses when the relative concentrations of sulphate equal 0.25, 0.50 and 0.75. Calculate the 95% confidence interval of each of these expected responses. The standard errors of the estimates are 0.49, 0.30 and 0.42, respectively. Back-transform the estimates obtained to counts of *Frustulia*.

Exercise 3.1.7 Calculate 95% prediction intervals when the relative sulphate concentrations are equal to 0.25, 0.50 and 0.75.

Exercise 3.2   Parabola, Gaussian response curve and weighted averaging

In a study aimed at reconstructing past temperatures of the sea-surface from fossil distributions of *Radiolaria*, Lozano & Hays (1976) investigated the relation between different taxa of *Radiolaria* and sea-surface temperature in present-day samples. The following data extracted from their Figure 11 concern the abundance (%) of *Spongotrochus glacialis* and February sea-surface temperature (temp., °C) at 34 sites in the Atlantic and Antarctic Oceans.

site	1	2	3	4	5	6	7	8	9	10	11	12
abundance	12	14	13	22	18	19	7	8	11	15	12	14
temp	0.8	1.1	1.6	1.8	1.7	2.0	1.6	1.9	2.0	2.5	3.7	4.2

site	13	14	15	16	17	18	19	20	21	22	23	24
abundance	16	21	35	30	34	48	47	63	54	62	56	52
temp.	4.1	5.8	6.1	6.6	7.9	10.2	11.0	11.9	12.8	14.8	15.9	18.1

site	25	26	27	28	29	30	31	32	33	34
abundance	41	38	30	18	25	35	37	38	42	41
temp.	16.9	17.1	18.0	18.5	20.0	21.0	19.4	19.8	19.0	21.6

Exercise 3.2.1   Construct a graph of the data, plotting the abundance on the vertical axis. Note that the relation looks unimodal. Plot also the logarithm of abundance against temperature.

Exercise 3.2.2   Use a computer program for least-squares regression to verify the following results. Fitting a parabola to the logarithm of the abundances gives:

		estimate	s.e.	t
constant	b_0	2.119	0.133	15.95
temp.	b_1	0.2497	0.0356	7.01
temp. squared	b_2	−0.00894	0.00164	−5.46

ANOVA table

	d.f.	s.s.	m.s.
regression	2	9.42	4.7101
residual	31	3.06	0.0988
total	33	12.48	0.3783

Exercise 3.2.3   Estimate the expected responses when the temperatures are 5, 10, 15 and 20 °C, calculate the optimum, tolerance and maximum of the fitted parabola and use the results to sketch the fitted parabola.

Exercise 3.2.4   What is the residual standard deviation and the fraction of variance accounted for?

68

Exercise 3.2.5 Calculate a 95% confidence interval for the regression coefficient b_2. Would a straight line be statistically acceptable for these data?

Exercise 3.2.6 Calculate a 95% confidence interval for the optimum using Equation 3.31. Here one needs to know also that covariance between the estimates of b_1 and b_2 equals -0.00005704; the variances required can be obtained from the table of regression coefficients. Hint: write a computer program for the calculations required in order to avoid lengthy hand-calculation.

Exercise 3.2.7 Back-transform the expected responses of Exercise 3.2.3 to abundance and sketch the fitted curve.

Exercise 3.2.8 Calculate (after reading Section 3.7) the weighted average of *Spongotrochus* with respect to temperature, using the abundances and, a second time, using log abundances. Explain the difference from the optimum estimated above. Is the difference large?

Exercise 3.3 *Logit link function*

Verify the equivalence of Equations 3.15 and 3.16 by showing that $\log_e [p/(1 - p)] = c$ if and only if $p = (\exp c)/(1 + \exp c)$.

Exercise 3.4 *Chi-square test and logit regression*

A sample of 160 fields of meadow is taken to investigate the occurrence of the grass species *Elymus repens* in relation to agricultural use (hayfield or pasture). The data, based on the study of Kruijne et al. (1967), are summarized in the following 2×2 table of number of fields.

E. repens	agricultural use		
	hayfield	pasture	total
present	12	96	108
absent	16	36	52
total	28	132	160

Exercise 3.4.1 Estimate the probability of occurrence of *E. repens* in hayfield and in pasture.

Exercise 3.4.2 Is there evidence that the probability of occurrence in hayfield differs from that in pasture? Apply here the chi-square test of Subsection 3.3.1, using a significance level of 5%.

Exercise 3.4.3   Instead of the chi-square test we can use logit regression of the presences and absences of *E. repens* in the 160 fields on the nominal explanatory variable agricultural use. Agricultural use has two classes in this problem and therefore we define a single dummy variable USE, which takes the value 1 if the field is a pasture and the value 0 if the field is a hayfield. A computer program for logit regression gave the following output with the response variable presence–absence of *E. repens*:

		estimate	s.e.	t
constant	c_0	-0.28	0.38	-0.74
USE	c_1	1.27	0.42	3.02
		d.f.	deviance	mean deviance
residual		158	192.9	1.221

The model corresponding to this output is $\log_e [p/(1 - p)] = c_0 + c_1 \times \text{USE}$.

Exercise 3.4.3.1   Calculate from the output the estimates for the probability of occurrence of *E. repens* in hayfield and in pasture. Hint: use Exercise 3.3. Compare the estimates with those of Exercise 3.4.1.

Exercise 3.4.3.2   Show by t test whether the probability of occurrence in hayfield differs from that in pasture. Compare the conclusion with that of Exercise 3.4.2.

Exercise 3.4.3.3   The deviance corresponding to the model $\log_e [p/(1 - p)] = c$ equals 201.7 with 159 degrees of freedom. Apply the deviance test instead of the t test of the previous exercise.

Exercise 3.5   Gaussian logit regression

The acidity (pH) of the fields was recorded also for the sample of the previous exercise. Spatial heterogeneity in acidity was disregarded; pH was the mean of several systematically located points in the field. To investigate the effect of acidity on the occurrence of *E. repens*, a Gaussian logit regression was carried out. The results were:

		estimate	s.e.	t
constant	b_0	-57.26	15.4	-3.72
pH	b_1	19.11	5.3	3.61
pH^2	b_2	-1.55	0.44	-3.52
		d.f.	deviance	mean deviance
residual		157	176.3	1.123

Exercise 3.5.1 At what pH did *E. repens* occur with the highest probability? Calculate also the tolerance and the maximum probability of occurrence.

Exercise 3.5.2 Calculate from the output the estimated probabilities of occurrence of *E. repens* at pH 4.5, 5.0, 5.5, 6.0, 6.5, 7.0 and 7.5 and use the results to sketch the response curve of *E. repens* against pH.

Exercise 3.5.3 Is the estimated Gaussian logit response curve significantly different ($P < 0.05$) from a sigmoid response curve; hence, is the optimum significant? Hint: use a one-tailed t test.

Exercise 3.6 *Multiple logit regression*

When considered separately, agricultural use and acidity appear to influence the occurrence of *E. repens* in fields (Exercises 3.4 and 3.5). Hayfield and pasture differ, however, in acidity; hayfields tend to be more acid than pastures. It is therefore of interest to investigate whether this difference in acidity between hayfields and pastures can explain the difference in probability of occurrence of *E. repens* between hayfields and pastures. This problem can be attacked by multiple (logit) regression. We fitted the model

$$\log_e [p/(1 - p)] = c_0 + c_1 \, \text{USE} + b_1 \, \text{pH} + b_2 \, \text{pH}^2$$

to the data and obtained the following results:

		estimate	s.e.	t
constant	c_0	−57.82	17.10	−3.38
USE	c_1	−0.04	0.57	−0.07
pH	b_1	19.30	5.81	3.32
pH^2	b_2	−1.56	0.49	−3.18
		d.f.	deviance	mean deviance
residual		156	176.2	1.129

Exercise 3.6.1 Calculate the estimated probabilities of occurrence in hayfields and pastures for pH 5 and for pH 6. Calculate also the optimum pH and the maximum probabilities of occurrence in hayfields and pastures, and the tolerance. Compare the results with those of Exercise 3.5.1 and 3.5.2, and sketch the response curves.

Exercise 3.6.2 Show by a t test whether the probability of occurrence in hayfields differs from that in pastures after correction for the effect of acidity. Can acidity account for the difference found in Exercise 3.4.2

71

Exercise 3.6.3 Use the deviance test instead of the t test in Exercise 3.6.2. Does the conclusion change?

Exercise 3.6.4 Show by a deviance test whether acidity has an effect on the probability of occurrence of *E. repens* after correction for the effect of agricultural use. Are the variables acidity and agricultural use substitutable in the sense of Subsection 3.5.3?

3.11 Solutions to exercises

Exercise 3.1 Straight-line regression

Exercise 3.1.3 The systematic part is $Ey = b_0 + b_1 S_{rel}$ and the error part is that the error $(y - Ey)$ follows a normal distribution with mean at zero and a variance that does not depend on S_{rel}. Pool B2 has a count of 14 (hence, $y = 2.71$) and $S_{rel} = 0.44$; hence, the fitted value is $5.848 - 5.96 \times 0.44 = 3.23$ and the residual is $2.71 - 3.23 = -0.52$. The fitted number of *Frustulia* frustules is thus $\exp(3.23) - 1 = 25 - 1 = 24$.

Exercise 3.1.4 From the ANOVA table, we obtain the residual sum of squares 19.11, the residual variance 1.365, the residual standard deviation $\sqrt{1.365} = 1.17$ and the fraction of variance accounted for is $1 - (1.365/2.897) = 0.529$. The residual sum of squares has 14 degrees of freedom.

Exercise 3.1.5 In Equation 3.2 with $t_{0.05}(14) = 2.145$, we insert the estimate for b_1 and its standard error and obtain a lower bound of $-5.96 - (2.145 \times 1.41) = -8.98$ and an upper bound of $-5.96 + (2.145 \times 1.41) = -2.94$. The 95% confidence interval for b_1 is therefore $(-8.98, -2.94)$. The value 0 does not lie in this interval. Alternatively, the t for b_1 (-4.22) is greater in absolute value than the critical t (2.145); hence, the estimate of b_1 is significantly $(P < 0.05)$ different from 0.

Exercise 3.1.6 In a pool with $S_{rel} = 0.25$ the expected response is estimated by $5.848 - 5.96 \times 0.25 = 4.36$. The standard error of this estimate is 0.49 and the 95% confidence interval is therefore $(4.36 - 2.145 \times 0.49, 4.36 + 2.145 \times 0.49) = (3.31, 5.41)$. For $S_{rel} = 0.50$ and 0.75 the estimates are 2.87 and 1.38, with confidence intervals of $(2.23, 3.50)$ and $(0.47, 2.29)$, respectively. Notice that the interval is shortest near the middle of the interval of the relative sulphate values actually sampled. For $S_{rel} = 0.25, 0.50, 0.75$ back-transformation to counts gives the estimates $\exp(4.36) - 1 = 77$, 17 and 3, respectively.

The latter values estimate the median number of frustules at the respective relative sulphate concentrations, and not the expected number of frustules. We assumed that the log-transformed data do follow a normal distribution. In the normal distribution, the mean is equal to the median (50-percentile) and transformations do not change percentiles of a distribution. Back-transforming the limits of the 95% confidence intervals gives 95% confidence intervals for the median counts. For $S_{rel} = 0.25$ this interval is $(26, 223)$.

72

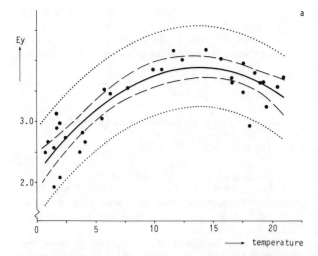

Figure 3.19a Parabola (solid line) fitted by least-squares regression of log-transformed relative abundance of *Spongotrochus glacialis* (●) on February sea-surface temperature (temp). 95% confidence intervals (dashed curve) and 95% prediction intervals (dotted line) are shown. Data from Lozano & Hays (1976).

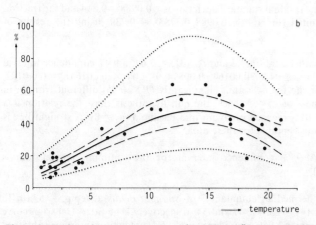

Figure 3.19b Gaussian response curve with 95% confidence and 95% prediction intervals obtained by back-transforming the curves of Figure 3.19a. Vertical axis: abundance (%) of *Spongotrochus glacialis*. Horizontal axis: February sea-surface temperature.

Exercise 3.1.7   With Equation 3.3 and $S_{rel} = 0.25$ we obtain the interval 4.36 $\pm\ 2.145 \times \sqrt{(1.17^2 + 0.49^2)} = 4.36 \pm 2.145 \times 1.27 = (1.63, 7.08)$.
Back-transforming to counts shows that 95% of the counts are expected to lie between 4 and 1187. The latter value is nonsensical as the maximum count is 400.

For $S_{rel} = 0.50$ and 0.75 we obtain 95% prediction intervals for the transformed counts of (0.28, 5.46) and (–1.28, 4.05), respectively.

Exercise 3.2   Parabola, Gaussian response curve and weighted averaging

Exercise 3.2.1   See Figure 3.19a,b.

Exercise 3.2.3   The expected response at temp. = 5 is estimated by
$2.119 + 0.2497 \times 5 - 0.00894 \times 5^2 = 3.14$. For temp. = 10, 15 and 20 the estimates are 3.72, 3.85 and 3.54, respectively. It is of interest to note that the standard errors of the estimates are 0.07, 0.10, 0.09 and 0.11 at temp. 5, 10, 15 and 20, respectively.

With Equations 3.11a and 3.11b, the optimum is estimated by $\hat{u} = -\hat{b}_1/(2\ \hat{b}_2) = -0.2497/(-2 \times 0.00894) = 14.0$, so that the optimum temperature is 14.0 °C and the tolerance by $\hat{t} = 1/\sqrt{(-2\ \hat{b}_2)} = 7.48$, so that the tolerance of temperature is 7.48 °C. The maximum of the parabola (Figure 3.9a) is estimated by $2.119 + 0.2497 \times 14.0 - 0.00894 \times 14.0^2 = 3.86$.

Exercise 3.2.4   The residual standard deviation is $\sqrt{0.0988} = 0.314$ and the fraction of variance accounted for is $1 - (0.0988/0.3783) = 0.739$, using the results of the ANOVA table.

Exercise 3.2.5   With Equation 3.2 and $t_{0.05}(31) = 2.04$, a 95% confidence interval for b_2 is $(-0.00894 - 2.04 \times 0.00164, 0.00894 + 2.04 \times 0.00164) = (-0.0122, -0.0056)$. The estimate for b_2 is thus significantly ($P < 0.05$) different from 0, in agreement with the t of –5.46; hence, the null hypothesis thast the relation is a straight line ($b_2 = 0$) is rejected in favour of a parabola ($b_2 \neq 0$). A straight line is thus statistically unacceptable for these data.

Exercise 3.2.6   A 95% confidence interval for the optimum temperature is (12.8 °C, 16.2 °C).

Exercise 3.2.7   The median abundances of *Spongotrochus* at temp. = 5, 10, 15 and 20 are exp (3.14) = 23, 41, 47 and 34, respectively. The fitted Gaussian curve with the data points and 95% confidence and 95% prediction intervals (obtained also by back-transformation) is plotted in Figure 3.19b.

Exercise 3.2.8 The weighted average is $(12 \times 0.8 + 14 \times 1.1 + \ldots + 41 \times 21.6)/(12 + 14 + \ldots + 41) = 12.7$, so that the weighted average temperature is 12.7 °C.

With log-transformed abundance data the weighted average temperature is smaller, namely 11.0 °C. Both values are smaller than the optimum (14.0 °C) estimated by regression, because the temperatures are not homogeneously distributed over the range of the species; in particular, the lower temperatures are over-represented and the optimum lies at the higher end of the temperature interval that was actually sampled. So the weighted average estimator is biased. The difference is large in a statistical sense: the weighted averages fall outside the 95% confidence interval for the optimum calculated in Exercise 3.2.6.

Exercise 3.3   Logit link function

$\log_e [p/(1 - p)] = c \rightarrow p/(1 - p) = \exp c$
$\rightarrow p = (\exp c)(1 - p) = \exp c - p \exp c \rightarrow p + p \exp c = \exp c.$
$\rightarrow p(1 + \exp c) = \exp c \rightarrow p = (\exp c)/(1 + \exp c).$

The arrows hold true also in the reverse direction; hence, the equivalence.

Exercise 3.4   Chi-square test and logit regression

Exercise 3.4.1 The estimated probability of occurrence is: in hayfield $12/28 = 0.43$; in pasture $96/132 = 0.73$.

Exercise 3.4.2 When the probability of occurrence in hayfield equals that in pasture, this probability is estimated by $108/160 = 0.675$. Then, we expect that out of 28 fields $0.675 \times 28 = 18.9$ fields contain *E. repens*, and $28 - 18.9 = 9.1$ fields do not contain *E. repens*.

With 132 fields (pastures) the expected numbers are: 89.1 with *E. repens* and 42.9 without *E. repens*. Inserting the observed and expected numbers in the equation for chi-square gives $(12 - 18.9)^2/18.9 + \ldots + (36 - 42.9)^2/42.9 = 9.39$ which is much greater than the critical value at the 5% significance level of a chi-square distribution with $(2 - 1) \times (2 - 1) = 1$ degree of freedom: $\chi^2_{0.05}(1) = 3.841$. The conclusion is that there is strong evidence ($P < 0.01$) that the probability of occurrence in hayfield differs from that in pasture.

Exercise 3.4.3.1 For hayfield the model reads: $\log_e [p/(1 - p)] = c_0$ because USE $= 0$ for hayfields. c_0 is estimated by -0.28; hence the estimated probability of occurrence is $\hat{p} = \exp(-0.28)/[1 + \exp(-0.28)] = 0.43$. For pastures, the model reads: $\log_e [p/(1 - p)] = c_0 + c_1$ because USE $= 1$ for pastures. $c_0 + c_1$ is estimated as $-0.28 + 1.27 = 0.99$, which gives $\hat{p} = \exp 0.99/(1 + \exp 0.99) = 0.73$. The estimates equal those of Exercise 3.4.1, because the regression model simply specifies two probabilities, one for hayfields and one for pastures.

Exercise 3.4.3.2 The estimate of the coefficient c_1 of USE differs significantly ($P < 0.05$) from 0, $t = 3.02$ being greater than $t_{0.05}(158) = 1.98$; hence, the estimated probabilities differ significantly. The conclusion is identical to that of Exercise 3.4.2; we applied a different test for the same purpose.

Exercise 3.4.3.3 The difference in deviance between the model with and without the variable USE is $201.7 - 192.9 = 8.8$, which is to be compared with a chi-square distribution with one degree of freedom.

Exercise 3.5   Gaussian logit regression

Exercise 3.5.1 From Equation 3.11a, the estimated optimum of pH is $\hat{u} = -19.11/(-2 \times 1.55) = 6.16$.

  With $u = 6.16$ in Equation 3.17, the maximum probability of occurrence is estimated by $\hat{p} = (\exp 1.641)/(1 + \exp 1.641) = 0.84$, because $-57.26 + 19.11 \times 6.16 - 1.55 \times 6.16^2 = 1.641$. The tolerance is $t = 0.57$ (Equation 3.11b).

Exercise 3.5.2 Inserting pH = 4.5, 5.0, 5.5, 6.0, 6.5, 7.0 and 7.5 in Equation 3.17, we obtain estimated probabilities of 0.07, 0.39, 0.72, 0.83, 0.81, 0.64 and 0.25.

Exercise 3.5.3 The estimate of b_2 is significantly ($P < 0.05$) smaller than 0, because the t (-3.52) is much greater in absolute value than the critical value of a one-tailed t test (1.65 at $P = 0.05$, one-tailed); hence, the estimated Gaussian logit response curve differs significantly from a sigmoid response curve, so that the optimum is significant.

Exercise 3.6   Multiple logit regression

Exercise 3.6.1 In hayfield (USE = 0) with pH = 5: $\log_e [p/(1 - p)] = -57.82 + (19.30 \times 5) - (1.56 \times 5^2) = -0.32$, which gives $\hat{p} = 0.421$. In pasture (USE = 1) with pH = 5: $\log_e [p/(1 - p)] = -0.32 - 0.04 = -0.36$, which gives $\hat{p} = 0.411$.

  For pH = 6, the estimated probabilities of occurrence are 0.860 and 0.855 in hayfield and pasture, respectively. The optimum pH is now estimated as $-19.30/(-2 \times 1.56) = 6.18$ and the tolerance as 0.57, identical for hayfields and pastures. The maximum probabilities of occurrence are 0.867 and 0.862 in hayfield and pasture, respectively. The difference between the estimated curves is small (Figure 3.20), the difference from the curve estimated in Exercise 3.5 is small too.

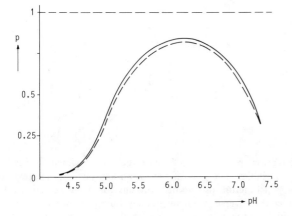

Figure 3.20 Gaussian logit curves of probability of occurrence of *Elymus repens* in hayfield (solid line) and pasture (broken line) against acidity (pH), as fitted by multiple logit regression. The probability of occurrence of *Elymus repens* at pH = 5 is estimated at 0.421 in hayfield and 0.411 in pasture; the difference is not statistically significant. Data from Kruijne et al. (1967).

Exercise 3.6.2 The t of the coefficient c_1 of USE is much smaller than the critical t at 5%. Therefore there is no evidence from these data that the probability of occurrence in fields with the same pH differs between hayfields and pastures. Acidity can therefore account for the overall difference between hayfields and pastures found in Exercise 3.4. The test result is not surprising after our observation in the previous exercise that the difference between the estimated response curves is small.

Exercise 3.6.3 The deviance of the model with acidity and agricultural use is 176.2; dropping agricultural use (variable USE) gives us the model with acidity only (Exercise 3.5), whose deviance is 176.3. The change in deviance (0.1) is much smaller than the critical value of a chi-square distribution with one degree of freedom, the change in the number of parameters between the models being one. The conclusion is the same as in Exercise 3.6.2.

Exercise 3.6.4 The deviance of the model with acidity and agricultural use is 176.2; dropping acidity (pH and pH^2) gives us the model with agricultural use only (Exercise 3.4), whose deviance is 192.9. The change in deviance is 16.7, which must be compared with a chi-square distribution with two degrees of freedom: $\chi^2_{0.05}$ (2) = 5.99.

The conclusion is that acidity has an effect after correction for the effect of agricultural use. Acidity and agricultural use are not substitutable in the sense of Subsection 3.5.3; agricultural use cannot replace acidity in explanatory power, as judged by the deviance tests.

4   Calibration

C.J.F. ter Braak

4.1   Introduction

In Chapter 3, we used regression analysis to analyse the way in which species respond to environmental variables. The goal of regression analysis is to express the response of a species as a function of one or more environmental variables. In this chapter, we consider the reverse problem: namely how to express values of an environmental variable as a function of species data. This function is termed the 'transfer function' or 'biotic index' and its construction is termed calibration. The calibration problem differs from the regression problem, because the causal and statistical relations between species and environment are asymmetric.

It might be thought easier to measure environmental variables at a site than to infer their values from the species that occur there. But often it is not. For example, total values over time may be required; repeated measurements are costly, while species automatically integrate environmental conditions over time. This is one of the ideas behind biological evaluation of water quality and bio-monitoring in general. There are also situations where it is impossible to measure environmental variables by direct means, whereas a biological record does exist. An example is the reconstruction of past changes in acidity (pH) in lakes from fossil diatoms from successive strata of the bottom sediment.

An indicator species is ideally a species that always occurs under a unique set of environmental conditions and does not occur elsewhere. Such an ideal indicator species indicates its unique set of environmental conditions without error. Ideal indicator species do not exist, however. Species with narrow ecological amplitudes exist, but such species are not always present in their specific environment and many of them have a low probability of occurrence there, partly because we do not know their specific environmental requirements fully. If such species occur somewhere, they indicate the environmental conditions at that place precisely, but their absence provides hardly any information about the environment. This is a major reason to use the whole community composition at a site for calibration purposes, including species with wider ecological amplitudes. In practice, 'community composition' is restricted to mean species of a particular taxonomic group, e.g. diatoms or vascular plants. Our definition of indicator species is broader than the one used in standard bioassay applications, where individuals of a single species are put on test to determine the amount of some drug or pollutant. Environmental calibration can, however, be considered as a multi-species form of bioassay.

In this chapter, we will introduce three calibration methods, one based on

78

response functions, one on indicator values and one on inverse regression. In the first method (Section 4.2), the response functions can be of any type, whereas in the other two methods particular response curves are assumed, unimodal curves in the one based on indicator values (Section 4.3) and straight lines in the inverse regression method (Section 4.4).

4.2 Maximum likelihood calibration using response functions

4.2.1 Introduction

Maximum likelihood calibration is based on response functions of species against environmental variables. We shall assume that these functions are known, i.e. they have already been estimated from an appropriate and sufficiently large set of data by regression analysis (Chapter 3). (This set of data is termed the training set.) For each set of values of environmental variables, we thus know what the probability is of observing a particular species composition. What we want is to predict the set of values of environmental variables at a particular site from its species composition. When the maximum likelihood principle is used, the prediction is the set of values that would give the maximum probability of observing that particular species composition, if that set of values were the true condition (cf. Subsection 3.3.2). This principle is illustrated in Subsection 4.2.2, together with the concept of a prior distribution and the loss in efficiency when ignoring possible correlations between species. In Subsection 4.2.2, we consider the problem of predicting a nominal environmental variable from presence–absence species data. This type of calibration is also known as discriminant analysis. How to discriminate between classes of a nominal variable by using abundance data will be discussed in the next chapter, in Subsection 5.5.5. In Subsection 4.2.3, the maximum likelihood principle is used to predict values of a quantitative environmental variable, first from presence–absence species data and then from abundance data.

One is commonly interested in a single environmental variable, whereas the species might respond to many more environmental variables. This problem can be solved in maximum likelihood calibration by using response functions of all the important environmental variables; the principles remain the same. But the response functions have first to be estimated from data by regression (Chapter 3), and the size of the training set of data will put a limit on the number of environmental variables that can be taken into account.

4.2.2 Predicting a nominal environmental variable

As an example, suppose we want to estimate the unknown value of soil type from the presence of a particular species. Let us assume that soil type has three classes, clay, peat and sand, and that the probabilities that the species occurs on a field of a given size are 0.1 for clay, 0.2 for peat and 0.4 for sand. If this species is encountered, then the maximum likelihood estimate of the soil type is sand, because sand is the soil type on which the species occurs with the highest probability. If the species is absent, the maximum likelihood estimate is clay, because clay is the soil type where the species is absent with the highest probability. These are the rules of assignment or classification. When the species is present, the proportion of wrong assignments is $(0.1 + 0.2)/(0.1 + 0.2 + 0.4) = 0.43$. If the species is absent, the proportion of wrong assignments is $(0.8 + 0.6)/(0.9 + 0.8 + 0.6) = 0.61$, a small reduction compared to random assignment, so then the assignment procedure is not very effective; note also that the rules defined above never assign to soil type peat.

In these rules, it was implicit that clay, peat and sand occurred equally frequently. This may not be so. If we know beforehand that soil type clay is encountered three times as often as peat or sand, then we could bet on the soil type being clay without any further information. This knowledge about the soil types a priori is termed the 'prior distribution', which is 0.6, 0.2, 0.2 in the example. If we also know that the species is present in a particular field, the probability that its soil type is clay is (apart from a normalizing constant) the product of the prior probability of clay and the probability that the species occurs on clay, that is: $0.6 \times 0.1 = 0.06$, compared to $0.2 \times 0.2 = 0.04$ for peat and $0.2 \times 0.4 = 0.08$ for sand. From these values, we obtain 'posterior probabilities' by dividing these values by their sum, $0.06 + 0.04 + 0.08 = 0.18$ in the example, so that the posterior probabilities for clay, peat and sand are 0.33, 0.22 and 0.44, respectively. The maximum of these probabilities is 0.44 for sand. The extra information that the species is present in the field changes our preference a priori from clay to sand. If the prior distribution is, however, 0.8, 0.1 and 0.1, then the maximum likelihood estimate is always clay, even if the species is present at the field. It is therefore important for the construction of the assignment rule for what frequencies the soil types are expected to be encountered on when the assignment rule will be used. The prior distribution is said to be uniform when the frequencies are equal. This distribution is often assumed when the true distribution is unknown. Many of the controversies in the statistical literature about calibration concern the question whether it is prudent to use the distribution of the fields in the training set as a prior distribution (Brown 1979). Rules based on the maximum likelihood principle have the attractive property that they minimize the number of wrong assignments (misclassifications). As a consequence, each wrong assignment is counted equally. There are, however, situations where one wrong assignment (e.g. assignment to peat instead of to clay) has more serious consequences than another (e.g. assignment to peat instead of to sand). This aspect of costs can be incorporated in the construction of assignment rules (e.g. Lachenbruch 1975).

In the following, we will assume equal costs for wrong assignments and a uniform prior distribution unless explicitly stated otherwise. So environmental conditions will be predicted on the basis of the response function of the species only.

We now extend the example. Apart from the species of that example, Species A, there is a second species, Species B, that only occurs rarely on clay or peat $(p = 0.01)$ but often on sand $(p = 0.98)$. If a field only contains Species A, then the absence of Species B indicates that its soil type is not likely to be sand; peat is then the most logical bet. Peat is also the maximum likelihood estimate if the responses of the species are independent; the probabilities of 'Species A present and Species B absent' for the three soil types are $0.1 \times 0.99 = 0.099$, 0.198 and 0.008, respectively, the maximum being for peat. The proportion of wrong assignment (0.35) is less than in the first example with Species A only. In this example (and also in the previous one), the absence of a species thus provides information on the environment.

In this example, an extra assumption was needed to calculate the probability of 'Species A present and Species B absent', namely that the responses of the two species were independent, so that the joint probability could simply be obtained by multiplication of the probability of 'Species A present' and the probability of 'Species B absent'. However the example was constructed in such a way that the best assignment rule would not change, even if the responses of the species were interdependent. In the next example, the assignment rule can be improved considerably if we account for known correlation between the responses of species.

For simplicity, this example includes only two soil types, clay and sand, with equal probabilities of occurrence of Species A ($p = 0.2$) and of Species B ($p = 0.4$). If the responses of Species A and Species B are independent, there is no way of discriminating between clay and sand on the basis of their responses; each assignment rule is wrong for half the cases. But suppose now that these species have preference for a different water-table when on sand, and are indifferent to the water-table when on clay. If both species are encountered in a field, its soil type is not likely to be sand. The probability of both species being present is close to zero on sand, whereas this probability is much larger on clay (0.2 $\times$ 0.4 = 0.08). It is therefore possible to improve the assignment rule by using the (negative) correlation between the species. To construct this improved rule, we must know four probabilities:
- the probability of A only
- the probability of B only
- the probability of A and B
- the probability of neither A nor B.

If there are m species, we need to know 2^m probabilities to construct the maximum likelihood assignment rule. All these probabilities must be estimated from the training set, an impossible task if the number of species exceeds 10, even if the training set is huge. Lachenbruch (1975, p. 41-46) described solutions to this problem when the dependence between species is simple. If the dependence between responses is caused by another environmental variable, it is most natural to incorporate this variable explicitly in the response function and to maximize the likelihood for both environmental variables jointly.

4.2.3 Predicting a quantitative environmental variable

Presence–absence species data

Assume that the response curve of the probability that a particular species is present is unimodal. Further assume that the environmental variable to be inferred takes the value x_0 for a particular field. If the species is present, the

81

maximum likelihood estimate of x_0 is then the optimum of the curve. At the optimum, the probability of occurrence of the species is clearly maximum. If the species is absent, there are two maximum likelihood estimates, $-\infty$ and $+\infty$.

Suppose now that there are m species that respond to a single quantitative environmental variable x only and suppose that the responses of the species are mutually independent for each fixed value of x. Denote the response curve of the probability of occurrence of the k-th species by $p_k(x)$. The probability that the k-th species is absent also depends on x and equals $1 - p_k(x)$.

The probability of a combination of species is, by their independence, the product of the probabilities of occurrence of the species that are present and the probabilities of absence of the species that are absent. The maximum likelihood estimate of x_0 is, again, the value for which the probability of the observed combination of species is maximum. In principle, we can calculate this probability for any value of x and determine the value of x that gives the highest probability. In practice, we need to write a computer program to do so.

The ratios of probabilities for different values of x, and not the absolute probabilities, are relevant in the estimation because a product of probabilities is calculated for every value of x. For rare species, whose maximum probability of occurrence is small, the ratio of the probabilities of occurrence for two values of x can still be very large. But the probability that a rare species is absent is always close to 1, irrespective of the value of x. The ratio of the probabilities of absence for different values of x is therefore always close to 1. Consequently, absences of rare species cannot influence the maximum likelihood estimate very much and so provide hardly any information on the environment at a site.

Quantitative abundance data

We now consider the estimation of an unknown value of a quantitative environmental variable (x) from a quantitative response (y) of a single species. If the response function is $Ey = f(x)$ and the error is normally distributed, we obtain the maximum likelihood estimate by solving the equation $y = f(x_0)$ for x_0. In a graph of the response curve, this simply means drawing a horizontal line at the level of the value y and reading off x where this line cuts the response curve. For the straight line (Figure 3.1), this gives the estimate

$$\hat{x}_0 = (y - b_0)/b_1.$$

If the response curve is unimodal, the horizontal line cuts the response curve twice so that we obtain two estimates. This problem has led de Wit et al. (1984) to suggest that an indicator species should have a monotonic relation with the environmental variable of interest. But, if more than one species is used for calibration, the problem generally disappears (Brown 1982).

For later reference (Subsection 5.3.2), we consider the case where each of m species shows a straight-line relation with x, and we want to predict x_0 from the m abundance values at the site. Reading off the graph for each species would give m possibly different estimates of x_0, and we want to combine them. The model for the data can be written as

$$Ey_k = a_k + b_k x \qquad\qquad \text{Equation 4.1}$$

where
y_k is the response of species k,
a_k its intercept and
b_k its slope parameter.

By minimizing the sum of squares of differences between the observed and expected responses, we obtain the combined estimate (as Equation 3.6):

$$\hat{x}_0 = \Sigma_{k=1}^m (y_k - a_k) b_k / \, \Sigma_{k=1}^m b_k^2 \qquad\qquad \text{Equation 4.2}$$

This is the maximum likelihood estimate only in the special case that the species are independent and have equal error variances. For the general case see Brown (1982).

4.3 Weighted averaging using indicator values

In this calibration method, the relation between a species and a (semi-) quantitative environmental variable (x) is summarized by a single quantity, the indicator value. Intuitively, the indicator value is the optimum, i.e. the value most preferred by a species. The value of the environmental variable at a site (x_0) is likely to be somewhere near the indicator values of the species that are present at that site. The method of weighted averaging takes it to be the average of these indicator values. If we have recorded abundances of the species, we may take a weighted average with weighting proportional to species' abundance and absent species carrying zero weight. The weighted average of indicator values is thus

$$\hat{x}_0 = (y_1 u_1 + y_2 u_2 + ... + y_m u_m) \,/\, (y_1 + y_2 + ... + y_m) \qquad \text{Equation 4.3}$$

where
$y_1, y_2, ..., y_m$ are the responses of the species at the site,
$u_1, u_2, ..., u_m$ are their indicator values.

For presence–absence data, the average of the indicator values of the species present is also called 'weighted' because absent species implicitly carry zero weight. Note that the method of weighted averaging is also used in Section 3.7 to estimate the indicator value of a species, in particular, by taking a weighted average of values of an environmental variable (Equation 3.28).

The weighted average was proposed as a biotic index for many types of organisms: for vascular plants by Ellenberg (1948) and by Whittaker (1956); for algae by Zelinka & Marvan (1961); and for faunal communities in streams and rivers by Chutter (1972). A typical example is Ellenberg's (1948; 1979) system for predicting soil acidity, reviewed by Böcker et al. (1983). Ellenberg has grouped Central European plants into nine preference groups of soil acidity and assigned the scores 1 to 9 to these groups, the score 1 to the group with species that preferred the most acid conditions and the score 9 to the group with species that preferred the most alkaline conditions. Ellenberg based this grouping on his field observations of the conditions under which particular species occurred and, to a lesser extent, on laboratory tests. The scores are thus the indicator values and are used to derive site scores by weighted averaging. In Ellenberg's system, the indicator values are ordinal and the resulting weighted average is a semiquantitative estimate of soil acidity. Ellenberg (1979), Rogister (1978) and Vevle & Aase (1980) demonstrated a strong relation between the weighted average for acidity based on plant composition and acidity actually measured in the field, thus confirming the empirical predictive value of the weighted average in Ellenberg's system.

From a theoretical viewpoint, it is surprising that the absent species have been disregarded in the weighted average. Apparently it is supposed that absent species do not provide information on the environment of a site (cf. Subsection 4.2.3). Further, each species is regarded as an equally good indicator in weighted averaging, whereas it is intuitively reasonable to give species with a narrow ecological amplitude more weight than species with a broader ecological amplitude. Ellenberg (1979) circumvented this problem by disregarding indifferent species; they were not assigned an indicator value. Zelinka & Marvan (1961) solved this problem in a heuristic way by assigning species not only an indicator value but also an indicator weight. Finally, because the indicator values are ordinal, calculating averages is a dangerous arithmetic operation; ordinal scale values are rather arbitrary, so they could be transformed monotonically without change of meaning. However the order of weighted averages calculated for different sites can be scrambled by such a transformation.

Ter Braak & Barendregt (1986) provided a theoretical justification of using the weighted average (Equation 4.3). For presence–absence data, the weighted average of indicator values is about as efficient as the maximum likelihood estimate of x_0 if the response curves of the species are Gaussian logit curves (Equation 3.17) with equal tolerances and the species presences are independent and if, in addition:

- either the maximum probability of occurrence is very small for any species so that absent species provide no information on the environment (Subsection 4.2.3)
- or as illustrated in Figure 4.1, the indicator values (optima) are homogeneously distributed over a large interval around x_0
- and the maxima of the response curves of species are equal.

If the condition of equal tolerances does not hold true, we must take a tolerance-weighted version of the weighted average

$$\hat{x}_0 = (\Sigma_{k=1}^m y_k \, u_k / t_k^2) / (\Sigma_{k=1}^m y_k / t_k^2) \qquad\qquad \text{Equation 4.4}$$

to retain high efficiency. Here, t_k is the tolerance of species k (Equation 3.17).

For quantitative abundance data, the method of weighted averaging can be justified analogously (ter Braak & Barendregt 1986). If the abundances follow a Poisson distribution and the response curves are Gaussian curves (Equation 3.8) with homogeneously distributed optima, equal tolerances and maxima, the weighted average again approximates the maximum likelihood estimate. This result may help to decide whether it is prudent to transform to presence–absence before the weighted average is calculated.

The conditions (homogeneously distributed optima, equal tolerances and maxima) together make a species packing model (Figure 4.1). This is an ecological model based on the idea that species evolve to occupy maximally separate niches with respect to a limiting resource. Christiansen & Fenchel (1977, Chapter 3) provide a lucid introduction here. This idea applies also to the occurrence of competing species along environmental variables (Whittaker et al. 1973). Response curves should therefore have minimum overlap.

Despite its theoretical basis, the species packing model is not likely to hold in real life. Nevertheless, the derivation of the weighted average provided above indicates the kind of situation in which the weighted average performs reasonably well. Species may not really be distributed according to the species packing model, but neither are they tightly clumped along environmental gradients; there is usually a fairly even turnover of species along gradients. In addition, Equation 4.4 shows how one can incorporate information on ecological amplitudes in the weighted average.

In lists of indicator values, the values are often expressed on an ordinal scale. For weighted averaging to be useful, the scale values (and, hence, the indicator values) must be chosen such that most species show fairly symmetric response curves. If this can be achieved, the weighted average is an informative semiquantitative biotic index. The method of weighted averaging of indicator values is

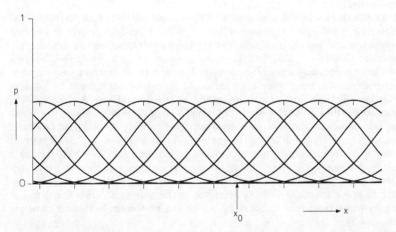

Figure 4.1 Species packing model: Gaussian logit curves of the probability (p) that a species occurs at a site, against environmental variable x. The curves shown have equispaced optima (spacing $= 1$), equal tolerances ($t = 1$) and equal maximum probabilities of occurrence ($p_{max} = 0.5$). x_0 is the value of x at a particular site.

85

also attractive to reveal a possible structure in data tables such as Table 0.1 of this book. We simply rearrange the species in order of their indicator value for a particular environmental variable and the sites in order of their weighted average, as in Section 3.7.

4.4 Inverse regression

In Subsection 4.2.3, we discussed a calibration method for when abundance values of a species show a linear relation with the environmental variable of interest. An attractive alternative method is then inverse regression. In inverse regression, the training set is not used to construct response curves by regressing the responses of the species on the environmental variable; instead the environmental variable is taken as the response variable and the responses of the species as the explanatory variable. The regression equation so constructed is then directly the transfer function that is used for prediction. This method has attractive properties if the prior distribution of the environmental variable equals the distribution in the training set (Brown 1979).

The method of inverse regression can easily be extended to prediction on the basis of the responses of more than one species. Each species then makes an explanatory variable, so that the inverse regression is a multiple (least-squares) regression of the environmental variable on the response variables of the species. Predictions are again derived directly from the multiple regression equation so obtained. This method is most efficient if the relation between each of the species and the environmental variable is a straight line with a normal distribution of error (Equation 4.1) and if the environmental variable too has a normal distribution (Brown 1982).

However species do not in general have monotonic relations with environmental variables. For example, response surfaces of pollen types with respect to summer temperature and annual precipitation over large geographic regions are strongly non-linear (Bartlein et al. 1986). Inverse regression could not therefore be used to build one generally applicable transfer function to reconstruct past climates from pollen data. But response curves could be made about linear by limiting the geographic area and transforming the pollen data (Howe & Webb 1983). Therefore Bartlein & Webb (1985) subdivided a large geographic area into regions and, for the actual climatic reconstruction, chose among the transfer functions obtained separately for different regions by using an analogue method (a method to decide to which training set of modern pollen data (i.e. to which region) a fossil pollen sample is most similar). Inverse regression was thus just one step in the whole calibration procedure. A simpler procedure would be to fit non-linear response functions first, as described by Bartlein et al. (1986), and to use these to reconstruct past climates by use of the maximum likelihood principle (Section 4.2).

4.5 Bibliographic notes

The history of the method of weighted averaging has been sketched in Section 4.3. Other biotic indices are listed in Sheenan (1984). Battarbee (1984) reviews various biotic indices for pH reconstruction from diatoms, including one based on inverse regression (see also Davis & Anderson 1985).

Much of the statistical literature on calibration is devoted to the prediction of a single quantitative variable on the basis of a single quantitative response variable, assuming a straight-line relation and a normal distribution of error. Brown (1979) compared the method of inverse regression with the Classical approach by first fitting response functions (Subsection 4.2.3). Calibration with polynomial response functions is treated, for instance, by Scheffé (1973), Schwartz (1977) and Brown (1982). Williams (1959, Chapter 9), Brown (1979), Brown (1982), and Naes & Martens (1984) discuss linear multivariate calibration, the prediction of one or more quantitative variables from more than one quantitative response variable, assuming a linear model.

Discrimination (calibration of a nominal explanatory variable) is treated by Lachenbruch (1975) in a general statistical context, by Titterington et al. (1981) in a medical context and by Kanal (1974) in electrical engineering.

4.6 Exercises

Exercise 4.1 Weighted averaging and maximum likelihood calibration with Gaussian logit curves

With data from Kruijne et al. (1967) on the occurrence of plant species and soil acidity (pH) in meadow fields, ter Braak & Looman (1986) fitted a Gaussian logit curve with respect to pH for each of the species. The curves of seven of the species are shown in Figure 4.2. Their parameters are:

Species name	Code	Optimum	Tolerance	Maximum
Agrostis canina	AC	3.4	1.1	0.84
Stellaria graminea	SG	5.7	0.4	0.38
Alopecurus geniculatus	AG	5.8	0.6	0.58
Plantago major	PM	6.2	0.7	0.34
Bellis perennis	BP	6.4	0.5	0.89
Hordeum secalinum	HS	7.1	0.7	0.57
Glechoma hederacea	GH	8.1	1.5	0.55

Although the parameters were estimated from only 100 fields, we treat them in this exercise as the true parameters. For three meadow fields with a unknown soil acidity, we want to predict the soil acidity from the presences and absences of these seven species. The species that are present are in Field 1 AC, SG and BP, in Field 2 AG and BP, and in Field 3 HS and BP (species not mentioned are absent). Predict the pH of each of these fields:

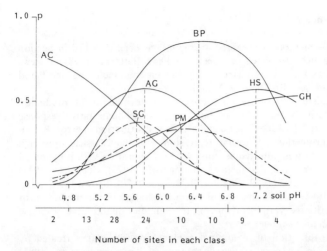

Figure 4.2 Probability of occurrence of seven contrasting species in relation to soil acidity (pH) in meadows, as fitted by logit regression. The curves can be identified by the code near their optimum indicated by dotted lines. The species arranged in order of their optima are: *Agrostis canina* (AC); *Stellaria graminea* (SG); *Alopecurus geniculatus* (AG); *Plantago major* (PM); *Bellis perennis* (BP); *Hordeum secalinum* (HS); *Glechoma hederacea* (GH). Nomenclature follows Heukels–van der Meijden (1983).

Exercise 4.1.1   By the method of weighted averaging using the optima as indicator values.

Exercise 4.1.2   By the tolerance-weighted version of the method of weighted averaging (Equation 4.4).

Exercise 4.1.3   By the method of maximum likelihood. Hint: calculate the likelihood for a limited number of pH values, for example, pH = 5.0, 5.5, 6.0, 6.5, 7.0, 7.5 and next for the most likely value of these plus and minus 0.1. Use Equation 3.17 of Chapter 3 to calculate probabilities of occurrence. In that equation: $c = \text{maximum}/(1 - \text{maximum})$.

Exercise 4.2   Calibration using a straight line

Predict, by using the results of Exercise 3.1, the relative sulphate concentration of a moorland pool in which *Frustulia rhomboides* var. *saxonica* occurs with 70 frustules.

88

Exercise 4.3   Calibration using a Gaussian response curve

Predict, by using the results of Exercise 3.2, the February sea-surface temperatures of two samples in which the abundances of *Spongotrochus glacialis* are 20% and 60%, respectively.

4.7   Solutions to exercises

Exercise 4.1   Weighted averaging and maximum likelihood calibration with Gaussian logit curves

Exercise 4.1.1   The weighted average (Equation 4.3) is for Field 1
$\hat{x}_0 = (1 \times 3.4 + 1 \times 5.7 + 0 \times 5.8 + 0 \times 6.2 + 1 \times 6.4 + 0 \times 7.1 + 0 \times 8.1)/(1 + 1 + 0 + 0 + 1 + 0 + 0) = 15.5/3 = 5.17$.
The prediction is thus pH 5.17. Analogously, the weighted average for Field 2 is 6.10 and for Field 3 is 6.75.

Exercise 4.1.2   The tolerance weighted version of the weighted average (Equation 4.4) gives for Field 1
$\hat{x}_0 = (1 \times 3.4/1.1^2 + 1 \times 5.7/0.4^2 + 0 \times 5.8/0.6^2 + ... + 0 \times 8.1/1.5^2)/(1/1.1^2 + 1/0.4^2 + ... + 0/1.5^2) = 64.03/11.08 = 5.78$. For Field 2 we obtain 6.15 and for Field 3 we obtain 6.64.

Exercise 4.1.3   With Equation 3.17, we obtain the probability of occurrence (p_k) at pH 5.0, which is for AC 0.646, for SG 0.117, for AG 0.362, for PM 0.106, for BP 0.138, for HS 0.015 and for GH 0.126. The probability that the k-th species is absent is $1 - p_k$. For pH 5.0, the likelihood of the species combination of Field 1 (AC, SG and BP present) is therefore $0.646 \times 0.117 \times (1 - 0.362) \times (1 - 0.106) \times 0.138 = (1 - 0.015) \times (1 - 0.126) = 0.0051$.
For pH 5.5, 6.0, 6.5, 7.0 and 7.5, we obtain likelihoods of 0.0244, 0.0094, 0.0008, 0.0000, 0.0000, respectively. The maximum of these likelihoods is 0.0244, at pH 5.5. The likelihoods at pH 5.4 and 5.6 are slightly lower and, within the precision of 0.1, 5.5 is the maximum likelihood prediction of the pH of Field 1.
For Field 2, the likelihood at pH 5.0 becomes 0.0121; the maximum (0.083) occurs at pH 6.0. Slightly lower likelihoods are obtained for pH 5.9 and 6.1. The maximum likelihood prediction is thus 6.0.
For Field 3 the likelihood at pH 5.0 becomes 0.0003; the maximum of the six likelihoods occurs at pH 7.0. pH 7.1 gives a slightly higher likelihood, whereas for pH 7.2 the likelihood decreases again. The maximum likelihood prediction is thus 7.1.

Exercise 4.2   Calibration using a straight line

In Exercise 3.1, the regression equation E $\log_e$ (*Frustulia* count $+ 1) = 5.848 - 5.96 S_{rel}$ was obtained. In the pool under study, the count is 70, so that $y = \log_e (70 + 1) = 4.263$. Replacing the left side of the regression equation by 4.263, we obtain $S_{rel} = (5.848 - 4.263)/5.96 = 0.27$.

Exercise 4.3   Calibration using a Gaussian response curve

For the sample with 20% *S. glacialis*, we have to solve the quadratic equation $-0.00894 \, temp^2 + 0.2497 \, temp + 2.119 = \log_e (20) = 2.996$. There are two solutions, temp = 4.1 °C and 23.8 °C. The temperatures on which the regression equation is based lies between 0.8 and 21.6 °C. If this range is relevant prior information, the prediction of 23.8 °C can be discarded and the remaining prediction is 4.1 °C.

For the sample with 60% *S.glacialis*, the quadratic equation for temperature has no solution. This is not surprising, because the maximum of the Gaussian curve was 48%, which was obtained at 14 °C. The most likely temperature is therefore 14 °C.

5 Ordination

C.J.F. ter Braak

5.1 Introduction

5.1.1 Aim and usage

Ordination is the collective term for multivariate techniques that arrange sites along axes on the basis of data on species composition. The term ordination was introduced by Goodall (1954) and, in this sense, stems from the German 'Ordnung', which was used by Ramensky (1930) to describe this approach.

The result of ordination in two dimensions (two axes) is a diagram in which sites are represented by points in two-dimensional space. The aim of ordination is to arrange the points such that points that are close together correspond to sites that are similar in species composition, and points that are far apart correspond to sites that are dissimilar in species composition. The diagram is a graphical summary of data, as in Figure 5.1, which shows three groups of similar sites. Ordination includes what psychologists and statisticians refer to as multidimensional scaling, component analysis, factor analysis and latent-structure analysis.

Figure 5.1 also shows how ordination is used in ecological research. Ecosystems are complex: they consist of many interacting biotic and abiotic components. The way in which abiotic environmental variables influence biotic composition is often explored in the following way. First, one samples a set of sites and records which species occur there and in what quantity (abundance). Since the number of species is usually large, one then uses ordination to summarize and arrange the data in an ordination diagram, which is then interpreted in the light of whatever is known about the environment at the sites. If explicit environmental data are lacking, this interpretation is done in an informal way; if environmental data have been collected, in a formal way (Figure 5.1). This two-step approach is indirect gradient analysis in the sense used by Whittaker (1967). By contrast, direct gradient analysis is impossible without explicit environmental data. In direct gradient analysis, one is interested from the beginning in particular environmental variables, i.e. either in their influence on the species as in regression analysis (Chapter 3) or in their values at particular sites as in calibration (Chapter 4).

Indirect gradient analysis has the following advantages over direct gradient analysis. Firstly, species compositions are easy to determine, because species are usually clearly distinguishable entities. By contrast, environmental conditions are difficult to characterize exhaustively. There are many environmental variables and even more ways of measuring them, and one is often uncertain of which variables the species react to. Species composition may therefore be a more informative

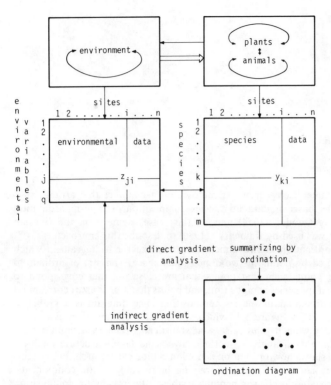

Figure 5.1 Outline of the role of ordination in community ecology, showing the typical format of data sets obtained by sampling ecosystems and their analysis by direct gradient and indirect gradient analysis. Also shown is the notation used in Chapter 5. Point of site in the ordination diagram (•).

indicator of environment than any given set of measured environmental variables. Ordination can help to show whether important environmental variables have been overlooked: an important variable has definitely been missed if their is no relation between the mutual positions of the sites in the ordination diagram and the measured environmental variables.

Secondly, the actual occurrence of any individual species may be too unpredictable to discover the relation of its occurrence to environmental conditions by direct means (Chapter 3) and therefore more general patterns of coincidence of several species are of greater use in detecting species–environment relations.

Thirdly, for example in landscape planning, interest may from the onset be focused more on the question of which combinations of species can occur, and less on the behaviour of particular species. Regression analysis of single species then provides too detailed an account of the relations between species and

92

environment. The ordination approach is less elaborate and gives a global picture, but – one hopes – with sufficient detail for the purpose in hand.

Between regression analysis and ordination (in the strict sense) stand the canonical ordination techniques. They are ordination techniques converted into multivariate direct gradient analysis techniques; they deal simultaneously with many species and many environmental variables. The aim of canonical ordination is to detect the main pattern in the relations between the species and the observed environment.

5.1.2   Data approximation and response models in ordination

Ordination techniques can be viewed in two ways (Prentice 1977). According to one view, the aim of ordination is to summarize multivariate data in a convenient way in scatter diagrams. Ordination is then considered as a technique for matrix approximation (as the data are usually presented in the two-way layout of a matrix). A second, more ambitious, view assumes from the beginning that there is an underlying (or latent) structure in the data, i.e. that the occurrences of all species under consideration are determined by a few unknown environmental variables (latent variables) according to a simple response model (Chapter 3). Ordination in this view aims to recover that underlying structure. This is illustrated in Figure 5.2 for a single latent variable. In Figure 5.2a, the relations of two species, A and B, with the latent variable are rectilinear. In Figure 5.2c they are unimodal. We now record species abundance values at several sites and plot the abundance of Species A against that of Species B. If relations with the latent variable were rectilinear, we would obtain a straight line in the plot of Species B against Species A (Figure 5.2b), but if relations were unimodal, we would obtain a complicated curve (Figure 5.2d). The ordination problem of indirect gradient analysis is to infer about the relations with the latent variable (Figures 5.2a,c) from the species data only (Figure 5.2b,d). From the second viewpoint, ordination is like regression analysis, but with the major difference that in ordination the explanatory variables are not known environmental variables, but 'theoretical' variables. These variables, the latent variables, are constructed in such a way that they best explain the species data. As in regression, each species thus constitutes a response variable, but in ordination these response variables are analysed simultaneously. (The distinction between these two views of ordination is not clear-cut, however. Matrix approximation implicitly assumes some structure in the data by the mere way the data are approximated. If the data structure is quite different from the assumed structure, the approximation is inefficient and fails.)

The ordination techniques that are most popular with community ecologists, are principal components analysis (PCA), correspondence analysis (CA), and techniques related to CA, such as weighted averaging and detrended correspondence analysis. Our introduction to PCA and CA will make clear that PCA and CA are suitable to detect different types of underlying data structure. PCA relates to a linear response model in which the abundance of any species either increases or decreases with the value of each of the latent environmental variables (Figure 5.2a). By contrast, CA is related, though in a less unequivocal way, to a unimodal response model (Figure 5.2c). In this model, any species occurs in a limited range

93

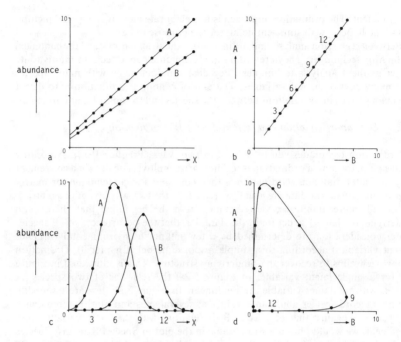

Figure 5.2 Response curves for two species A and B against a latent variable x (a, c) and the expected abundances of the species plotted against each other (b, d), for the straight line model (a, b) and a unimodal model (c, d). The numbers refer to sites with a particular value for x. The ordination problem is to make inferences about the relations in Figures a and c from species data plotted in Figures b and d.

of values of each of the latent variables. PCA and CA both provide simultaneously an ordination for the sites and an ordination for the species. The two ordinations may be plotted in the same diagram to yield 'joint plots' of site and species points, but the interpretation of the species points is different between PCA and CA.

PCA and CA operate directly on the species data. By contrast, multidimensional scaling is a class of ordination techniques that operate on a table of dissimilarity values between sites. To apply these techniques, we must therefore first choose an appropriate dissimilarity coefficient to express the dissimilarity in species composition between any two sites (Subsection 6.2.2). After choosing one, we can calculate the dissimilarity values of all pairs of sites required as input for multidimensional scaling. CA and PCA may also be considered as multidimensional scaling techniques, but ones that use a particular dissimilarity coefficient.

5.1.3   Outline of Chapter 5

Section 5.2 introduces CA and related techniques and Section 5.3 PCA. Section
5.4 discusses methods of interpreting ordination diagrams with external (environmental) data. It is also a preparation for canonical ordination (Section 5.5).
After a discussion of multidimensional scaling (Section 5.6), Section 5.7 evaluates
the advantages and disadvantages of the various ordination techniques and
compares them with regression analysis and calibration. After the bibliographic
notes (Section 5.8) comes an appendix (Section 5.9) that summarizes the ordination
methods described in terms of matrix algebra.

5.2   Correspondence analysis (CA) and detrended correspondence analysis (DCA)

5.2.1   From weighted averaging to correspondence analysis

Correspondence analysis (CA) is an extension of the method of weighted
averaging used in the direct gradient analysis of Whittaker (1967) (Section 3.7).
Here we describe the principles in words; the mathematical equations will be
given in Subsection 5.2.2.

Whittaker, among others, observed that species commonly show bell-shaped
response curves with respect to environmental gradients. For example, a plant
species may prefer a particular soil moisture content, and not grow at all in places
where the soil is either too dry or too wet. In the artificial example shown in
Figure 5.3a, Species A prefers drier conditions than Species E, and the Species
B, C and D are intermediate. Each of the species is therefore largely confined
to a specific interval of moisture values. Figure 5.3a also shows presence–absence
data for Species D: the species is present at four of the sites.

We now develop a measure of how well moisture explains the species data.
From the data, we can obtain a first indication of where a species occurs along
the moisture gradient by taking the average of the moisture values of the sites
in which the species is present. This average is an estimate of the optimum of
the species (the value most preferred), though not an ideal one (Section 3.7).
The average is here called the species score. The arrows in Figure 5.3a point
to the species scores so calculated for the five species. As a measure of how well
moisture explains the species data, we use the dispersion ('spread') of the species
scores. If the dispersion is large, moisture neatly separates the species curves and
moisture explains the species data well. If the dispersion is small, then moisture
explains less. To compare the explanatory power of different environmental
variables, each environmental variable must first be standardized; for example
by subtracting its mean and dividing by its standard deviation.

Suppose that moisture is the 'best' single environmental variable measured in
the artificial example. We might now wish to know whether we could in theory
have measured a variable that explains the data still better. CA is now the technique
that constructs the theoretical variable that best explains the species data. CA
does so by choosing the best values for the sites, i.e. values that maximize the
dispersion of the species scores (Figure 5.3b). The variable shown gives a larger

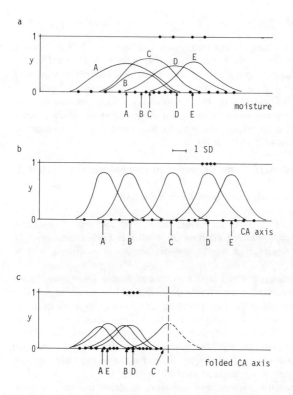

Figure 5.3 Artificial example of unimodal response curves of five species (A-E) with respect to standardized variables, showing different degrees of separation of the species curves. a: Moisture. b: First axis of CA. c: First axis of CA folded in this middle and the response curves of the species lowered by a factor of about 2. Sites are shown as dots at $y = 1$ if Species D is present and at $y = 0$ if Species D is absent. For further explanation, see Subsections 5.2.1 and 5.2.3.

dispersion than moisture; and consequently the curves in Figure 5.3b are narrower, and the presences of Species D are closer together than in Figure 5.3a.

The theoretical variable constructed by CA is termed the first ordination axis of CA or, briefly, the first CA axis; its values are the site scores on the first CA axis.

A second and further CA axes can also be constructed; they also maximize the dispersion of the species scores but subject to the constraint of being uncorrelated with previous CA axes. The constraint is intended to ensure that new information is expressed on the later axes. In practice, we want only a few axes in the hope that they represent most of the variation in the species data.

So we do not need environmental data to apply CA. CA 'extracts' the ordination

96

axes from the species data alone. CA can be applied not only to presence–absence data, but also to abundance data; for the species scores, we then simply take a weighted average of the values of the sites (Equation 3.28).

5.2.2 Two-way weighted averaging algorithm

Hill (1973) introduced CA into ecology by the algorithm of reciprocal averaging. This algorithm shows once more that CA is an extension of the method of weighted averaging.

If we have measured an environmental variable and recorded the species composition, we can estimate for each species its optimum or indicator value by averaging the values of the environmental variable over the sites in which the species occurs, and can use the averages so obtained to rearrange the species (Table 3.9). If the species show bell-shaped curves against the environmental variable, the rearranged table will have a diagonal structure, at least if the optima of the curves differ between the species (Table 3.9). Conversely, if the indicator values of species are known, the environmental variable at a site can be estimated from the species that it contains, by averaging the indicator values of these species (Section 4.3) and sites can be arranged in order of these averages. But, these methods are only helpful in showing a clear structure in the data if we know in advance which environmental variable determines the occurrences of the species. If this is not known in advance, the idea of Hill (1973) was to discover the 'underlying environmental gradient' by applying this averaging process both ways in an iterative fashion, starting from arbitrary initial values for sites or from arbitrary initial (indicator) values for species. It can be shown mathematically that this iteration process eventually converges to a set of values for sites and species that do not depend on the initial values. These values are the site and species scores of the first CA axis.

We illustrate now the process of reciprocal averaging. For abundance data, it is rather a process of two-way weighted averaging. Table 5.1a shows the Dune Meadow Data (Table 0.1), arranged in arbitrary order. We take as initial values for the sites the numbers 1 to 20, as printed vertically below Table 5.1a. As before, we shall use the word 'score', instead of 'value'. From the site scores, we derive species scores by calculating the weighted average of the site scores for each species. If we denote the abundance of species k at site i by y_{ki}, the score of site i by x_i and the score of species k by u_k, then the score of species k becomes the weighted average of site scores (Section 3.7)

$$u_k = \sum_{i=1}^{n} y_{ki} x_i / \sum_{i=1}^{n} y_{ki} \qquad \qquad \text{Equation 5.1}$$

For *Achillea millefolium* in Table 5.1a, we obtain $u_1 = (1 \times 1 + 3 \times 2 + 2 \times 5 + 2 \times 6 + 2 \times 7 + 4 \times 10 + 2 \times 17)/(1 + 3 + 2 + 2 + 2 + 4 + 2) = 117/16 = 7.31$. The species scores thus obtained are also shown in Table 5.1a. From these species scores, we derive new site scores by calculating for each site the weighted average of the species scores, i.e.

Table 5.1a

k	Species	Sites (i) 0000000001111111112 1234567890123456789 0	u_k
1	Ach mil	13 222 4 45447 5	7.31
2	Agr sto	48 43 2 3	11.33
3	Air pra	272 53 85 4 4 4	18.20
4	Alo gen	432 4 4	9.03
5	Ant odo	3222 2 2	11.24
6	Bel per	4 32 2 4	6.62
7	Bro hor		5.60
8	Che alb	1	13.00
9	Cir arv	2	4.00
10	Ele pal	2 458 4	14.84
11	Ely rep	44444 6	4.38
12	Emp nig	2	19.00
13	Hyp rad	44 2 2 5	16.78
14	Jun art	44 33 4	13.39
15	Jun buf	2 4 43	10.54
16	Leo aut	5223332352222 2562	10.94
17	Lol per	7565266426 7 2	6.31
18	Pla lan	555 33 23	9.27
19	Poa pra	44542344444 2 13	7.25
20	Poa tri	2765645454 49 2	7.25
21	Pot pal	22	14.50
22	Ran fla	2 2222 4	15.14
23	Rum ace	563 2	6.89
24	Sag pro	5 22 242 3	10.35
25	Sal rep	335	19.18
26	Tri pra	252	6.00
27	Tri rep	5212522363261 22	9.47
28	Vic lat	12 1	12.50
29	Bra rut	2226222244 44 634	12.02
30	Cal cus	4 3 3	16.40

x_i 1111111112
1234567890123456789 0

Table 5.1 Two-way weighted averaging algorithm of CA applied to the Dune Meadow Data presented in a preliminary section of this book. The site numbers and site scores are printed vertically. a: Original data table with at the bottom the initial site scores. b: Species and sites rearranged in order of their scores obtained after one cycle of two-way weighted averaging. c: Species and sites arranged in order of their final scores (CA scores). Note the minus signs in the site scores; for example, the score of Site 17 is −1.46.

98

Table 5.1b

```
Species              Sites (i)                       u_k
                     0000001001110111112
 k                   1253470693128876459 0
 9 Cir arv                     2                      4.00
11 Ely rep           44444   6                        4.38
 7 Bro hor           42 324                            5.60
26 Tri pra            2  2 5                           6.00
17 Lol per           752656662 7 42                   6.31
 6 Bel per           3222 2   2                        6.62
23 Rum ace           5 3 62  2                         6.89
19 Poa pra           44254443424 431                  7.25
20 Poa tri           2766554459 44  2                 7.25
 1 Ach mil           132 242   2                       7.31
 4 Alo gen           2 72  35 85  4                    9.03
18 Pla lan           5 535 3 32                        9.27
27 Tri rep           52216532322  612                 9.47
24 Sag pro           5 22242      3                   10.35
15 Jun buf              2 43 4                        10.54
16 Leo aut           53223332252352  2262            10.94
 5 Ant odo           4 243      4  4                  11.24
 2 Agr sto           48  35 44 744 5                  11.33
29 Bra rut           222262 4426 4 434               12.02
28 Vic lat                 1 2 1                      12.50
 8 Che alb                  1   3 3 4                 13.00
14 Jun art                4   4  3 3 4                13.39
21 Pot pal                   22                       14.50
10 Ele pal                 4 845 4                    14.84
22 Ran fla                2 2 222 4                   15.14
30 Cal cus                     34 3                   16.40
13 Hyp rad              2  2 5                        16.78
 3 Air pra           2 2 3                            18.20
12 Emp nig           2 3                              19.00
25 Sal rep                 3  35                      19.18

                            1111111
x_i     6778888889900122234
        2490113469773786983
        5634887804384912479 6
```

Table 5.1c

```
Species              Sites (i)                       u_k
                     1010001101010001101121
 k                   75076191283492385406
 3 Air pra           2  3                            -0.99
 5 Ant odo           44423 4                         -0.96
 1 Ach mil           224221  3                       -0.91
26 Tri pra           2 25                            -0.88
13 Hyp rad           25355  3  3                     -0.84
18 Pla lan           2                               -0.84
12 Emp nig           242    4   3                    -0.67
 7 Bro hor           5 36     22                     -0.66
23 Rum ace           1   2 1                         -0.65
28 Vic lat           22   3222                       -0.62
 6 Bel per           26667 752652   4                -0.62
17 Lol per           124434 443544 24                -0.50
19 Poa pra           4   4 446                       -0.50
11 Ely rep           23333 6555222223222             -0.39
16 Leo aut           64542  7 655494     2           -0.37
20 Poa tri           2625 23522133221 6              -0.19
27 Tri rep                                           -0.18
 9 Cir arv           32  52422                       -0.18
24 Sag pro                443                        -0.06
15 Jun buf           2                               0.00
29 Bra rut           2226 34 62224 24  44            0.08
 4 Alo gen             2 723855   4                  0.08
 8 Che alb              3  3      1                  0.18
25 Sal rep                  3   5                    0.40
 2 Agr sto                4834544457                 0.42
14 Jun art                    4  43 43               0.62
22 Ran fla                       222242              0.93
10 Ele pal                        45448              1.28
21 Pot pal                           22              1.56
30 Cal cus                          433              1.77
                                                     1.92
                                                     1.96

                     -------------------------------
x_i    10000011010001100001112
       :988866310024 7990
       4988866631002479990
       6587628441698262250
```

$x_i = \sum_{k=1}^{m} y_{ki} u_k / \sum_{k=1}^{m} y_{ki}$ \hfill Equation 5.2

For Site 1 in Table 5.1a, we obtain $x_1 = (1 \times 7.31 + 4 \times 4.38 + 7 \times 6.31 + 4 \times 7.25 + 2 \times 7.25)/(1 + 4 + 7 + 4 + 2) = 112.5/18 = 6.25$. In Table 5.1b, the species and sites are arranged in order of the scores obtained so far. The new site scores are also printed vertically underneath. There is already some diagonal structure, i.e. the occurrences of each species tend to come together along the rows. We can improve upon this structure by calculating new species scores from the site scores that we have just calculated, and so on.

A practical numerical problem with this technique is that, by taking averages, the range of the scores gets smaller and smaller. For example, we started off with a range of 19 (site scores from 1 to 20) and after one cycle the site scores have a range of $14.36 - 6.25 = 8.11$ (Table 5.1b). To avoid this, either the site scores or the species scores must be rescaled. Here the site scores have been rescaled. There are several ways of doing so. A simple way is to rescale to a range from 0 to 100 by giving the site with the lowest score the value 0 and the site with the highest score the value 100 and by calculating values for the remaining sites in proportion to their scores; in the example, the rescaled scores would be obtained with the formula $(x_i - 6.25)/0.0811$.

We shall use another way in which the site scores are standardized to (weighted)

Table 5.2 Two-way weighted averaging algorithm of CA.

---

a: Iteration process

Step 1. Take arbitrary, but unequal, initial site scores (x_i).
Step 2. Calculate new species scores (u_k) by weighted averaging of the site scores   (Equation 5.1).
Step 3. Calculate new site scores (x_i) by weighted averaging of the species scores   (Equation 5.2).
Step 4. For the first axis, go to Step 5. For second and higher axes, make the site scores (x_i) uncorrelated with the previous axes by the orthogonalization procedure described below.
Step 5. Standardize the site scores (x_i). See below for the standardization procedure.
Step 6. Stop on convergence, i.e. when the new site scores are sufficiently close to the site scores of the previous cycle of the iteration; ELSE go to Step 2.

b: Orthogonalization procedure

Step 4.1. Denote the site scores of the previous axis by f_i and the trial scores of the present axis by x_i.
Step 4.2. Calculate $v = \sum_{i=1}^{n} y_{+i} x_i f_i / y_{++}$
where $y_{+i} = \sum_{k=1}^{m} y_{ki}$
and $y_{++} = \sum_{i=1}^{n} y_{+i}$.
Step 4.3 Calculate $x_{i,\text{new}} = x_{i,\text{old}} - v f_i$.
Step 4.4 Repeat Steps 4.1-4.3 for all previous axes.

c: Standardization procedure

Step 5.1 Calculate the centroid, z, of site scores (x_i) $z = \sum_{i=1}^{n} y_{+i} x_i / y_{++}$.
Step 5.2 Calculate the dispersion of the site scores $s^2 = \sum_{i=1}^{n} y_{+i} (x_i - z)^2 / y_{++}$.
Step 5.3 Calculate $x_{i,\text{new}} = (x_{i,\text{old}} - z)/s$.
Note that, upon convergence, s equals the eigenvalue.

---

mean 0 and variance 1 as described in Table 5.2c. If the site scores are so standardized, the dispersion of the species scores can be written as

$$\delta = \Sigma_{k=1}^{m} y_{k+} u_k^2 / y_{++} \qquad\qquad\qquad \text{Equation 5.3}$$

where
y_{k+} is the total abundance of species k
y_{++} the overall total.

The dispersion will steadily increase in each iteration cycle until, after about 10 cycles, the dispersion approaches its maximum value. At the same time, the site and species scores stabilize. The resulting scores have maximum dispersion and thus constitute the first CA axis.

If we had started from a different set of initial site scores or from a set of arbitrary species scores, the iteration process would still have resulted in the same ordination axis. In Table 5.1c, the species and sites are rearranged in order of their scores on the first CA axis and show a clear diagonal structure.

A second ordination axis can also be extracted from the species data. The need for a second axis may be illustrated in Table 5.1c; Site 1 and Site 19 lie close together along the first axis and yet differ a great deal in species composition. This difference can be expressed on a second axis. The second axis is extracted by the same iteration process, with one extra step in which the trial scores for the second axis are made uncorrelated with the scores of the first axis. This can be done by plotting in each cycle the trial site scores for the second axis against the site scores of the first axis and fitting a straight line by a (weighted) least-squares regression (the weights are y_{+i}/y_{++}). The residuals from this regression (i.e. the vertical deviations from the fitted line: Figure 3.1) are the new trial scores. They can be obtained more quickly by the orthogonalization procedure described in Table 5.2b. The iteration process would lead to the first axis again without the extra step. The intention is thus to extract information from the species data in addition to the information extracted by the first axis. In Figure 5.4, the final site scores of the second axis are plotted against those of the first axis. Site 1 and Site 19 lie far apart on the second axis, which reflects their difference in species composition. A third axis can be derived in the same way by making the scores uncorrelated with the scores of the first two axes, and so on. Table 5.2a summarizes the algorithm of two-way weighted averaging. A worked example is given in Exercise 5.1 and its solution.

In mathematics, the ordination axes of CA are termed eigenvectors (a vector is a set of values, commonly denoting a point in a multidimensional space and 'eigen' is German for 'self'). If we carry out an extra iteration cycle, the scores (values) remain the same, so the vector is transformed into itself, hence, the term eigenvector. Each eigenvector has a corresponding eigenvalue, often denoted by λ (the term is explained in Exercise 5.1.3). The eigenvalue is actually equal to the (maximized) dispersion of the species scores on the ordination axis, and is thus a measure of importance of the ordination axis. The first ordination axis has the largest eigenvalue (λ_1), the second axis the second largest eigenvalue (λ_2),

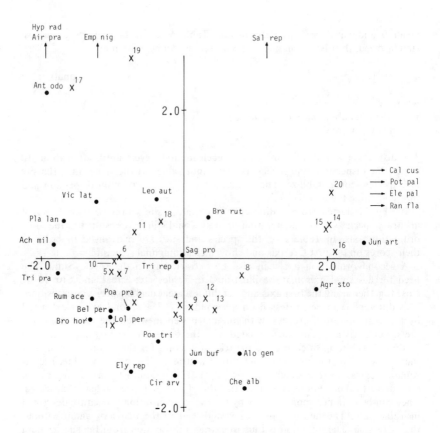

Figure 5.4 CA ordination diagram of the Dune Meadow Data in Hill's scaling. In this and the following ordination diagrams, the first axis is horizontal and the second axis vertical; the sites are represented by crosses and labelled by their number in Table 5.1; species names are abbreviated as in Table 0.1.

and so on. The eigenvalues of CA all lie between 0 and 1. Values over 0.5 often denote a good separation of the species along the axis. For the Dune Meadow Data, $\lambda_1 = 0.53$; $\lambda_2 = 0.40$; $\lambda_3 = 0.26$; $\lambda_4 = 0.17$. As λ_3 is small compared to λ_1 and λ_2, we ignore the third and higher numbered ordination axes, and expect the first two ordination axes to display the biologically relevant information (Figure 5.4).

When preparing an ordination diagram, we plot the site scores and the species scores of one ordination axis against those of another. Because ordination axes differ in importance, one would wish the scores to be spread out most along the most important axis. But our site scores do not do so, because we standardized them to variance 1 for convenience in the algorithm (Table 5.2). An attractive standardization is obtained by requiring that the average width of the species curves is the same for each axis. As is clear from Figure 5.3b, the width of the curve for Species D is reflected in the spread among its presences along the axis. Therefore, the average curve width along an axis can be estimated from the data. For example, Hill (1979) proposed to calculate, for each species, the variance of the scores of the sites containing the species and to take the (weighted) average of the variances so obtained, i.e. Hill proposed to calculate

$$\Sigma_k \, y_{k+} \, [\Sigma_i \, y_{ki} \, (x_i - u_k)^2 / y_{k+}] / y_{++}.$$

To equalize the average curve width among different axes, we must therefore divide all scores of an axis by its average curve width (i.e. by the square root of the value obtained above). This method of standardization is used in the computer program DECORANA (Hill 1979a). Other than in Table 5.2, the program further uses the convention that site scores are weighted averages of species scores; so we must iterate Step 3 of our algorithm once more, before applying the standardization procedure just described. This scaling has already been used in preparing Figure 5.4 and we shall refer to it as Hill's scaling. A short cut to obtain Hill's scaling from the scores obtained from our algorithm is to divide the site scores after convergence by $\sqrt{(1 - \lambda)/\lambda}$ and the species scores by $\sqrt{\lambda(1 - \lambda)}$. The scores so obtained are expressed in multiples of one standard deviation (s.d.) and have the interpretation that sites that differ by 4 s.d. in score tend to have few species in common (Figure 5.3b). This use of s.d. will be discussed further in Subsection 5.2.4.

CA cannot be applied on data that contain negative values. So the data should not be centred or standardized (Subsection 2.4.4). If the abundance data of each species have a highly skew distribution with many small values and a few extremely large values, we recommend transforming them by taking logarithms: $\log_e (y_{ki} + 1)$, as in Subsection 3.3.1. By doing so, we prevent a few high values from unduly influencing the analysis. In CA, a species is implicitly weighted by its relative total abundance y_{k+}/y_{++} and, similarly, a site is weighted by y_{+i}/y_{++}. If we want to give a particular species, for example, triple its weight, we must multiply all its abundance values by 3. Sites can also be given greater or smaller weight by multiplying their abundance values by constants (ter Braak 1987b).

5.2.3 Diagonal structures: properties and faults of correspondence analysis

Table 5.3a shows artificial data in which the occurrences of species across sites appear rather chaotic and Table 5.3b shows the same data after arranging the species and sites in order of their score on the first CA axis. The data are rearranged into a perfectly diagonal table, also termed a two-way Petrie matrix. (A Petrie matrix is an incidence matrix that has a block of consecutive ones in every row; the matrix is two-way Petrie if the matrix also has a block of consecutive ones in every column, the block in the first column starting in the first row and the block of the last column ending in the last row.) For any table that permits such a rearrangement, we can discover the correct order of species and sites from

103

the scores of the first axis of CA. This property of CA can be generalized to quantitative data (Gifi 1981) and to (one-way) Petrie matrices (Heiser 1981; 1986). For two-way Petrie matrices with many species and sites and with about equal numbers of occurrences per species and per site, the first eigenvalue is close to 1; e.g. for Table 5.3, $\lambda_1 = 0.87$.

Note that CA does not reveal the diagonal structure if the ones and zeros are interchanged. Their role is asymmetrical, as is clear from the reciprocal averaging algorithm. The ones are important; the zeros are disregarded. Many ecologists feel the same sort of asymmetry between presences and absences of species.

The ordination of Table 5.3 illustrates two 'faults' of CA (Figure 5.5). First, the change in species composition between consecutive sites in Table 5.3, Column b is constant (one species appears; one disappears) and one would therefore wish that this constant change were reflected in equal distances between scores of neighbouring sites along the first axis. But the site scores at the ends of the first axis are closer together than those in the middle of the axis (Figure 5.5b). Secondly, the species composition is explained perfectly by the ordering of the sites and species along the first axis (Table 5.3, Column b) and the importance of the second axis should therefore be zero. However $\lambda_2 = 0.57$ and the site scores on the second axis show a quadratic relation with those on the first axis (Figure 5.5a). This fault is termed the arch effect. The term 'horseshoe' is also in use but is less appropriate, as the ends do not fold inwards in CA.

Table 5.3 CA applied to artificial data (– denotes absence). Column a: The table looks chaotic. Column b: After rearrangement of species and sites in order of their scores on the first CA axis (u_k and x_i), a two-way Petrie matrix appears: $\lambda_1 = 0.87$.

Column a		Column b		u_k
Species	Sites 1 2 3 4 5 6 7	Species	Sites 1 7 2 4 6 5 3	
A	1 – – – – – –	A	1 – – – – – –	–1.40
B	1 – – – – – 1	B	1 1 – – – – –	–1.24
C	1 1 – – – – 1	C	1 1 1 – – – –	–1.03
D	– – – 1 1 1 –	E	– 1 1 1 – – –	–0.56
E	– 1 – 1 – – 1	F	– – 1 1 1 – –	0.00
F	– 1 – 1 – 1 –	D	– – – 1 1 1 –	0.56
G	– – 1 – 1 1 –	G	– – – – 1 1 1	1.03
H	– – 1 – 1 – –	H	– – – – – 1 1	1.24
I	– – 1 – – – –	I	– – – – – – 1	1.40
			– – –	
			1 1 0 0 0 1 1	
			· · · · · · ·	
		x_i	4 0 6 0 6 0 4	
			0 8 0 0 0 8 0	

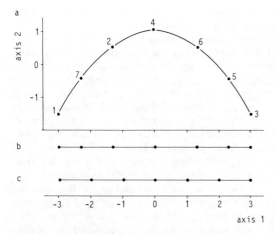

Figure 5.5 Ordination by CA of the two-way Petrie matrix of Table 5.3. a: Arch effect in the ordination diagram (Hill's scaling; sites labelled as in Table 5.3; species not shown). b: One-dimensional CA ordination (the first axis scores of Figure a, showing that sites at the ends of the axis are closer together than sites near the middle of the axis. c: One-dimensional DCA ordination, obtained by nonlinearly rescaling the first CA axis. The sites would not show variation on the second axis of DCA.

Let us now give a qualitative explanation of the arch effect. Recall that the first CA axis maximally separates the species curves by maximizing the dispersion (Equation 5.3) and that the second CA axis also tries to do so but subject to the constraint of being uncorrelated with the first axis (Subsection 5.2.1). If the first axis fully explains the species data in the way of Figure 5.3b, then a possible second axis is obtained by folding the first axis in the middle and bringing the ends together (Figure 5.3c). This folded axis has no linear correlation with the first axis. The axis so obtained separates the species curves; at least Species C from Species B and D, and these from Species A and E, and is thus a strong candidate for the second axis of CA. Commonly CA will modify this folded axis somewhat, to maximize its dispersion, but the order of the site and species scores on the second CA axis will essentially be the same as that of the folded axis. Even if there is a true second underlying gradient, CA will not take it to be the second axis if its dispersion is less than that of the modified folded first axis. The intention in constructing the second CA axis is to express new information, but CA does not succeed in doing so if the arch effect appears.

5.2.4 Detrended correspondence analysis (DCA)

Hill & Gauch (1980) developed detrended correspondence analysis (DCA) as a heuristic modification of CA, designed to correct its two major 'faults': (1) that the ends of the axes are often compressed relative to the axes middle; (2) that

105

the second axis frequently shows a systematic, often quadratic relation with the first axis (Figure 5.5). The major of these is the arch effect.

The arch effect is 'a mathematical artifact, corresponding to no real structure in the data' (Hill & Gauch 1980). They eliminate it by 'detrending'. Detrending is intended to ensure that, at any point along the first axis, the mean value of the site scores on the subsequent axes is about zero. To this end, the first axis is divided into a number of segments and within each segment the site scores on Axis 2 are adjusted by subtracting their mean (Figure 5.6). In the computer program DECORANA (Hill 1979a), running segments are used for this purpose. This process of detrending is built into the two-way weighted averaging algorithm, and replaces the usual orthogonalization procedure (Table 5.2). Subsequent axes are derived similarly by detrending with respect to each of the existing axes. Detrending applied to Table 5.3 gives a second eigenvalue of 0, as required.

The other fault of CA is that the site scores at the end of the first axis are often closer together than those in the middle of the axis (Figure 5.5b). Through this fault, the species curves tend to be narrower near the ends of the axis than in the middle. Hill & Gauch (1980) remedied this fault by nonlinearly rescaling the axis in such a way that the curve widths were practically equal. Hill & Gauch (1980) based their method on the tolerances of Gaussian response curves for the species, using the term standard deviation (s.d.) instead of tolerance. They noted that the variance of the optima of species present at a site (the 'within-site variance') is an estimate of the average squared tolerance of those species. Rescaling must therefore equalize the within-site variances as nearly as possible. For rescaling, the ordination axis is divided into small segments; the species ordination is expanded in segments with sites with small within-site variance and contracted in segments with sites with high within-site variance. Subsequently, the site scores are calculated by taking weighted averages of the species scores and the scores of sites and species are standardized such that the within-site variance equals 1. The tolerances of the curves of species will therefore approach 1. Hill & Gauch (1980) further define the length of the ordination axis to be the range of the site scores. This length is expressed in multiples of the standard deviation, abbreviated as s.d.

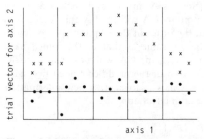

axis 1

Figure 5.6 Method of detrending by segments (simplified). The crosses indicate site scores before detrending; the dots are site scores after detrending. The dots are obtained by subtracting, within each of the five segments, the mean of the trial scores of the second axis (after Hill & Gauch 1980).

106

The use of s.d. is attractive: a Gaussian response curve with tolerance 1 rises and falls over an interval of about 4 s.d. (Figure 3.6). Because of the rescaling, most species will about have this tolerance. Sites that differ 4 s.d. in scores can therefore be expected to have no species in common. Rescaling of the CA axis of Table 5.3 results in the desired equal spacing of the site scores (Figure 5.5c); the length of the axis is 6 s.d.

DCA applied to the Dune Meadow Data gives, as always, the same first eigenvalue (0.53) as CA and a lower second eigenvalue (0.29 compared to 0.40 in CA). The lengths of the first two axes are estimated as 3.7 and 3.1 s.d., respectively. Because the first axis length is close to 4 s.d., we predict that sites at opposite ends of the first axis have hardly any species in common. This prediction can be verified in Table 5.1c (the order of DCA scores on the first axis is identical to that of CA); Site 17 and Site 16 have no species in common, but closer sites have one or more species in common. The DCA ordination diagram (Figure 5.7) shows the same overall pattern as the CA diagram of Figure 5.4. There are, however,

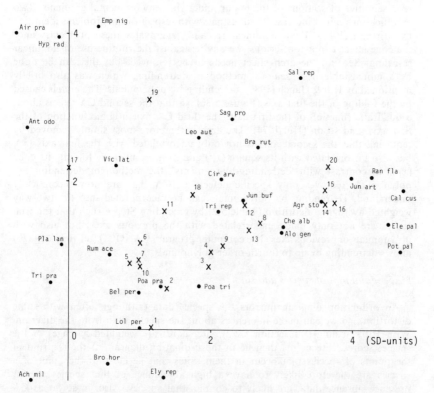

Figure 5.7 DCA ordination diagram of the Dune Meadow Data. The scale marks are in multiples of the standard deviation (s.d.).

differences in details. The arch seen in Figure 5.4 is less conspicuous, the position of Sites 17 and 19 is less aberrant. Further, *Achillea millefolium* is moved from a position close to Sites 2, 5, 6, 7 and 10 to the bottom left of Figure 5.7 and is then closest to Site 1; this move is unwanted, as this species is most abundant in the former group of sites (Table 5.1).

In an extentive simulation study, Minchin (1987) found that DCA, as available in the program DECORANA, can flatten out some of the variation associated with one of the underlying gradients. He ascribed this loss of information to an instability in either, or both, detrending and rescaling. Pielou (1984, p. 197) warned that DCA is 'overzealous' in correcting the 'defects' in CA and that it 'may sometimes lead to the unwitting destruction of ecologically meaningful information'.

DCA is popular among practical field ecologists; presumably because it provides an effective approximate solution to the ordination problem for a unimodal response model in two or more dimensions – given that the data are reasonably representative of sections of the major underlying environmental gradients. Two modifications might increase its robustness with respect to the problems identified by Minchin (1987). First, nonlinear rescaling aggravates these problems; since the edge effect is not too serious, we advise against the routine use of nonlinear rescaling. Second, the arch effect needs to be removed, but this can be done by a more stable, less 'zealous' method of detrending, which was also briefly mentioned by Hill & Gauch (1980): detrending-by-polynomials. The arch is caused by the folding of the first axis (Figure 5.3c), so that the second CA axis is about a quadratic function of the first axis, the third CA axis a cubic function of the first axis, and so on (Hill 1974). The arch is therefore most simply removed by requiring that the second axis is not only uncorrelated with the first axis (x_i), but also uncorrelated with its square (x_i^2) and, to prevent more folding, its cube (x_i^3). In contrast with 'detrending-by-segments', the method of detrending-by-polynomials removes only specific defects of CA that are now theoretically understood. Detrending by polynomials can be incorporated into the two-way weighted averaging algorithm (Table 5.2) by extending Step 4 such that the trial scores are not only made uncorrelated with the previous axes, but also with polynomials of previous axes. The computer program CANOCO (ter Braak 1987b) allows detrending by up to fourth-order polynomials.

5.2.5 Joint plot of species and sites

An ordination diagram mirrors the species data (although often with some distortion), so we can make inferences about the species data from the diagram. With Hill's scaling (Subsection 5.2.2), site scores are weighted averages of the species scores. Site points then lie in the ordination diagram at the centroid of the points of species that occur in them. Sites that lie close to the point of a species are therefore likely to have a high abundance of that species or, for presence–absence data, are likely to contain that species. Also, in so far as CA and DCA are a good approximation to fitting bell-shaped response surfaces to the species data (Subsection 5.2.1 and Section 5.7), the species points are close

108

to the optima of these surfaces; hence, the expected abundance or probability of occurrence of a species decreases with distance from its position in the plot (Figure 3.14).

Using these rules to interpret DCA diagrams, we predict as an example the rank order of species abundance for three species from Figure 5.7 and compare the order with the data in Table 5.1. The predicted rank order for *Juncus bufonius* is Sites 12, 8, 13, 9, 18 and 4; in the data *Juncus bufonius* is present at four sites, in order of abundance Sites 9, 12, 13 and 7. The predicted rank order for *Rumex acetosa* is Sites 5, 7, 6, 10, 2 and 11; in the data *R. acetosa* occurs in five sites, in order of abundance Sites 6, 5, 7, 9 and 12. *Ranunculus flammula* is predicted to be most abundant at Sites 20, 14, 15, 16 and less abundant, if present at all, at Sites 8, 12 and 13; in the data, *R. flammula* is present in six sites, in order of abundance Sites 20, 14, 15, 16, 8 and 13. We see some agreement between observations and predictions but also some disagreement. What is called for is a measure of goodness of fit of the ordination diagram. Such a measure is, however, not normally available in CA and DCA.

In interpreting ordination diagrams of CA and DCA, one should be aware of the following aspects. Species points on the edge of the diagram are often rare species, lying there either because they prefer extreme (environmental) conditions or because their few occurrences by chance happen to be at sites with extreme conditions. One can only decide between these two possibilities by additional external knowledge. Such species have little influence on the analysis; if one wants to enlarge the remainder of the diagram, it may be convenient not to display them at all. Further, because of the shortcomings of the method of weighted averaging, species at the centre of the diagram may either be unimodal with optima at the centre, or bimodal, or unrelated to the ordination axes. Which possibility is most likely can be decided upon by table rearrangement as in Table 5.1c or by plotting the abundance of a species against the axes. Species that lie between the centre and the outer edge are most likely to show a clear relation with the axes.

5.2.6 *Block structures and sensitivity to rare species*

CA has attractive properties in the search for block structures. A table is said to have block structure if its sites and species can be divided into clusters, with each cluster of species occurring in a single cluster of sites (Table 5.4). For any table that allows such a clustering, CA will discover it without fail. With the four blocks in Table 5.4, the first three eigenvalues of CA equal 1 and sites from the same cluster have equal scores on the three corresponding axes. An eigenvalue close to 1 can therefore point to an almost perfect block structure or to a diagonal structure in the data (Subsection 5.2.3). The search for block structures or 'near-block structures' by CA forms the basis of the cluster-analysis program TWINSPAN (Chapter 6).

This property of CA is, however, a disadvantage in ordination. If a table contains two disjoint blocks, one of which consists of a single species and a single site, then the first axis of CA finds this questionably uninteresting block. For a similar

Table 5.4 Data table with block structure. Outside the Sub-tables A_1, A_2, A_3 and A_4, there are no presences, so that there are four clusters of sites that have no species in common ($\lambda_1 = 1$, $\lambda_2 = 1$, $\lambda_3 = 1$).

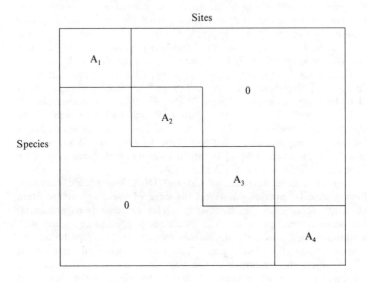

reason, CA is sensitive to species that occur only in a few species-poor sites. In the 'down-weighting' option of the program DECORANA (Hill 1979a), species that occur in a few sites are given a low weight, so minimizing their influence, but this does not fully cure CA's sensitivity to rare species at species-poor sites.

5.2.7 Gaussian ordination and its relation with CA and DCA

In the introduction to CA (Subsection 5.2.1), we assumed that species show unimodal response curves to environmental variables, intuitively took the dispersion of the species scores as a plausible measure of how well an environmental variable explains the species data, and subsequently defined CA to be the technique that constructs a theoretical variable that explains the species data best in the sense of maximizing the dispersion. Because of the shortcomings of CA noted in the subsequent sections, the dispersion of the species scores is not ideal to measure the fit to the species data. We now take a similar approach but with a better measure of fit and assume particular unimodal response curves. We will introduce ordination techniques that are based on the maximum likelihood principle (Subsections 3.3.2 and 4.2.1), in particular Gaussian ordination, which is a theoretically sound but computationally demanding technique of ordination. We also show that the simpler techniques of CA and DCA give about the same result if particular additional conditions hold true. This subsection may now be skipped at first reading; it requires a working knowledge of Chapters 3 and 4.

In maximum likelihood ordination, a particular response model (Subsection 3.1.2) is fitted to the species data by using the maximum likelihood principle. In this approach, the fit is measured by the deviance (Subsection 3.3.2) between the data and the fitted curves. Recall that the deviance is inversely related to the likelihood, namely deviance $= -2 \log_e$ (likelihood). If we fit Gaussian (logit) curves (Figure 3.9) to the data, we obtain Gaussian ordination. In Subsection 3.3.3, we fitted a Gaussian logit curve of pH to the presence–absence data of a particular species (Figure 3.10). In principle, we can fit a separate curve for each species under consideration. A measure of how badly pH explains the species data is then the deviance (Table 3.6) summed over all species. Gaussian ordination of presence–absence data is then the technique that constructs the theoretical variable that best explains the species data by Gaussian logit curves, i.e. that minimizes the deviance between the data and the fitted curves.

A similar approach can be used for abundance data by fitting Gaussian curves to the data, as in Section 3.4, with the assumption that the abundance data follow a Poisson distribution. A Gaussian curve for a particular species has three parameters: optimum, tolerance and maximum (Figure 3.6), for species k denoted by u_k, t_k and c_k, respectively. In line with Equation 3.8, the Gaussian curves can now be written as

$$Ey_{ki} = c_k \exp [-0.5(x_i - u_k)^2/t_k^2] \qquad \text{Equation 5.4}$$

where x_i is the score of site i on the ordination axis (the value of the theoretical variable at site i).

To fit this response model to data we can use an algorithm akin to that to obtain the ordination axis in CA (Table 5.2).

Step 1: Start from initial site scores x_i.

Step 2: Calculate new species scores by (log-linear) regression of the species data on the site scores (Section 3.4). For each species, we so obtain new values for u_k, t_k and c_k.

Step 3: Calculate new site scores by maximum likelihood calibration (Subsection 4.2.1).

Step 4: Standardize the site scores and check whether they have changed and, if so, go back to Step 2, otherwise stop.

In this algorithm, the ordination problem is solved by solving the regression problem (Chapter 3) and the calibration problem (Chapter 4) in an iterative fashion so as to maximize the likelihood. In contrast to the algorithm for CA, this algorithm may give different results for different initial site scores because of local maxima in the likelihood function for Equation 5.4. It is therefore not guaranteed that the algorithm actually leads to the (overall) maximum likelihood estimates; hence, we must supply 'good' initial scores, which are also needed to reduce the computational burden. Even for modern computers, the algorithm requires heavy computation. In the following, we show that a good choice for initial scores are the scores obtained by CA.

The CA algorithm can be thought of a simplification of the maximum likelihood algorithm. In CA, the regression and calibration problems are both solved by weighted averaging. Recall that in CA the species score (u_k) is a position on the ordination axis x indicating the value most preferred by that particular species (its optimum) and that the site score (x_i) is the position of that particular site on the axis.

We saw in Section 3.7 that the optimum or score of a species (u_k) can be estimated efficiently by weighted averaging of site scores provided that (Figure 3.18b):

A1. the site scores are homogeneously distributed over the whole range of occurrence of the species along the axis x.

In Section 4.3, we saw that the score (x_i) of a site is estimated efficiently by weighted averaging of species optima provided the species packing model holds, i.e. provided (Figure 4.1):

A2. the species' optima (scores) are homogeneously distributed over a large interval around x_i.

A3. the tolerances of species t_k are equal (or at least independent of the optima; ter Braak 1985).

A4. the maxima of species c_k are equal (or at least independent of the optima; ter Braak 1985).

Under these four conditions the scores obtained by CA approximate the maximum likelihood estimates of the optima of species and the site values in Gaussian ordination (ter Braak 1985). For presence–absence data, CA approximates similarly the maximum likelihood estimates of the Gaussian logit model (Subsection 3.3.3). CA does not, however, provide estimates for the maximum and tolerance of a species.

A problem is that assumptions A1 and A2 cannot be satisfied simultaneously for all sites and species: the first assumption requires that the range of the species optima is amply contained in the range of the site scores whereas the second assumption requires the reverse. So CA scores show the edge effect of compression of the end of the first axis relative to the axis middle (Subsection 5.2.3). In practice, the ranges may coincide or may only partly overlap. CA does not give any clue about which possibility is likely to be true. The algorithm in Table 5.2 results in species scores that are weighted averages of the site scores and, consequently, the range of the species scores is contained in the range of the site scores. But it is equally valid mathematically to stop at Step 3 of the algorithm, so that the site scores are weighted averages of the species scores and thus all lie within the range of the species scores; this is done in the computer program DECORANA (Hill 1979). The choice between these alternatives is arbitrary. It may help interpretation of CA results to go one step further in the direction of the maximum likelihood estimates by one regression step in which the data of each species are regressed on the site scores of CA by using the Gaussian response model. This can be done by methods discussed in Chapter 3. The result is new species scores (optima) as well as estimates for the tolerances and maxima. As an example, Figure 5.8 shows Gaussian response curves along the first CA axis fitted to the

112

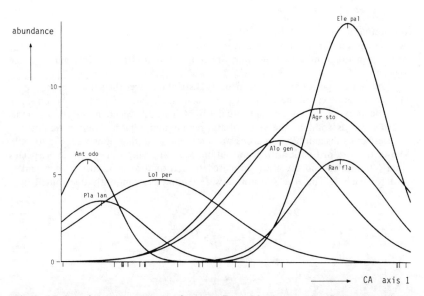

Figure 5.8 Gaussian response curves for some Dune Meadow species, fitted by log-linear regression of the abundances of species (Table 5.1) on the first CA axis. The sites are shown as small vertical lines below the horizontal axis.

Dune Meadow Data in Table 5.1. The curve of a particular species was obtained by a log-linear regression (Section 3.4) of the data of the species on the site scores of the first CA axis by using $b_0 + b_1 x + b_2 x^2$ in the linear predictor (Equation 3.18).

Two dimensions

In two dimensions, Gaussian ordination means fitting the bivariate Gaussian surfaces (Figure 3.14)

$$Ey_{ki} = c_k \exp (-0.5[(x_{i1} - u_{k1})^2 + (x_{i2} - u_{k2})^2]/t_k^2) \qquad \text{Equation 5.5}$$

where
(u_{k1}, u_{k2}) are the coordinates of the optimum of species k in the ordination diagram
c_k is the maximum of the surface
t_k is the tolerance
(x_{i1}, x_{i2}) are the coordinates of site i in the diagram.

These Gaussian surfaces look like that of Figure 3.14, but have circular contours because the tolerances are taken to be the same in both dimensions.
    One cannot hope for more than that the two-axis solution of CA provides an approximation to the fitting of Equation 5.5 if the sampling distribution of

113

the abundance data is Poisson and if:

A1. site points are homogeneously distributed over a rectangular region in the ordination diagram with sides that are long compared to the tolerances of the species,

A2. optima of species are homogeneously distributed over the same region,

A3. the tolerances of species are equal (or at least independent of the optima),

A4. the maxima of species are equal (or at least independent of the optima).

However as soon as the sides of the rectangular region differ in length, the arch effect (Subsection 5.2.3) crops up and the approximation is bad. Figure 5.9b shows the site ordination diagram obtained by applying CA to artificial species data (40 species and 50 sites) simulated from Equation 5.5 with $c_k = 5$ and $t_k = 1$ for each k. The true site points were completely randomly distributed over

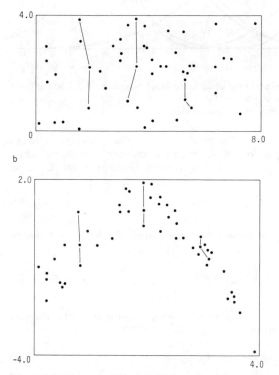

Figure 5.9 CA applied to simulated species data. a: True configuration of sites (•). b: Configuration of sites obtained by CA, showing the arch effect. The data were obtained from the Gaussian model of Equation 5.5 with Poisson error, $c_k = 5$, $t_k = 1$ and optima that were randomly distributed in the rectangle $[-1,9] \times [-0.5,4.5]$. The vertical lines in Figures a and b connect identical sites.

114

a rectangular region with sides of 8 and 4 s.d. (Figure 5.9a). The CA ordination diagram is dominated by the arch effect, although the actual position of sites within the arch still reflects their position on the second axis in Figure 5.9a. The configuration of site scores obtained by DCA was much closer to the true configuration. DCA forcibly imposes Conditions A1, A2 and A3 upon the solution, the first one by detrending and the second and third one by rescaling of the axes.

We also may improve the ordination diagram of DCA by going one step further in the direction of maximum likelihood ordination by one extra regression step. We did so for the DCA ordination (Figure 5.7) of Dune Meadow Data in Table 0.1. For each species with more than 4 presences, we carried out a log-linear regression of the data of the species on the first two DCA axes using the response model

$$\log_e E y_{ki} = b_0 + b_{1k} x_{i1} + b_{3k} x_{i2} + b_{4k} (x_{i1}^2 + x_{i2}^2) \qquad \text{Equation 5.6}$$

where x_{i1} and x_{i2} are the scores of site i on the DCA axes 1 and 2, respectively.

If $b_{4k} < 0$, this model is equivalent to Equation 5.5 (as in Subsection 3.3.3). The new species scores are then obtained from the estimated parameters in Equation 5.6 by $u_{k1} = -b_{1k}/(2b_{4k})$, $u_{k2} = -b_2/(2b_{4k})$ and $t_k = 1/\sqrt{(-2b_{4k})}$.

If $b_{4k} > 0$, the fitted surface shows a minimum and we have just plotted the DCA scores of the species. Figure 5.10a shows how the species points obtained by DCA change by applying this regression method to the 20 species with four or more presences. A notable feature is that *Achillea millefolium* moves towards its position in the CA diagram (Figure 5.4). In Figure 5.10b, circles are drawn with centres at the estimated species points and with radius t_k. The circles are contours where the expected abundance is 60% of the maximum expected abundance c_k. Note that $\exp(-0.5) = 0.60$.

From Figure 5.10b, we see, for example, that *Trifolium repens* has a high tolerance (a large circle, thus a wide ecological amplitude) whereas *Bromus hordaceus* has a low tolerance (a small circle, thus a narrow ecological amplitude). With regression, the joint plot of DCA can be interpreted with more confidence. This approach also leads to a measure of goodness of fit. A convenient measure of goodness of fit is here

$$V = 1 - (\Sigma_k D_{k1})/ (\Sigma_k D_{k0}) \qquad \text{Equation 5.7}$$

where D_{k0} and D_{k1} are the residual deviances of the kth species for the null model (the model without explanatory variables) and the model depicted in the diagram (Equation 5.6), respectively. These deviances are obtained from the regressions (as in Table 3.7). We propose to term V the fraction of deviance accounted for by the diagram. For the two-axis ordination (only partially displayed in Figure 5.10b) $V = (1 - 360/987) = 0.64$. For comparison, $V = 0.51$ for the one-axis ordination (partially displayed in Figure 5.8).

115

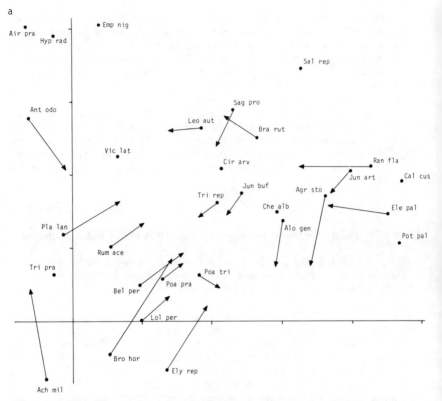

a

Figure 5.10 Gaussian response surfaces for several Dune Meadow species fitted by log-linear regression of the abundances of species on the site scores of the first two DCA axes (Figure 5.7). a: Arrows for species running from their DCA scores (Figure 5.7) to their fitted optimum. b: Optima and contours for some of the species. The contour indicates where the abundance of a species is 60% of the abundance at its optimum.

The regression approach can of course be extended to more complicated surfaces (e.g. Equation 3.24), but this will often be impractical, because these surfaces are more difficult to represent graphically.

5.3 Principal components analysis (PCA)

5.3.1 From least-squares regression to principal components analysis

Principal components analysis (PCA) can be considered to be an extension of fitting straight lines and planes by least-squares regression. We will introduce

116

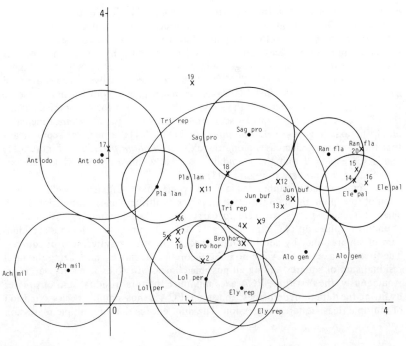

Figure 5.10b

PCA, assuming the species data to be quantitative abundance values.

Suppose we want to explain the abundance values of several species by a particular environmental variable, say moisture, and suppose we attempt to do so by fitting straight lines to the data. Then, for each species, we have to carry out a least-squares regression of its abundance values on the moisture values and obtain, among other things, the residual sum of squares, i.e. the sum of squared vertical distances between the observed abundance values and the fitted line (Figure 3.1; Subsection 3.2.2). This is a measure of how badly moisture explains the data of a single species. To measure how badly moisture explains the data of all species, we now use the total of the separate residual sums of squares over all species, abbreviated the total residual sum of squares. If the total residual sum of squares is small, moisture can explain the species data well.

Now, suppose that, among a set of environmental variables, moisture is the variable that best explains the species data in the sense of giving the least total residual sum of squares. As in all ordination techniques, we now wish to construct a theoretical variable that explains the species data still better. PCA is the ordination technique that constructs the theoretical variable that minimizes the total residual sum of squares after fitting straight lines to the species data. PCA does so by

117

choosing best values for the sites, the site scores. This is illustrated in Figure 5.11 for the Dune Meadow Data. The site scores are indicated by ticks below the horizontal axis. The fitted lines are shown for six of the 30 species and the observed abundance values and residuals for one of them. Any other choice of site scores would result in a larger sum of squared residuals. Note that Figure 5.11 shows only 20 out of all 20 × 30 = 600 residuals involved. The horizontal axis in Figure 5.11 is the first PCA axis, or first principal component. The score of a species in PCA is actually the slope of the line fitted for the species against the PCA axis. A positive species score thus means that the abundance increases along the axis (e.g. *Agrostis stolonifera* in Figure 5.11); a negative score means that the abundance decreases along the axis (e.g *Lolium perenne* in Figure 5.11) and a score near 0 that the abundance is poorly (linearly) related to the axis (e.g. *Sagina procumbens* in Figure 5.11).

If a single variable cannot explain the species data sufficiently well, we may attempt to explain the data with two variables by fitting planes (Subsection 3.5.2). Then, for each species we have to carry out a least-squares regression of its abundance values on two explanatory variables (Figure 3.11), obtain its residual sum of squares and, by addition over species, the total residual sum of squares. The first two axes of PCA are now the theoretical variables minimizing the total residual sum of squares among all possible choices of two explanatory variables. Analogously, the first three PCA axes minimize the total residual sum of squares by fitting the data to hyperplanes, and so on. PCA is thus a multi-species extension of multiple (least-squares) regression. The difference is that in multiple regression

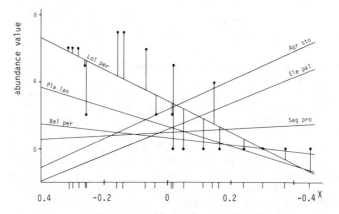

Figure 5.11 Straight lines for several Dune Meadow species, fitted by PCA to the species abundances of Table 5.1. Also shown are the abundances of *Lolium perenne* and their deviations from the fitted straight line. The horizontal axis is the first principal component. Fitting straight lines by least-squares regression of the abundances of species on the site scores of the first PCA axis gives the same results. The slope equals the species score of the first axis. The site scores are shown by small vertical lines below the horizontal axis.

118

the explanatory variables are supplied environmental variables whereas in PCA the explanatory variable are theoretical variables estimated from the species data alone. It can be shown (e.g. Rao 1973) that the same result as above is obtained by defining the PCA axes sequentially as follows. The first PCA axis is the variable that explains the species data best, and second and later axes also explain the species data best but subject to the constraint of being uncorrelated with previous PCA axes. In practice, we ignore higher numbered PCA axes that explain only a small proportion of variance in the species data.

5.3.2 Two-way weighted summation algorithm

We now describe an algorithm that has much in common with that of CA and that gives the ordination axes of PCA. The algorithm also shows PCA to be a natural extension of straight-line regression.

If the relation between the abundance of a species and an environmental variable is rectilinear, we can summarize the relation by the intercept and slope of a straight line. The error part of the model is taken to consist of independent and normally distributed errors with a constant variance. The parameters (intercept and slope) are then estimated by least-squares regression of the species abundances on the values of the environmental variable (Subsection 3.2.2). Conversely, when the intercepts and slopes are known, we can estimate the value of the environmental variable from the species abundances at a site by calibration (Subsection 4.2.3). If it is not known in advance which environmental variable determines the abundances of the species, the idea is as in CA (Subsection 5.2.2) to discover the 'underlying environmental gradient' by applying straight-line regression and calibration alternately in an iterative fashion, starting from arbitrary initial values for sites or from arbitrary initial values for the intercepts and slopes of species. As in CA, the iteration process eventually converges to a set of values for species and sites that does not depend on the initial values.

The iteration process reduces to simple calculations when we first centre the abundances of each species to mean 0 and standardize the site scores to $\bar{x} = 0$ and $\Sigma_i (x_i - \bar{x})^2 = 1$. Then, the equations to estimate the intercept and the slope of a straight line (Equations 3.6a,b) reduce to $b_0 = 0$ and $b_1 = \Sigma_i y_i x_i$, because in the notation of Subsection 3.2.2 $\bar{y} = 0$, $\bar{x} = 0$ and $\Sigma_i (x_i - \bar{x})^2 = 1$. Hence we ignore the intercepts and concentrate on the slope parameters. From now on, b_k will denote the slope parameter for species k and y_{ki} the centred abundance of species k at site i (i.e. $y_{k+} = 0$). In this notation, the slope parameter of species k is calculated by

$$b_k = \Sigma_{i=1}^n y_{ki} x_i \qquad\qquad \text{Equation 5.8}$$

As an example, Table 5.5a shows the Dune Meadow Data used before with an extra column of species means and, as arbitrary initial scores for the sites, values obtained by standardizing the numbers 1 to 20 (bottom row). For *Achillea millefolium*, the mean abundance is 0.80 and we obtain

Table 5.5a

```
                              Sites (i)
                              00000000011111111112
  Species                     12345678901234567890     mean    b_k
  k

 1  Ach mil   13 222  4 45447  2        5              0.80   -1.98
 2  Agr sto        48  43          2  3                2.40    1.55
 3  Air pra   272    53 85  4                          0.25    1.49
 4  Alo gen          432 4      4 4                    1.80   -2.06
 5  Ant odo          3222   2        2                 1.05    0.60
 6  Bel per          4 32 2  4                         0.65   -1.96
 7  Bro hor               1                            0.75   -2.85
 8  Che alb    2                                       0.05    0.10
 9  Cir arv                   458        4             0.05   -0.50
10  Ele pal   44444   6                                0.10    4.21
11  Ely rep                    2                       1.25   -6.17
12  Emp nig               2      2 5                   1.30    0.66
13  Hyp rad                44  33        4             0.10    2.19
14  Jun art               2 4  43                      0.45    2.02
15  Jun buf   5223333352222 2562                       0.90    0.02
16  Leo aut   7652664267     2                         0.65    0.93
17  Lot per       555  33     23                       2.70   -9.42
18  Pla lan   445423444444 2     13                    2.90   -1.24
19  Poa pra   2785645454 49 2                          1.30   -6.05
20  Poa tri                22                          2.40   -7.93
21  Pot pal            2  2222      4                  3.15    0.62
22  Ran fla        563 2  2                            0.20    2.52
23  Rum ace        5  22 242       3                   0.70   -2.52
24  Sag pro                    335                     0.90   -0.12
25  Sal rep        252                                 1.00    3.70
26  Tri pra   5212522363261     22                     0.55   -1.57
27  Tri rep              12                            2.35   -1.88
28  Vic lat   2226222244   44 634                      0.20    0.31
29  Bra rut                4 3     3                    2.45    2.89
30  Cal cus                                            0.50    2.29
                              --------
                              00000000000000000000
  x_i                         33222111000011122233
                              73951740622604715937
```

Table 5.5 Two-way weighted summation algorithm of PCA applied to the Dune Meadow Data. a: The original data table with at the bottom the initial site scores. b: The species and sites rearranged in order of their scores obtained after one cycle of two-way weighted summation. c: The species arranged in order of their final scores (PCA scores).

Table 5.5b

```
Species      Sites (i)                      b_k
k            0001000011011111112
             2370415693182874365 0

17 Lol per   566657262 74 2                -9.42
20 Poa tri   7654526459 44  2              -7.93
11 Ely rep   44  444 6                      -6.17
19 Poa pra   45444234244 31                -6.05
7  Bro hor   4 243 2                        -2.85
23 Rum ace   3 562  2                       -2.52
4  Alo gen   27 2  35 58    4              -2.06
1  Ach mil   3 24 122  2                    -1.98
6  Bel per   32 22  2                       -1.96
27 Tri rep   52261 25323232 62 1           -1.88
26 Tri pra   2 25                           -1.57
18 Pla lan   53 55   3  32                  -1.24
9  Cir arv   2                              -0.50
24 Sag pro   2 5   22224     3              -0.12
15 Jun buf     43 4                         -0.02
8  Che alb      1                            0.10
28 Vic lat   2    1  2                       0.31
5  Ant odo   24 43       4 4                 0.60
21 Pot pal            2 2                     0.62
12 Emp nig            2                       0.66
16 Leo aut   52332 33253325226 22            0.93
3  Air pra         2 3                        1.49
2  Agr sto   4  8  35 44 4 745               1.55
                  4 4  334                    2.02
14 Jun art   4  4                             2.02
13 Hyp rad   2 2  5                           2.19
30 Cal cus       2  4 3 3                     2.29
22 Ran fla   2 2  2224                        2.52
29 Bra rut   2222 262 4246 3444              2.89
25 Sal rep          3 3 5                     3.70
10 Ele pal       4  4 854                     4.21

x_i          000000000000000000000
             3222111100001122334
             3810998745417486922 0
```

Table 5.5c

```
Species      Sites (i)                      b_k
k            0100010010011011121
             6072511389479382450 6

17 Lol per   6665277622 5   4              -9.21
18 Pla lan   535 53  3  2                  -5.77
19 Poa pra   344424453441 24              -5.69
20 Poa tri   44576 26 55 944   2           -4.80
1  Ach mil   24232 1   2                    -3.81
23 Rum ace   6 3 5    2  2                  -3.68
27 Tri rep   562523 2231 222361            -3.67
5  Ant odo   342 4    44                    -3.52
7  Bro hor   4242   3                       -3.31
16 Leo aut   333535 25222623222 2          -2.86
11 Ely rep   44 44 64                       -2.86
26 Tri pra   5 2 2                          -2.63
6  Bel per   2 32  22 2                     -2.11
28 Vic lat   1  2   1    25                 -0.67
13 Hyp rad        2                         -0.08
9  Cir arv   2                               0.01
12 Emp nig             1                      0.09
8  Che alb         23                         0.11
3  Air pra            4  3 4                  0.15
15 Jun buf   2    4   3 4 444                 0.40
29 Bra rut   622 24 2622 3 24 444            0.94
24 Sag pro   2    25 3224                     0.98
21 Pot pal                22                  1.07
25 Sal rep   2  3  3  5                       1.86
4  Alo gen   2 7 32 558   4                   3.33
30 Cal cus            4 33                    3.40
22 Ran fla           22 2242                  3.95
14 Jun art       4   4 343                    4.29
10 Ele pal       4  38 4548                   8.08
2  Agr sto        4 38 5444457                8.67

x_i          0000000000000000000
             3322110000011112334
             1086664740115146407 5
```

121

$b_1 = (1 - 0.80) \times (-0.37) + (3 - 0.80) \times (-0.33) + (0 - 0.80) \times (-0.29) + ...$
$+ (0 - 0.80) \times (0.37) = -1.98.$

From the slopes thus obtained (Table 5.5a, last column), we derive new site scores by least-squares calibration (Equation 4.2 with $a_k = 0$). The site scores so obtained are proportional to

$$x_i = \Sigma_{k=1}^{m} y_{ki} b_k \qquad\qquad \text{Equation 5.9}$$

because the denominator of Equation 4.1 has the same value for each site. This denominator is unimportant in PCA, because the next step in the algorithm is to standardize the site scores, as shown in Table 5.6c. For Site 1 in Table 5.5a, we get from Equation 5.9 the site score $x_1 = (1 - 0.80) \times (-1.98) + (0 - 2.40)$ $\times (1.55) + (0 - 0.25) \times (1.49) + ... + (0 - 0.50) \times (2.29) = -0.19$. Note that the species mean abundance is subtracted each time from the abundance value. In Table 5.5b, the species and sites are arranged in order of the scores obtained so far, in which the slopes (b_k) form the species scores. The abundance of the species in the top row (*Lolium perenne*) has the tendency to decrease along the row, whereas the abundance of the species in the bottom row (*Eleocharis palustris*) has the tendency to increase across the row. The next cycle of the iteration is to calculate new species scores (b_k), then new site scores, and so on. As in CA, the scores stabilize after several iterations and the resulting scores (Table 5.5c) constitute the first ordination axis of PCA. In Table 5.5c, the species and sites are arranged in order of their scores on the first axis. Going from top row to bottom row, we see first a decreasing trend in abundance across the columns (e.g. for *Lolium perenne*), then hardly any trend (e.g. for *Sagina procumbens*) and finally an increasing trend (e.g. for *Agrostis stolonifera*). A graphical display of the trends has already been shown in Figure 5.11. The order of species in Table 5.5c is quite different from the order in the table arranged by CA (Table 5.1c), but the difference in ordering of the sites is more subtle.

In the above iteration algorithm of PCA (Table 5.6), weighted sums (Equations 5.8 and 5.9) replace the weighted averages in CA (Table 5.2; Equations 5.1 and 5.2). For this analogy to hold, let us consider the data y_{ki} as weights (which can be negative in PCA), so that the species scores are a weighted sum of the site scores and, conversely, the site scores are a weighted sum of the species scores (Table 5.6). The standard terminology used in mathematics is that x_i is a linear combination of the variables (species) and that b_k is the loading of species k.

After the first axis, a second axis can be extracted as in CA, and so on. (There is a subtle difference in the orthogonalization procedure, which need not concern us here.) The axes are also eigenvectors to which correspond eigenvalues as in CA (Subsection 5.2.2). The meaning of the eigenvalues in PCA is given below. The axes are also termed principal components.

So PCA decomposes the observed values into fitted values and residuals (Equations 3.1 and 3.2). In one dimension, we have the decomposition

$$y_{ki} = b_k x_i + \text{residual} \qquad\qquad \text{Equation 5.10}$$

Table 5.6 Two-way weighted summation algorithm of PCA.

---

a: Iteration process

Step 1. Take arbitrary initial site scores (x_i), not all equal to zero.
Step 2. Calculate new species scores (b_k) by weighted summation of the site scores (Equation 5.8).
Step 3. Calculate new site scores (x_i) by weighted summation of the species scores (Equation 5.9).
Step 4. For the first axis go to Step 5. For second and higher axes, make the site scores (x_i) uncorrelated with the previous axes by the orthogonalization procedure described below.
Step 5. Standardize the site scores (x_i). See below for the standardization procedure.
Step 6. Stop on convergence, i.e. when the new site scores are sufficiently close to the site scores of the previous cycle of the iteration; ELSE go to Step 2.

b: Orthogonalization procedure

Step 4.1. Denote the site scores of the previous axis by f_i and the trial scores of the present axis by x_i.
Step 4.2. Calculate $v = \Sigma_{i=1}^{n} x_i f_i$.
Step 4.3 Calculate $x_{i,new} = x_{i,old} - v f_i$.
Step 4.4 Repeat Steps 4.1-4.3 for all previous axes.

c: Standardization procedure

Step 5.1 Calculate the sum of squares of the site scores $s^2 = \Sigma_{i=1}^{n} x_i^2$.
Step 5.2 Calculate $x_{i,new} = x_{i,old}/s$.
Note that, upon convergence, s equals the eigenvalue.

---

where y_{ki} is the (mean corrected) observed value and $b_k x_i$ the fitted value.

As an example, the values fitted by the first PCA axis (Table 5.5c) for the centred abundances of *Agrostis stolonifera* ($b_2 = 8.67$) at Site 6 ($x_6 = -0.31$) and Site 16 ($x_{16} = 0.45$) are: $8.67 \times (-0.31) = -2.75$ and $8.67 \times 0.45 = 3.99$, respectively. Adding the mean value of *A. stolonifera* (2.40), we obtain the values -0.35 and 6.39, respectively, which are close to the observed abundance values of 0 and 7 at Site 6 and Site 16. In PCA, the sum of squared residuals in Equation 5.10 is minimized (Subsection 5.3.1). Analogously, one can say that PCA maximizes the sum of squares of fitted values and the maximum is the eigenvalue of the first axis. In two dimensions (Figure 5.12), we have the decomposition

$$y_{ki} = (b_{k1} x_{i1} + b_{k2} x_{i2}) + \text{residual} \hspace{2cm} \text{Equation 5.11}$$

where
b_{k1} and b_{k2} are the scores of species k
x_{i1} and x_{i2} are the scores of site i on Axis 1 and Axis 2, respectively.

On the second axis, the score of *A. stolonifera* is 6.10 and the scores of Sites 6 and 16 are -0.17 and 0.033 (Figure 5.12), so that the fitted values become $8.67 \times (-0.31) + 6.10 \times (-0.17) = -3.72$ and $8.67 \times 0.45 + 6.10 \times 0.033 = 4.10$.

123

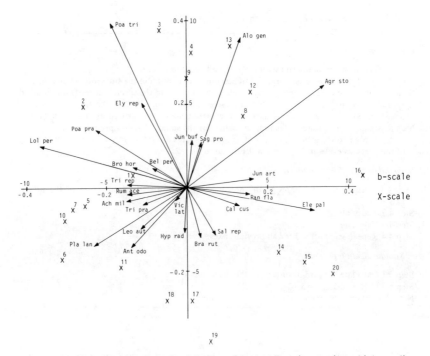

Figure 5.12 PCA-ordination diagram of the Dune Meadow Data in covariance biplot scaling with species represented by arrows. The *b* scale applies to species, the *x* scale to sites. Species not represented in the diagram lie close to the origin (0,0).

The first two PCA axes thus give approximate abundance values of $-3.72 + 2.40 = -1.3$ and $4.10 + 2.40 = 6.5$, slightly worse values than those obtained from the first axis, but most of the remaining abundance values in the data table will be approximated better with two axes than with a single axis. The sum of squares of fitted values now equals $\lambda_1 + \lambda_2$. Further, the total sum of squares ($\Sigma_k \Sigma_i \, y_{ki}^2$) equals the sum of all eigenvalues. (This equality means that we can reconstruct the observed values exactly from the scores of species and sites on all eigenvectors and the mean abundance values.) The fraction of variance accounted for (explained) by the first two axes is therefore $(\lambda_1 + \lambda_2)/$ (sum of all eigenvalues). This measure is the equivalent of R^2 in Section 3.2. For the Dune Meadow Data, $\lambda_1 = 471$, $\lambda_2 = 344$ and the total sum of squares = 1598. So the two-axes solution explains $(471 + 344)/1598 = 51\%$ of the variance. The first axis actually explains $471/1598 = 29\%$ of the variance and the second axis $344/1598 = 22\%$.

124

5.3.3 Best lines and planes in m-dimensional space

Here we present a geometric approach to PCA. In this approach, the aim of PCA is seen as being to summarize multivariate data in a graphical way. The approach is best illustrated with data on two species only. Figure 5.13a displays the abundances of Species A and B at 25 sites in the form of a scatter diagram, with axes labelled by the species names. The simplest summary of data is by the mean abundances of A (25) and B (15). Knowing the means, we may shift the axes to the centroid of the data points, i.e. to the point with the coordinates (25,15), provided we remember that the origin (0,0) of the new coordinate system is the point (25,15) in the old coordinate system. Next we draw a line through the new origin in the direction of maximum variance in the plot. This line is the first principal component (PC1), or first PCA axis, and perpendicularly we draw PC2. Next we rotate the plot, so that PC1 is horizontal (Figure 5.13b). Figure 5.13b is an ordination diagram with arrows representing the species. These arrows are the shifted and rotated axes of the species in the original diagram. PC2 shows so much less variation than PC1 that PC2 can possibly be neglected. This is done in Figure 5.13c showing a one-dimensional ordination; the points in Figure 5.13c were obtained from Figure 5.13b by drawing perpendicular lines from each point on the horizontal axis (projection onto PC1). In this way, the first coordinate of the points in Figure 5.13b is retained in Figure 5.13c; this coordinate is the site score on PC1. The first coordinate of the arrows in Figure 5.13b is the species loading on PC1, which is also represented by an arrow in Figure 5.13c. These arrows indicate the direction in which Species A and Species B increase in abundance; hence Figure 5.13c still shows which sites have high abundances of Species A and of Species B (those on the right side) and which sites have low abundance (those on the left side).

The example is, of course, artificial. Usually there are many species ($m \geqslant 3$), so that we need an m-dimensional coordinate system, and we want to derive a two-dimensional or three-dimensional ordination diagram. Yet the principle remains the same: PCA searches for the direction of maximum variance; this is PC1, the best line through the data points. It is the best line in the sense that it minimizes the sum of squares of perpendicular distances between the data points and the line (as is illustrated in Figure 5.13a for $m = 2$). So the first component in Figure 5.13a is neither the regression line of Species B on Species A nor that of Species A on Species B, because regression minimizes the sum of squares of vertical distances (Figure 3.1). But, as we have seen in Subsection 5.3.1, PCA does give the best regression of Species A on PC1 and of Species B on PC1 (Figure 5.11). After the first component, PCA seeks the direction of maximum variance that is perpendicular onto the first axis; that is PC2, which with PC1 forms the best plane through the data points, and so on. In general, the site scores are obtained by projecting each data point from the m-dimensional space onto the PCA axes and the species scores are obtained by projecting the unit vectors: for the first species (1,0,0,...); for the second species (0,1,0,0,...), etc., onto the PCA axes (Figure 5.13).

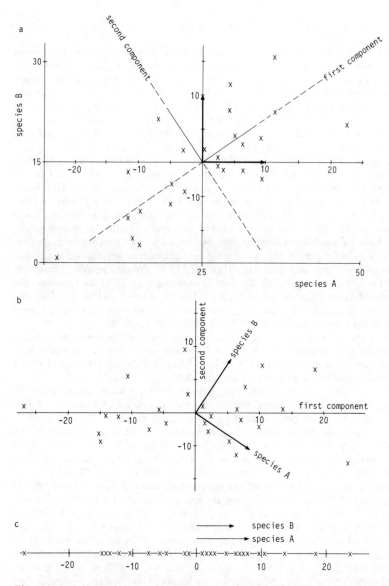

Figure 5.13 Artificial abundance data for two species A and B at 25 sites. a: First principal component, running through the centroid of the sites in the direction of the greatest variance (at 34° of the axis of Species A). b: Rotated version of Figure a with the first principal component horizontally. c: One-dimensional PCA ordination with species represented by arrows. The scores are simply those of the first axis of b.

126

5.3.4   Biplot of species and site scores

The scores obtained from a PCA for species and sites can be used to prepare a biplot (Gabriel 1971). The biplot serves the same function as the joint plot in CA (Subsection 5.2.5), but the rules to interpret the biplot are rather different. We limit the discussion to the two-dimensional biplot as it is more difficult to visualize three-dimensional or higher ones. The prefix 'bi' in biplot refers to the joint representation of sites and species, and not to the dimension of the plot; for example, Figure 5.13c shows a one-dimensional biplot.

The ranges of the scores for sites and for species (scores and loadings) in PCA are often of a different order of magnitude. For example in Table 5.3c, the range of the species scores is 17.9 whereas the range of the site scores is 0.8. A biplot is therefore constructed most easily by drawing separate plots of sites and of species on transparent paper, each one with its own scaling. In each of the plots, the scale unit along the vertical axis must have the same physical length as the scale unit along the horizontal axis, as in CA. A biplot is obtained by superimposing the plots with the axes aligned. A biplot may therefore have different scale units for the sites (x scale) and species (b scale). Figures 5.12 and 5.15 provide examples for the Dune Meadow Data.

In Subsection 5.3.1, we showed that for each species PCA fits a straight line in one dimension to the (centred) abundances of the species (Figure 5.11; Equation 5.10) and in two dimensions a plane with respect to the PCA axes (Figure 3.11; Equation 5.11). The abundance of a species as fitted by PCA thus changes linearly across the biplot. We represent the fitted planes in a biplot by arrows as shown in Figure 5.12. The direction of the arrow indicates the direction of steepest ascent of the plane, i.e. the direction in which the abundance of the corresponding species increases most, and the length of the arrow equals the rate of change in that direction. In the perpendicular direction, the fitted abundance is constant. The arrows are obtained by drawing lines that join the species points to the origin, the point with coordinates (0,0).

The fitted abundances of a species can be read from the biplot in very much the same way as from a scatter diagram, i.e. by projecting each site onto the axis of the species. (This is clear from Figure 5.13a.) The axis of a species in a biplot is in general, however, not the horizontal axis or the vertical axis, as in Figure 5.13a, but an oblique axis, the direction of which is given by the arrow of the species. As an example of how to interpret Figure 5.12, some of the site points are projected onto the axis of *Agrostis stolonifera* in Figure 5.14. Without doing any calculations, we can see the ranking of the fitted abundances of *A. stolonifera* among the sites from the order of the projection points of the sites along the axis of that species. From Figure 5.14, we thus infer that the abundance of *A. stolonifera* is highest at Site 16, second highest at Site 13, and so on to Site 6, which has the lowest inferred abundance. The inferred ranking is not perfect when compared with the observed ranking, but not bad either.

Another useful rule to interpret a biplot is that the fitted value is positive if the projection point of a site lies, along the species' axis, on the same side of the origin as the species point does, and negative if the origin lies between the

127

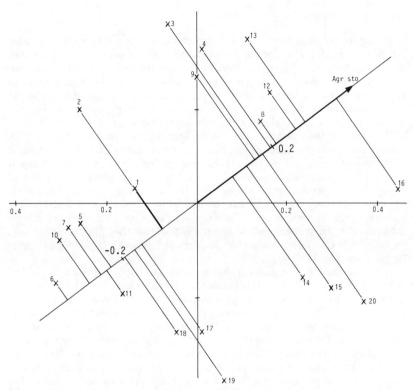

Figure 5.14 Biplot interpretation of Figure 5.12 for *Agrostis stolonifera*. For explanation, see text.

projection point and the species point. As we have centred the abundance data, the fitted abundance is higher than the species mean in the former case and lower than the species mean in the latter case. For example, Site 3 and Site 20 are inferred to have a higher than average abundance of *A. stolonifera*, whereas Sites 2 and 19 are inferred to have a lower than average abundance of this species. These inferences are correct, as can be seen from Table 5.5. One can also obtain quantitative values for the abundances as represented in the biplot, either algebraically with Equation 5.11 or geometrically as follows (ter Braak 1983). For this, we need the distance of the species point from the origin. In Figure 5.12, we see from the *b* scale that *A. stolonifera* lies at a distance of about 10 from the origin. We need further the projection points of sites onto the species' axis (Figure 5.14). From the *x* scale, we see that, for example, the projection point of Site 20 lies a distance of about 0.2 from the origin. The fitted value is now about $10 \times 0.2 = 2$. Adding the mean of *A. stolonifera* (2.4), we obtain

128

4.4 as the fitted abundance for *A. stolonifera* at Site 20; the observed value is 5. This biplot accounts in this way for 51% of the variance in abundance values of all species. This value was computed at the end of Subsection 5.3.2. Note, however, that the fraction of variance accounted for usually differs among species. In general, the abundances of species that are far from the origin are better represented in the biplot than the abundances of species near the origin. For example, the fractions accounted for are 80% for *Agrostis stolonifera*, 78% for *Poa trivialis*, 25% for *Bromus hordaceus*, 4% for *Brachythecium rutabulum* and 3% for *Empetrum nigrum*.

The scaling of the species and site scores in the biplot requires attention. From Equation 5.11, we deduce that scaling is rather arbitrary; for example, the fitted values remain the same if we jointly plot the species points $(3b_{k1}, 5b_{k2})$ and the site points $(x_{i1}/3, x_{i2}/5)$. Yet, there are two types of scaling that have special appeal.

In the first type of scaling, the site scores are standardized to unit sum of squares and the species scores are weighted sums of the site scores (Table 5.6). The sum of squared scores of species is then equal to the eigenvalue of the axis. In this scaling, the angle between arrows of each pair of species (Figure 5.12) provides an approximation of their pair-wise correlation, i.e.

$$r \approx \cos \theta$$

with r the correlation coefficient and θ the angle.

Consequently, arrows that point in the same direction indicate positively correlated species, perpendicular arrows indicate lack of correlation and arrows pointing in the opposite direction indicate negatively correlated species. This biplot is termed the covariance biplot and is considered in detail by Corsten & Gabriel (1976).

In the second type of scaling, the species scores are standardized to unit sum of squares and the site scores are standardized, so that their sum of squares equals the eigenvalue of each axis. Then, the site scores are the weighted sum of the species scores. This scaling was used implicitly in Subsection 5.3.3 and is intended to preserve Euclidean Distances between sites (Equation 5.16), i.e. the length of the line segment joining two sites in the biplot then approximates the length of the line segment joining the sites in *m*-dimensional space, the axes of which are formed by the species. When scaled in this way, the biplot is termed a Euclidean Distance biplot (ter Braak 1983). Figure 5.15 shows this biplot for the Dune Meadow Data.

The Euclidean Distance biplot is obtained from the covariance biplot by simple rescaling of species and site scores. Species k with coordinates (b_{k1}, b_{k2}) in the covariance biplot gets the coordinates $(b_{k1}/\sqrt{\lambda_1}, b_{k2}/\sqrt{\lambda_2})$ in the Euclidean Distance biplot, and site i with coordinates (x_{i1}, x_{i2}) gets coordinates $(x_{i1}\sqrt{\lambda_1}, x_{i2}\sqrt{\lambda_2})$ in the Euclidean Distance biplot. Figure 5.15 does not look very different from Figure 5.12, because the ratio of $\sqrt{\lambda_1}$ and $\sqrt{\lambda_2}$ is close to 1.

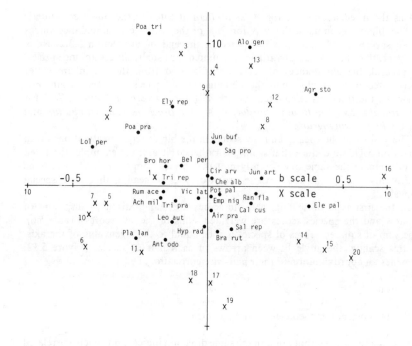

Figure 5.15 Euclidean Distance biplot of Dune Meadow Data.

5.3.5 Data transformation

We have so far described the standard form of PCA as treated in statistical textbooks (e.g. Morrison 1967). In ecology, this form is known as 'species-centred PCA'. In a variant of this, 'standardized PCA', abundances of each species are also divided by its standard deviation. In species-centred PCA, each species is implicitly weighted by the variance of its abundance values. Species with high variance, often the abundant ones, therefore dominate the PCA solution, whereas species with low variance, often the rare ones, have only minor influence on the solution. This may be reason to apply standardized PCA, in which all species receive equal weight. However the rare species then unduly influence the analysis if there are a lot of them, and chance can dominate the results. We therefore recommend species-centred PCA, unless there is strong reason to use standardized PCA. Standardization is necessary if we are analysing variables that are measured in different units, for example quantitative environmental variables such as pH, mass fraction of organic matter or ion concentrations. Noy Meir et al. (1975) fully discuss the virtues and vices of various data transformations in PCA.

The fraction of variance accounted for by the first few axes is not a measure of the appropriateness of a particular data transformation. By multiplying the abundances of a single species by a million, the first axis of a species-centred PCA will in general account for nearly all the variance, just because nearly all the variance after this transformation is due to this species and the first axis almost perfectly represents its abundances.

If some environmental variables are known to influence the species data strongly, the axes of a PCA will probably show what is already known. To detect unknown variation, one can for each species first apply a regression on the known environmental variables, collect the residuals from these regressions in a two-way table and apply PCA to this table of residuals. This analysis is called partial PCA and is standardly available in the computer program CANOCO (ter Braak 1987b). The analysis is particularly simple if, before sampling, groups of sites are recognized. Then, the deviations of the group means should be analysed instead of the deviations from the general mean. An example is the analysis of vegetation change in permanent plots by Swaine & Greig-Smith (1980).

5.3.6 R-mode and Q-mode algorithms

The iteration algorithm in Table 5.6 is a general-purpose algorithm to extract eigenvectors and eigenvalues from an $m \times n$ matrix Y with elements y_{ki}. The algorithm is used in the computer program CANOCO (ter Braak 1987b) to obtain the solution to species-centred PCA if the rows are centred and to standardized PCA if the rows are standardized, but also to non-centred PCA (Noy Meir 1973) if the data are neither centred nor standardized. However many computer programs for PCA use other algorithms, most of which implicitly transform the data. Centring by variables is done implicitly when PCA is carried out on the matrix of covariances between the variables. Also, standardization by variables is implicit in an analysis of the correlation matrix. The role of species in our discussion therefore corresponds to the role of variables in a general-purpose computer program for PCA. The rest of Subsection 5.3.6 may be skipped at a first reading.

Algorithms that are based on the covariance matrix or correlation matrix are termed R-mode algorithms. More generally, R-mode algorithms extract eigenvectors from the species-by-species cross-product matrix A with elements

$$a_{kl} = \Sigma_i y_{ki} y_{li} \ (k, l = 1, ..., m)$$

where, as before, y_{ki} is the data after transformation.

By contrast, Q-mode algorithms extract eigenvectors from the site-by-site cross-product matrix C with elements

$$c_{ij} = \Sigma_k y_{ki} y_{kj} \ (i, j = 1, ..., n).$$

A particular Q-mode algorithm is obtained from Table 5.6 by inserting Equation 5.8 in Equation 5.9. In this way, Steps 2 and 3 are combined into a single step, in which

131

new $x_i = \sum_{j=1}^{n} c_{ij} x_j$.

It can be shown that the eigenvalues of the Matrix **A** equal those of the Matrix **C**, and further that the eigenvectors of **C** can be obtained from those of **A** by applying Equation 5.9 to each eigenvector and, conversely, that the eigenvectors of **A** can be obtained from those of **C** by applying Equation 5.8 to each eigenvector of **C**. The terms R-mode and Q-mode therefore refer to different algorithms and not to different methods. If the number of species is smaller than the number of sites, R-mode algorithms are more efficient than Q-mode algorithms, and conversely.

5.4 Interpretation of ordination with external data

Once data on species composition have been summarized in an ordination diagram, the diagram is typically interpreted with help of external knowledge on sites and species. Here we discuss methods that facilitate interpretation when data on environmental variables are collected at different sites. Analogous methods exist when there is external data on the species, for example growth form of plant species or indicator values for environmental variables from previous studies or from the literature (Table 5.7).

Simple interpretative aids include:
- writing the values of an environmental variable in the order of site scores of an ordination axis below the arranged species data table (Table 5.7)
- writing the values of an environmental variable near the site points in the ordination diagram (Figure 5.16)
- plotting the site scores of an ordination axis against the values of an environmental variable (Figure 5.17)
- calculating (rank) correlation coefficients between each of the quantitative environmental variables and each of the ordination axes (Table 5.8)
- calculating mean values and standard deviations of ordination scores for each class of a nominal environmental variable (ANOVA, Subsection 3.2.1) and plotting these in the ordination diagram (Figure 5.16).

An ordination technique that is suited for the species composition data extracts theoretical environmental gradients from these data. We therefore expect straight line (or at least monotonic) relations between ordination axes and quantitative environmental variables that influence species. Correlation coefficients are therefore often adequate summaries of scatter plots of environmental variables against ordination axes.

Three of these simple interpretative aids are directed to the interpretation of axes instead of to the interpretation of the diagram as a whole. But the ordination axes do not have a special meaning. Interpretation of other directions in the diagram is equally valid. A useful idea is to determine the direction in the diagram that has maximum correlation with a particular environmental variable (Dargie 1984). For the jth environmental variable, z_j, that direction can be found by multiple (least-squares) regression of z_j on the site scores of the first ordination axis (x_1) and the second ordination axis (x_2), i.e. by estimating the parameters b_1 and

132

Table 5.7 Values of environmental variables and Ellenberg's indicator values of species written alongside the ordered data table of the Dune Meadow Data, in which species and sites are arranged in order of their scores on the second DCA axis. A1: thickness of A1 horizon (cm), 9 meaning 9 cm or more; moisture: moistness in five classes from 1 = dry to 5 = wet; use: type of agricultural use, 1 = hayfield, 2 = a mixture of pasture and hayfield, 3 = pasture; manure: amount of manure applied in five classes from 0 = no manure to 5 = heavy use of manure. The meadows are classified by type of management: SF, standard farming; BF, biological farming; HF, hobby farming; NM, nature management; F, R, N refer to Ellenberg's indicator values for moisture, acidity and nutrients, respectively.

Species		Site (i)	F	R	N
k		`00001000010111111121` `12350749638162485709`			
1	Ach mil	`13 242  2         2`	4		5
11	Ely rep	`4444  46`	5		8
7	Bro hor	` 4 2423`			3
17	Lol per	`756266526  47    2`	5		7
6	Bel per	`3222 2         2`			5
19	Poa pra	`445244443244   3 1`	5		
20	Poa tri	`27664555494 24`	7		7
26	Tri pra	`   2 2  5`			
23	Rum ace	` 5 3 26    2`			5
21	Pot pal	`          2 2`	10	3	2
18	Pla lan	`       535  5  3`	3		2
4	Alo gen	`27   23 55 48`	9	7	7
8	Che alb	`      1`	4		7
10	Ele pal	`       4 8 4 5  4`	10		
27	Tri rep	`52262135223 3621  2`			7
2	Agr sto	`4  83 54 744 4 5`	6		5
15	Jun buf	`  2 4 3   4`	7	3	
30	Cal cus	`        3 4   3`			
14	Jun art	`  4  4 3   3 4`	8		2
9	Cir arv	`2`			7
22	Ran fla	`     22 2 2 2 4`	9	3	2
28	Vic lat	`1    2   1`	2	3	2
29	Bra rut	`2222226 2444 64 43`			
16	Leo aut	`52333223235 2252226`	5		5
5	Ant odo	`442  3        4 4`		5	
24	Sag pro	`  52 222 4     3`	6	7	6
25	Sal rep	`          3 53`			
13	Hyp rad	`      2      2 5`	5	4	3
3	Air pra	`             2 3`	2	3	1
12	Emp nig	`        2`	6	4	

---

j		
1	A1	`34463344464466959444`
2	moisture	`11212124155154515255`
3	use	`22211321223332312111`
4	manure	`42421341233132000000`
5	SF	`10100010010011000000`
6	BF	`01001000000100000000`
7	HF	`00010101101000000000`
8	NM	`00000000000000111111`

133

Table 5.8 Correlation coefficients ($100 \times r$) of environmental variables with the first four DCA axes for the Dune Meadow Data.

Variable	Axes			
	1	2	3	4
1 Al	58	24	7	9
2 moisture	76	57	7	-7
3 use	35	-21	-3	-5
4 manure	6	-68	-7	-64
5 SF	22	-29	5	-60
6 BF	-28	-24	39	22
7 HF	-22	-26	-55	-14
8 NM	21	73	17	56
Eigenvalue	0.54	0.29	0.08	0.05

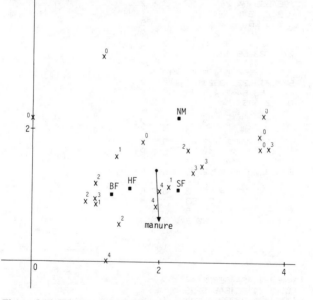

Figure 5.16 The amount of manure written on the DCA ordination of Figure 5.7. The trend in the amount across the diagram is shown by an arrow, obtained by a multiple regression of manure on the site scores of the DCA axes. Also shown are the mean scores for the four types of management, which indicate, for example, that the nature reserves (NM) tend to lie at the top of the diagram.

134

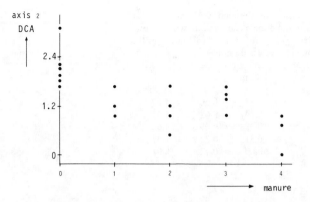

Figure 5.17 Site scores of the second DCA axis plotted against the amount of manure.

b_2 of the regression equation (as in Subsection 3.5.2)

$$Ez_j = b_0 + b_1 x_1 + b_2 x_2 \qquad \text{Equation 5.12}$$

The direction of maximum correlation makes an angle of θ with the first axis where $\theta = \arctan (b_2/b_1)$ and the maximum correlation equals the multiple correlation coefficient (Subsection 3.2.1). This direction can be indicated in the ordination diagram by an arrow running from the centroid of the plot, for instance with coordinates (0,0), to the point with coordinates (b_1,b_2), as illustrated for manure in Figure 5.16. This is an application of the biplot idea; the environmental variable is represented in the diagram by an arrow that points in the direction of maximum change (Subsection 5.3.4). Several environmental variables can be accommodated in this way in a single ordination diagram.

In Chapter 3, presence and abundance of a single species represented the response variable to be explained by the environmental variables. By applying an ordination technique to the abundances of many species, we have reduced many response variables to a few ordination axes. It is therefore natural to consider the ordination axes as new derived response variables and to attempt to explain each of them by use of multiple regression analysis. For example, we can fit for the first axis (x_1) the response model

$$Ex_1 = c_0 + c_1 z_1 + c_2 z_2 + ... + c_q z_q \qquad \text{Equation 5.13}$$

where z_j is the jth (out of q) environmental variables and c_j is the corresponding regression coefficient. The multiple correlation coefficient and the fraction of variance accounted for by the regression (Subsection 3.2.1) indicate whether the environmental variables are sufficient to predict the variation in species composition that is represented by the first ordination axis. Table 5.9 shows an example.

135

Table 5.9 Multiple regression of the first CA axis on four environmental variables of the dune meadow data, which shows that moisture contributes significantly to the explanation of the first axis, whereas the other variables do not.

Term	Parameter	Estimate	s.e.	t
constant	c_0	−2.32	0.50	−4.62
A1	c_1	0.14	0.08	1.71
moisture	c_2	0.38	0.09	4.08
use	c_3	0.31	0.22	1.37
manure	c_4	−0.00	0.12	−0.01

ANOVA table

	d.f.	s.s.	m.s.	F
Regression	4	17.0	4.25	10.6
Residual	15	6.2	0.41	
Total	19	23.2	1.22	

$R^2 = 0.75$
$R^2_{adj} = 0.66$

There are good reasons not to include the environmental variables in the ordination analysis itself, nor to reverse the procedure by applying ordination to the environmental data first and by adding the species data afterwards: the main variation in the environmental data is then sought, and this may well not be the major variation in species composition. For example, if a single environmental variable is important for the species and many more variables are included in the analysis, the first few axes of the environmental ordination mainly represent the relations among the unimportant variables and the relation of the important variable with the species' data would not be discovered. It is therefore better to search for the largest variation in the species' data first and to find out afterwards which of the environmental variables is influential.

5.5 Canonical ordination

5.5.1 Introduction

Suppose we are interested in the effect on species composition of a particular set of environmental variables. What can then be inferred from an indirect gradient analysis (ordination followed by environmental gradient interpretation)? If the ordination of the species data can be readily interpreted with these variables, the environmental variables are apparently sufficient to explain the main variation in the species' composition. But, if the environmental variables cannot explain the main variation, they may still explain some of the remaining variation, which

136

can be substantial, especially in large data sets. For example, a strong relation of the environmental variables with the fifth ordination axis will go unnoticed, when only four ordination axes are extracted, as in some of the computer programs in common use. This limitation can only be overcome by canonical ordination.

Canonical ordination techniques are designed to detect the patterns of variation in the species data that can be explained 'best' by the observed environmental variables. The resulting ordination diagram expresses not only a pattern of variation in species composition but also the main relations between the species and each of the environmental variables. Canonical ordination thus combines aspects of regular ordination with aspects of regression.

We introduce, consecutively, the canonical form of CA, the canonical form of PCA (redundancy analysis) and two other linear canonical techniques, namely canonical correlation analysis and canonical variate analysis. After introducing these particular techniques, we discuss how to interpret canonical ordination axes and the possible effect of data transformations.

5.5.2 Canonical correspondence analysis (CCA)

To introduce canonical correspondence analysis (CCA), we consider again the artificial example by which we have introduced CA (Subsection 5.2.1). In this example (reproduced in Figure 5.18a), five species each preferred a slightly different moisture value. The species score was defined to be the value most preferred and was calculated by averaging the moisture values of the sites in which the species is present. Environmental variables were standardized to mean 0 and variance 1 (Table 5.2c) and the dispersion of the species scores after standardization was taken to express how well a variable explains the species data.

Now suppose, as before, that moisture is the best single variable among the environmental variables measured. In Subsection 5.2.1, we proceeded by constructing the theoretical variable that best explains the species data and, in Section 5.4, we attempted to explain the variable so obtained by a combination of measured environmental variables (Equation 5.13). But, as discussed in Subsection 5.5.1, such attempts may fail, even if we measure environmental variables influencing the species. So why not consider combinations of environmental variables from the beginning? In the example, someone might suggest considering a combination of moisture and phosphate, and Figure 5.18b actually shows that, after standardization, the combination (3 $\times$ moisture + 2 $\times$ phosphate) gives a larger dispersion than moisture alone. So it can be worthwhile to consider not only the environmental variables singly but also all possible linear combinations of them, i.e. all weighted sums of the form

$$x_i = c_0 + c_1 z_{1i} + c_2 z_{2i} + ... + c_q z_{qi} \qquad \text{Equation 5.14}$$

where
z_{ji} is the value of environmental variable j at site i
c_j is the weight (not necessary positive) belonging to that variable
x_i is the value of the resulting compound environmental variable at site i.

137

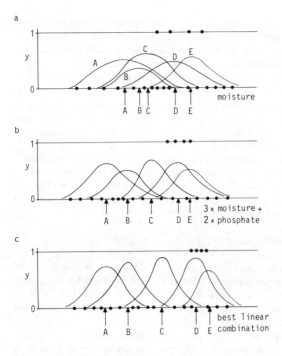

Figure 5.18 Artificial example of unimodal response curves of five species (A-E) with respect to standardized environmental variables showing different degrees of separation of the species curves. a: Moisture. b: Linear combination of moisture and phosphate, chosen a priori. c: Best linear combination of environmental variables, chosen by CCA. Sites are shown as dots, at $y = 1$ if Species D is present and at $y = 0$ if Species D is absent.

CCA is now the technique that selects the linear combination of environmental variables that maximizes the dispersion of the species scores (Figure 5.18c; ter Braak 1987a). In other words, CCA chooses the best weights (c_j) for the environmental variables. This gives the first CCA axis.

The second and further CCA axes also select linear combinations of environmental variables that maximize the dispersion of the species scores, but subject to the constraint of being uncorrelated with previous CCA axes (Subsection 5.2.1). As many axes can be extracted as there are environmental variables.

CA also maximizes the dispersion of the species scores, though irrespective of any environmental variable; that is, CA assigns scores (x_i) to sites such that the dispersion is absolutely maximum (Subsection 5.2.1). CCA is therefore 'restricted correspondence analysis' in the sense that the site scores are restricted to be a linear combination of measured environmental variables (Equation 5.14). By incorporating this restriction in the two-way weighted averaging algorithm

138

of CA (Table 5.2), we obtain an algorithm for CCA. More precisely, in each iteration cycle, a multiple regression must be carried out of the site scores obtained in Step 3 on the environmental variables (for technical reasons with y_{+i}/y_{++} as site weights). The fitted values of this regression are by definition a linear combination of the environmental variables (Equation 5.14) and are thus the new site scores to continue with in Step 4 of Table 5.2a. As in CA, the scores stabilize after several iterations and the resulting scores constitute an ordination axis of CCA. The corresponding eigenvalue actually equals the (maximized) dispersion of the species scores along the axis. The eigenvalues in CCA are usually smaller than those in CA because of the restrictions imposed on the site scores in CCA.

The parameters of the final regression in the iteration process are the best weights, also called canonical coefficients, and the multiple correlation of this regression is called the species–environment correlation. This is the correlation between the site scores that are weighted averages of the species scores and the site scores that are a linear combination of the environmental variables. The species–environment correlation is a measure of the association between species and environment, but not an ideal one; axes with small eigenvalues may have misleadingly high species–environment correlations. The importance of the association is expressed better by the eigenvalue because the eigenvalue measures how much variation in the species data is explained by the axis and, hence, by the environmental variables.

CCA is restricted correspondence analysis but the restrictions become less strict the more environmental variables are included in the analysis. If $q \geq n - 1$, then there are actually no restrictions any more; CCA is then simply CA. The arch effect may therefore crop up in CCA, as it does in CA (Gauch 1982). The method of detrending (Hill & Gauch 1980) can be used to remove the arch and is available in the computer program CANOCO (ter Braak 1987b). But in CCA, the arch can be removed more elegantly by dropping superfluous environmental variables. Variables that are highly correlated with the arched axis (often the second axis) are most likely to be superfluous. So a CCA with the superfluous variables excluded does not need detrending.

In Subsection 5.2.7, we saw that CA approximated the maximum likelihood solution of Gaussian ordination when Conditions A1-A4 hold true. If we change the Gaussian ordination model by stating that the site scores must be a linear combination of the environmental variables, the maximum likelihood solution of the model so obtained is again approximated by CCA when these conditions hold true (ter Braak 1986a). The data on species composition are thus explained by CCA through a Gaussian response model in which the explanatory variable is a linear combination of the environmental variables. Furthermore, tests of real data showed that CCA is extremely robust when these assumptions do not hold. The vital assumption is that the response model is unimodal. For a simpler model where relations are monotonic, the results can still be expected to be adequate in a qualitative sense, but for more complex models the method breaks down.

As an example, we use the Dune Meadow Data, which concerns the impact of agricultural use on vegetation in dune meadows on the Island of Terschelling (the Netherlands). The data set consists of 20 relevés, 30 plant species (Table

139

0.1) and 5 environmental variables (Table 0.2), one of which is the nominal variable 'type of management' consisting of four classes. CCA can accommodate nominal explanatory variables by defining dummy variables as in multiple regression (Subsection 3.5.5). For instance, the dummy variable 'nature management' (Table 5.7) indicates that meadows received that type of management. The first eigenvalue of CCA is somewhat lower than that of CA (0.46 compared to 0.54). Multiple regression of the site scores of the first CA axis on the environmental variables, as we proposed in Section 5.4, resulted in a multiple correlation of 0.87. If the multiple regression is carried out within the iteration algorithm, as in CCA, the multiple correlation increases to 0.96, which is the species–environment correlation. The CCA scores for species and sites look similar to those of CA: not surprisingly, since the multiple correlation obtained with CA is already high. We conclude that, in this example, the measured environmental variables account for the main variation in the species composition. This is true for the second axis also. The second eigenvalue of CCA is 0.29, compared to 0.40 for CA and the second species–environment correlation is 0.89, compared to a multiple correlation of 0.83 in CA. Table 5.10 shows the canonical coefficients that define the first two axes and the correlations of the environmental variables with these axes. These correlations are termed intra-set correlations to distinguish them from the inter-set correlations, which are the correlations between the environmental variables and the site scores that are derived from the species scores. (The inter-set correlation is R times the intra-set correlation; R is the species–environment correlation of the axis). From the correlations in Table 5.10, we infer that the first axis is a moisture gradient and that the second axis is a manuring axis, separating the meadows managed as a nature reserve from the standardly farmed meadows. This can be seen also from the CCA ordination diagram (Figure 5.19a).

Table 5.10 Canonical correspondence analysis: canonical coefficients ($100 \times c$) and intra-set correlations ($100 \times r$) of environmental variables with the first two axes of CCA for the Dune Meadow Data. The environmental variables were standardized first to make the canonical coefficients of different environmental variables comparable. The class SF of the nominal variable 'type of management' was used as reference class in the analysis (Subsection 3.5.5).

Variable	Coefficients		Correlations	
	Axis 1	Axis 2	Axis 1	Axis2
A1	9	-37	57	-17
moisture	71	-29	93	-14
use	25	5	21	-41
manure	-7	-27	-30	-79
SF	–	–	16	-70
BF	-9	16	-37	15
HF	18	19	-36	-12
NM	20	92	56	76

The species and sites are positioned as points in the CCA diagram as in CA and their joint interpretation is also as in CA; sites with a high value of a species tend to be close to the point for that species (Subsection 5.2.5). The environmental variables are represented by arrows and can be interpreted in conjunction with the species points as follows. Each arrow determines an axis in the diagram and the species points must be projected onto this axis. As an example, the points of a few species are projected on to the axis for manuring in Figure 5.19b. The order of the projection points now corresponds approximately to the ranking of the weighted averages of the species with respect to amount of manure. The weighted average indicates the 'position' of a species curve along an environmental variable (Figure 5.18a) and thus the projection point of a species also indicates this position, though approximately. Thus *Cirsium arvense*, *Alopecurus geniculatus*, *Elymus repens* and *Poa trivialis* mainly occur in these data in the highly manured meadows, *Agrostis stolonifera* and *Trifolium repens* in moderately manured meadows and *Ranunculus flammula* and *Anthoxanthum odoratum* in meadows with little manuring. One can interpret the other arrows in a similar way. From Figure 5.19a, one can see at a glance which species occur mainly in wetter conditions (those on the right of the diagram) and which prefer drier conditions (those on the left of the diagram).

The joint plot of species points and environmental arrows is actually a biplot that approximates the weighted averages of each of the species with respect to each of the environmental variables. The rules for quantitative interpretation of the CCA biplot are the same as for the PCA biplot described in Subsection 5.3.4. In the diagram, the weighted averages are approximated as deviations from the grand mean of each environmental variable; the grand mean is represented by the origin (centroid) of the plot. A second useful rule to interpret the diagram is therefore that the inferred weighted average is higher than average if the projection point lies on the same side of the origin as the head of an arrow and is lower than average if the origin lies between the projection point and the head of an arrow. As in Subsection 5.3.2, a measure of goodness of fit is $(\lambda_1 + \lambda_2)/$(sum of all eigenvalues), which expresses the fraction of variance of the weighted averages accounted for by the diagram. In the example, Figure 5.19a accounts for 65% of the variance of the weighted averages. (The sum of all canonical eigenvalues is 1.177.)

The positions of the heads of the arrows depend on the eigenvalues and on the intra-set correlations. In Hill's scaling (Subsection 5.2.2), the coordinate of the head of the arrow for an environmental variable on axis s is r_{js} $\sqrt{\lambda_s} (1 - \lambda_s)$, with r_{js} the intra-set correlation of environmental variable j with axis s and λ_s is the eigenvalue of axis s. The construction of biplots for detrended canonical correspondence analysis is described by ter Braak (1986a). Environmental variables with long arrows are more strongly correlated with the ordination axes than those with short arrows, and therefore more closely related to the pattern of variation in species composition shown in the ordination diagram.

Classes of nominal environmental variables can also be represented by arrows (ter Braak 1986a). The projection of a species on such an arrow approximates the fraction of the total abundance of that species that is achieved at sites of

141

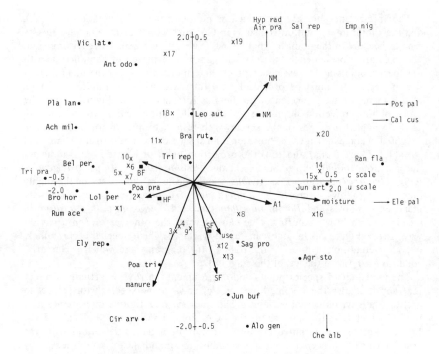

Figure 5.19 CCA of the Dune Meadow Data. a: Ordination diagram with environmental variables represented by arrows. The c scale applies to environmental variables, the u scale to species and sites. The types of management are also shown by closed squares at the centroids of the meadows of the corresponding types of management. b: Inferred ranking of the species along the variable amount of manure, based on the biplot interpretation of Part a of this figure.

that class. However it is sometimes more natural to represent each class of a nominal variable by a point at the centroid (the weighted average) of the sites belonging to that class (Figure 5.19a). Classes consisting of sites with high values for a species are then positioned close to the point of that species. In Figure 5.19a, the meadows managed as a nature reserve are seen to lie at the top-right of the diagram; the meadows of standard farms lie at the bottom.

A second example (from ter Braak 1986a) concerns the presence or absence of 133 macrophytic species in 125 freshwater ditches in the Netherlands. The first four axes of detrended correspondence analyses (DCA) were poorly related (multiple correlation $R < 0.60$) to the measured environmental variables, which were: electrical conductivity (κ), orthophosphate concentration (PHOSPHATE), both transformed to logarithms, chloride ratio (CHLORIDE, the share of chloride ions in κ) and soil type (clay, peaty soil, sand). By choosing the axes in the

142

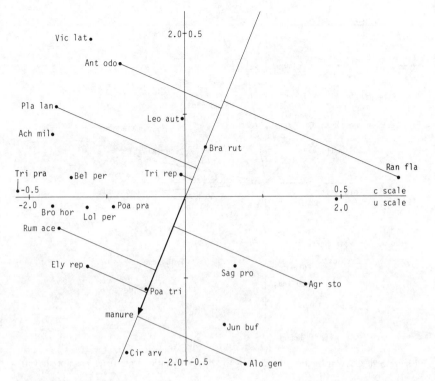

Figure 5.19b

light of these environmental variables by means of CCA, the multiple correlations increased considerably, R being 0.82 and 0.81 for the first two axes. The eigenvalues dropped somewhat – for the first two axes, from 0.34 and 0.25 in DCA to 0.20 and 0.13 in CCA. Apparently, the environmental variables are not sufficient to predict the main variation in species composition extracted by DCA, but they do predict a substantial part of the remaining variation. From the CCA ordination diagram (Figure 5.20), it can be seen that κ and PHOSPHATE are strongly correlated (> 0.8) with the first CCA axis. Species with a high positive score on that axis are therefore almost restricted to ditches with high κ and PHOS-PHATE, and species with a large negative score to ditches with low κ and PHOSPHATE. Species with intermediate scores are either unaffected by κ and PHOSPHATE or restricted to intermediate values of κ and PHOSPHATE. The second CCA axis is strongly correlated ($r = 0.9$) with CHLORIDE. The arrow for PEAT shows that species whose distribution is the most restricted to peaty soils lie in the top-left corner of the diagram. The arrows for SAND and CLAY are to be interpreted analogously.

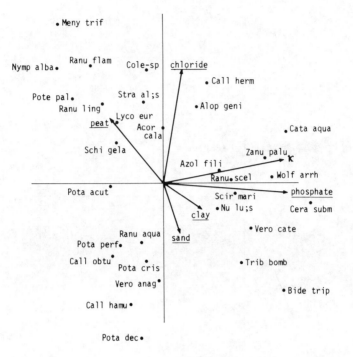

Figure 5.20 CCA ordination diagram of the ditch vegetation data (sites are not shown).

5.5.3 Redundancy analysis (RDA)

Redundancy analysis (RDA) is the canonical form of PCA and was invented by Rao (1964). RDA has so far been neglected by ecologists, but appears attractive when used in combination with PCA.

As in PCA (Subsection 5.3.1), we attempt to explain the data of all species by fitting a separate straight line to the data of each species. As a measure of how badly a particular environmental variable explains the species data, we take the total residual sum of squares, as in PCA (Figure 5.11). The best environmental variable is then the one that gives the smallest total residual sum of squares. From this, we can derive a canonical ordination technique, as in Subsection 5.5.2, by considering also linear combinations of environmental variables. RDA is the technique selecting the linear combination of environmental variables that gives the smallest total residual sum of squares.

PCA also minimizes the total residual sum of squares, but it does so without looking at the environmental variables (Subsection 5.3.1). We can obtain the RDA axes by extending the algorithm of PCA (Table 5.6) in a similar fashion to how

144

we modified the CA algorithm in Subsection 5.5.2; in each iteration cycle, the site scores calculated in Step 3 are regressed on the environmental variables with Equation 5.13 and the fitted values of the regression are taken as the new site scores to continue in Step 4 of the algorithm. (In contrast to CCA, we must now use equal site weights in the regression.) So the site scores are restricted to a linear combination of the environmental variables and RDA is simply PCA with a restriction on the site scores. The species–environment correlation is obtained in the same way as for CCA; but, in RDA, this correlation equals the correlation between the site scores that are weighted sums of the species scores and the site scores that are a linear combination of the environmental variables.

We illustrate RDA with the Dune Meadow Data, using the same environmental variables as in Subsection 5.5.2. The first two axes of PCA explained 29% and 21% of the total variance in the species data, respectively. RDA restricts the axes to linear combinations of the environmental variables and the RDA axes explain therefore less, namely 26% and 17% of the total variance. The first two species–environment correlations are 0.95 and 0.89, both a little higher than the multiple correlations resulting from regressing the first two PCA axes on the environmental variables. We conclude, as with CCA, that the environmental variables account for the main variation in the species composition. From the canonical coefficients and intra-set correlations (Table 5.11), we draw the same conclusions as with CCA, namely that the first axis is mainly a moisture gradient and the second axis a manuring gradient.

The RDA ordination diagram (Figure 5.21) can be interpreted as a biplot (Subsection 5.3.4). The species points and site points jointly approximate the species abundance data as in PCA, and the species points and environmental arrows

Table 5.11 Redundancy analysis: canonical coefficients (100 $\times c$) and intra-set correlations (100 $\times r$) of environmental variables with the first two axes of RDA for the Dune Meadow Data. The environmental variables were standardized first to make the canonical coefficients of different environmental variables comparable. The class SF of the nominal variable 'type of management' was used as reference class in the analysis (as in Table 5.10).

Variable	Coefficients		Correlations	
	Axis 1	Axis 2	Axis 1	Axis 2
A1	–1	–5	54	–6
moisture	15	9	92	12
use	5	–6	15	29
manure	–8	16	–26	86
SF	–	–	25	76
BF	–10	0	–48	–11
HF	–10	–2	–40	13
NM	–4	–13	51	–79

145

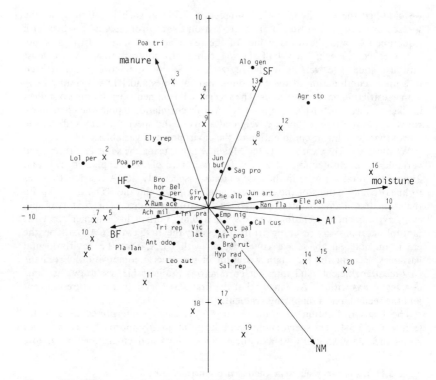

Figure 5.21 RDA ordination diagram of the Dune Meadow Data with environmental variables represented by arrows. The scale of the diagram is: 1 unit in the plot corresponds to 1 unit for the sites, to 0.067 units for the species and to 0.4 units for the environmental variables.

jointly approximate the covariances between species and environmental variables. If species are represented by arrows as well (a natural representation in a PCA biplot), the cosine of the angle between the arrows of a species and an environmental variable is an approximation of the correlation coefficient between the species and the environmental variable. One gets a qualitative idea of such correlations from the plot by noting that arrows pointing in roughly the same direction indicate a high positive correlation, that arrows crossing at right angles indicate near-zero correlation, and that arrows pointing in roughly opposite directions indicate a high negative correlation. If arrows are drawn for *Poa trivialis*, *Elymus repens* and *Cirsium arvense* in Figure 5.21, they make sharp angles with the arrow for manuring; hence, the abundances of these species are inferred to be positively correlated with the amount of manure. We can be more confident about this inference for *Poa trivialis* than for *Cirsium arvense* because the former species lies much further from the centre of the diagram than the latter species. As in

146

PCA, species at the centre of the diagram are often not very well represented and inferences from the diagram about their abundances and correlations are imprecise. From Figure 5.21, we infer also that, for instance, *Salix repens*, *Hypochaeris radicata* and *Aira praecox* are negatively correlated with the amount of manure.

A measure of goodness of fit of the biplot of species and environmental variables is $(\lambda_1 + \lambda_2)/$(sum of all eigenvalues), which expresses the fraction of variance of all covariances between species and environment accounted for by the diagram. For the example, Figure 5.21 accounts for 71% of this variance.

The scaling of Figure 5.21 conforms to that of the Euclidean distance biplot (Subsection 5.3.4): the sum of squares of the species scores is unity and site points are obtained by weighted summation of species scores. The positions of the heads of arrows of the environmental variables depend on the intra-set correlations (Table 5.11) and the eigenvalues. With this scaling, the coordinate of the head of the arrow for an environmental variable on axis s must be $r_{js} \sqrt{(\lambda_s/n)}$ where r_{js} is the intra-set correlation of environmental variable j with axis s, n is the number of sites, and λ_s the eigenvalue of axis s. The diagram scaled in this way gives not only a least-squares approximation of the covariances between species and environment, but also approximations of the (centred) abundances values, of the Euclidean Distances among the sites as based on the species data (Equation 5.15), and of covariances among the environmental variables, though the latter two approximations are not least-squares approximations. Other types of scaling are possible (ter Braak 1987b).

5.5.4 Canonical correlation analysis (COR)

The species–environment correlation was a by-product in CCA and RDA, but is central in canonical correlation analysis (COR). The idea of COR is to choose coefficients (scores) for species and coefficients for environmental variables so as to maximize the species–environment correlation. In COR, the species–environment correlation is defined as in RDA, as the correlation between site scores (x_i^*) that are weighted sums of species scores: $(x_i^* = \Sigma_k b_k y_{ki})$ and site scores (x_i) that are a linear combination of the environmental variables $(x_i = c_0 + \Sigma_j c_j z_{ji})$. An algorithm to obtain the COR axes is given in Table 5.12. The resulting species–environment correlation is termed the canonical correlation, and is actually the squareroot of the first eigenvalue of COR. Step 2 of the algorithm makes the difference from RDA: in RDA, the species scores are simply a weighted sum of the site scores, whereas in COR the species scores are parameters estimated by a multiple regression of the site scores on the species variables. This regression has the practical consequence that, in COR, the number of species must be smaller than the number of sites. It can be shown that the restriction on the number of species is even stronger than that: the number of species *plus* the number of environmental variables must be smaller than the number of sites. This requirement is not met in our Dune Meadow Data and is generally a nuisance in ecological research. By contrast, RDA and CCA set no upper limit to the number of species that can be analysed. Examples of COR can be found in Gittins

Table 5.12 An iteration algorithm for canonical correlation analysis (COR).

Step 1. Start with arbitrary initial site scores (x_i), not all equal to zero.
Step 2. Calculate species scores by multiple regression of the site scores on the species variables. The species scores (b_k) are the parameter estimates of this regression.
Step 3. Calculate new site scores (x_i^*) by weighted summation of the species scores (Equation 5.9). The site scores in fact equal the fitted values of the multiple regression of Step 2.
Step 4. Calculate coefficients for the environmental variables by multiple regression of the site scores (x_i^*) on the environmental variables. The coefficients (c_j) are the parameter estimates of this regression.
Step 5. Calculate new site scores (x_i) by weighted summation of the coefficients of the environmental variables, i.e. by $x_i = \sum_{j=1}^{q} c_j z_{ji}$. The site scores in fact equal the fitted values of the multiple regression of Step 4.
Step 6. For second and higher axes, orthogonalize the site scores (x_i) as in Table 5.6.
Step 7. Standardize the site scores (x_i) as in Table 5.6.
Step 8. Stop on convergence, i.e. when the new site scores are sufficiently close to the site scores of the previous cycle of the iteration process; ELSE go to Step 2.

(1985). COR allows a biplot to be made, from which the approximate covariances between species and environmental variables can be derived in the same way as in RDA (Subsection 5.5.3). The construction of the COR biplot is given in Subsection 5.9.3.

In our introduction to COR, species and environmental variables enter the analysis in a symmetric way (Table 5.12). Tso (1981) presented an asymmetric approach in which the environmental variables explain the species data. In this approach COR is very similar to RDA, but differs from it in the assumptions about the error part of the model (Equations 5.10 and 5.14): uncorrelated errors with equal variance in RDA and correlated normal errors in COR. The residual correlations between errors are therefore additional parameters in COR. When the number of species is large, there are so many of them that they cannot be estimated reliably from data from few sites. This causes practical problems with COR that are absent in RDA and CCA.

5.5.5 Canonical variate analysis (CVA)

Canonical variate analysis (CVA) belongs to the classical linear multivariate techniques along with PCA and COR. CVA is also termed linear discriminant analysis.

If sites are classified into classes or clusters, we may wish to know how the species composition differs among sites of different classes. If we have recorded the abundance values of a single species only, this question reduces to how much the abundance of the species differs between classes, a question studied in Subsection 3.2.1 by analysis of variance. If there are more species, we may wish to combine the abundance values of the species to make the differences between classes clearer than is possible on the basis of the abundance values of a single species. CVA does so by seeking a weighted sum of the species abundances; however not one

148

that maximizes the total variance along the first ordination axis, as PCA does, but one that maximizes the ratio of the between-class sum of squares and the within-class sum of squares of the site scores along the first ordination axis. (These sums of squares are the regression sum of squares and residual sum of squares, respectively, in an ANOVA of the site scores, cf. Subsection 3.2.1.).

Formally, CVA is a special case of COR in which the set of environmental variables consists of a single nominal variable defining the classes. So from Subsection 3.5.5, the algorithm of Table 5.12 can be used to obtain the CVA axes. We deduce that use of CVA makes sense only if the number of sites is much greater than the number of species and the number of classes (Schaafsma & van Vark 1979; Varmuza 1980). Consequently, many ecological data sets cannot be analysed by CVA without dropping many species. Examples of CVA can be found in Green (1979), Pielou (1984) and Gittins (1985).

In contrast to CVA, CCA and RDA can be used to display differences in species composition between classes without having to drop species from the analysis.

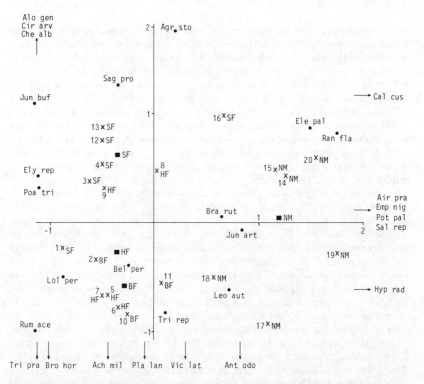

Figure 5.22 CCA ordination diagram of the Dune Meadow Data optimally displaying differences in species composition among different types of management.

149

For this, we must code classes as dummy environmental variables, as in Subsection 3.5.5. Such an analysis by CCA is equivalent to the analysis of concentration proposed by Feoli & Orlóci (1979). As an example, Figure 5.22 displays the differences in vegetation composition between the meadows receiving different types of management in our Dune Meadow Data. The first axis ($\lambda_1 = 0.32$) is seen to separate the meadows receiving nature management from the remaining meadows and second axis ($\lambda_2 = 0.18$) separates the meadows managed by standard farming from those managed by hobby farming and biological farming, although the separations are not perfect. The species displayed on the right side of the diagram occur mainly in the meadows receiving nature management and those on the upper left in the meadows managed by standard farming, and so on. Figure 5.22 displays almost the same information as Figure 5.19a, as can be seen by joining Site Points 16 and 18 in both diagrams. Moisture and manuring are presumably the major factors bringing about vegetation differences between types of management.

5.5.6 Interpreting canonical axes

To interpret the ordination axes, one can use the canonical coefficients and the intra-set correlations. The canonical coefficients define the ordination axes as linear combinations of the environmental variables by means of Equation 5.14 and the intra-set correlations are the correlation coefficients between the environmental variables and these ordination axes. As before, we assume that the environmental variables have been standardized to a mean of 0 and a variance of 1 before the analysis. This standardization removes arbitrariness in the units of measurement of the environmental variables and makes the canonical coefficients comparable among each other, but does not influence other aspects of the analysis.

By looking at the signs and relative magnitudes of the intra-set correlations and of the canonical coefficients standardized in this way, we may infer the relative importance of each environmental variable for prediction of species composition. The canonical coefficients give the same information as the intra-set correlations, if the environmental variables are mutually uncorrelated, but may provide rather different information if the environmental variables are correlated among one another, as they usually are in field data. Both a canonical coefficient and an intra-set correlation relate to the rate of change in species composition by changing the corresponding environmental variable. However it is assumed that other environmental variables are being held constant in the former case, whereas the other environmental variables are assumed to covary with that one environmental variable in the particular way they do in the data set in the latter case. If the environmental variables are strongly correlated with one another, for example simply because the number of environmental variables approaches the number of sites, the effects of different environmental variables on the species composition cannot be singled out and, consequently, the canonical coefficients will be unstable. This is the multicollinearity problem discussed in the context of multiple regression in Subsection 3.5.3. The algorithms to obtain the canonical axes show that canonical coefficients are actually coefficients of a multiple regression (Subsection 5.5.2), so both suffer identical problems. If the multicollinearity problem arises (the program CANOCO (ter Braak 1987b) provides statistics to help detecting it), one should abstain from attempts to interpret the canonical coefficients. But the intra-set correlations do not suffer from this problem. They are more like simple correlation coefficients. They can still be interpreted. One can also remove environmental variables from the analysis, keeping at least one variable per set of strongly correlated environmental variables. Then, the eigenvalues and species–environment correlations will usually only decrease slightly. If the eigenvalues and species–environment correlations drop considerably, one has removed too many (or the wrong) variables.

Their algorithms indicate that COR and CVA are hampered also by strong correlations among species, whereas CCA and RDA are not. So in CCA and RDA, the number of species is allowed to exceed the number of sites.

5.5.7 Data transformation

As in CA and PCA, any kind of transformation of the species abundances may affect the results of CCA and RDA. We refer to Subsections 5.2.2 and 5.3.5 for recommendations about this. The results of COR and CVA are affected by non-linear transformations of the species data, but not by linear transformations. Canonical ordination techniques are not influenced by linear transformations of the environmental variables, but non-linear transformation of environmental variables can be considered if there is some reason to do so. Prior knowledge about the possible impact of the environmental variables on species composition may suggest particular non-linear transformations and particular non-linear combinations, i.e. environmental scalars in the sense of Loucks (1962) and Austin et al. (1984). The use of environmental scalars can also circumvent the multicollinearity problem described in Subsection 5.5.6.

5.6 Multidimensional scaling

In Section 5.1, ordination was defined as a method that arranges site points in the best possible way in a continuum such that points that are close together correspond to sites that are similar in species composition, and points which are far apart correspond to sites that are dissimilar. A particular ordination technique is obtained by further specifying what 'similar' means and what 'best' is. The definition suggests that we choose a measure of (dis)similarity between sites (Subsection 6.2.2), replace the original species composition data by a matrix of dissimilarity values between sites and work further from the dissimilarity matrix to obtain an ordination diagram. This final step is termed multidimensional scaling.

In general, it is not possible to arrange sites such that the mutual distances between the sites in the ordination diagram are equal to the calculated dissimilarity values. Therefore a measure is needed that expresses in a single number how well or how badly the distances in the ordination diagram correspond to the dissimilarity values. Such a measure is termed a loss function or a stress function. In metric ordination techniques such as CA and PCA, the loss function depends on the actual numerical values of the dissimilarities, whereas, in non-metric techniques, the loss function depends only on the rank order of the dissimilarities.

In CA and PCA, one need not calculate a matrix of dissimilarity values first, yet those techniques use particular measures of dissimilarity. In CA, the implied measure of dissimilarity is the chi-squared distance and, in PCA, the Euclidean Distance, as follows immediately from Subsection 5.3.3. The chi-square distance δ_{ij} between site i and site j is defined as

$$\delta_{ij}^2 = y_{++} \, \Sigma_{k=1}^m (y_{ki}/y_{+i} - y_{kj}/y_{+j})^2 \, /y_{k+} \qquad\qquad \text{Equation 5.15}$$

and the Euclidean Distance δ_{ij} between these sites is

$$\delta_{ij}^2 = \Sigma_{k=1}^m (y_{ki} - y_{kj})^2 \qquad\qquad \text{Equation 5.16}$$

The chi-squared distance involves proportional differences in abundances of species between sites, whereas the Euclidean Distance involves absolute differences. Differences in site and species totals are therefore less influential in CA than in PCA, unless a data transformation is used in PCA to correct for this effect.

A simple metric technique for multidimensional scaling is principal coordinate analysis (PCO), also called classical scaling (Gower 1966; Pielou 1977, p.290-395). PCO is based on PCA, but is more general than PCA, in that other measures of dissimilarity may be used than Euclidean Distance. In PCO, the dissimilarity values δ_{ij} are transformed into similarity values by the equation

$$c_{ij} = -0.5 \, (\delta_{ij}^2 - \delta_{i+}^2/n - \delta_{+j}^2/n + \delta_{++}^2/n^2) \qquad\qquad \text{Equation 5.17}$$

where the index $+$ denotes a sum of squared dissimilarities. The matrix with elements c_{ij} is then subjected to the Q-mode algorithm of PCA (Subsection 5.3.6). If the original dissimilarities were computed as Euclidean Distances, PCO is identical to species-centred PCA calculated by the Q-mode algorithm.

In most techniques for (non-metric) multidimensional scaling, we must specify a priori the number of ordination axes and supply an initial ordination of sites. The technique then attempts to modify the ordination iteratively to minimize the stress. In contrast to the iterative algorithms for CA, PCA and PCO, different initial ordinations may lead to different results, because of local minima in the stress function (Subsection 5.2.7); hence, we must supply a 'good' initial ordination or try a series of initial ordinations. From such trials, we then select the ordination with minimum stress.

The best known technique for non-metric multidimensional scaling is ascribed to Shepard (1962) and Kruskal (1964). The stress function, which is minimized in their technique, is based on the Shepard diagram. This is a scatter diagram of the dissimilarities (δ_{ij}) calculated from the species data against the distances d_{ij} between the sites in the ordination diagram.

The ordination fits perfectly (stress $= 0$), if the dissimilarities are monotonic with the distances, i.e. if the points in the Shepard plot lie on a monotonically increasing curve. If they do not, we can fit a monotonic curve through the points by least-squares. This is called monotonic or isotonic regression (Barlow et al. 1972). We then use as stress, a function of the residual sum of squares (for example, Kruskal's stress formula 1, which is the residual sum of squares divided by the total sum of squared distances). The algorithm to seek the ordination that minimizes the stress proceeds further as described above. Note that the method can work equally well with similarities, the only modifications being that a monotonically decreasing curve is fitted in the Shepard diagram. There are two methods to deal with equal dissimilarity values (ties). In the primary approach to ties, the corresponding fitted distances need not be equal, whereas they must be equal in the secondary approach. The primary approach to ties is recommended, because equal dissimilarity values do not necessarily imply equal habitat differences, in part·cular if the equalities arise between pairs of sites that have no species in common (Prentice 1977).

The Shepard–Kruskal method is based on the rank order of all dissimilarities. But calculated dissimilarities may not be comparable in different parts of a gradient, for example if there is a trend in species richness. This potential problem can be overcome by making a separate Shepard diagram for each site, in which we plot the dissimilarities and distances between the particular site and all remaining sites. Each Shepard diagram the distances leads to a stress value and the total stress is taken to be a combination of the separate stress values. This is the local non-metric technique proposed by Sibson (1972). Prentice (1977; 1980) advocated a particular similarity coefficient for use in Sibson's technique. This coefficient is

$$s_{ij} = \Sigma_k \min(y_{ki}, y_{kj}) \hspace{4cm} \text{Equation 5.18}$$

Kendall (1971) proved that this coefficient contains all the information required to reconstruct the order of sites when abundances of species follow arbitrary unimodal response curves.

5.7 Evaluation of direct gradient and indirect gradient analysis techniques

Table 5.13 summarizes the techniques described in Chapters 3, 4 and 5 by type of response model and types of variables. We can classify response models as linear and non-linear. Each linear technique (from multiple regression to COR) has non-linear counterparts. A non-linear model that has special relevance in community ecology is the unimodal model. In principle, unimodal models can be fitted to data by the general methods used for non-linear models (in particular by maximum likelihood methods). For regression analysis, these methods are available (GLM, Chapter 3) but, in ordination, they are not so readily available and tend to require excessive computing. Therefore we have also introduced much simpler methods for analysing data for which unimodal models are appropriate. These simple methods start from the idea that the optima of species response curves can be estimated roughly by weighted averaging and we have shown (Section 3.7) that under particular conditions the estimates are actually quite good. This idea resulted in CA, DCA and CCA.

Multidimensional scaling is left out of Table 5.13, because it is unclear what

153

Table 5.13 Summary of gradient analysis techniques classified by type of response model and types of variables involved. MR, normal multiple regression; IR, inverse regression; PCA, principal components analysis; RDA, redundancy analysis; COR, canonical correlation analysis; WAE, weighted averaging of environmental values; GLM, generalized linear modelling; ML, maximum likelihood; WAI, weighted averaging of indicator values; CA, correspondence analysis; DCA, detrended correspondence analysis; CCA, canonical correspondence analysis; DCCA, detrended canonical correspondence analysis; env.vars, environmental variables; comp. gradients, composite gradients of environmental variables, either measured or theoretical.

		Response model	Number of variables		
	linear	unimodal	response (species)	explanatory (env. vars)	composite (comp.gradients)
Regression	MR	WAE, GLM, ML	one at a time	$\geqslant 1$*	one per species
Calibration	IR	WAI, ML	$\geqslant 1$*	rarely >1	none
Ordination	PCA	CA, DCA, ML	many	none	a few for all species
Canonical ordination	RDA	CCA, DCCA, ML	many	many*	a few for all species
	COR	variants of CCA,ML	many*	many*	a few

* less than number of sites, except for WAE, WAI and some applications of ML.

response models multidimensional scaling can cope with. Whether (non-metric) multidimensional scaling may detect a particular underlying data structure depends in an unknown way on the chosen dissimilarity coefficient and on the initial ordinations supplied. Non-metric multidimensional scaling could sometimes give better ordinations than DCA does, but the question is whether the improvements are worth the extra effort in computing power and manpower (Clymo 1980; Gauch et al. 1981).

Unimodal models are more general than monotonic ones (Figure 3.3), so it makes sense to start by using unimodal models and to decide afterwards whether one could simplify the model to a monotonic one. Statistical tests can help in this decision (Subsection 3.2.3). In ordination, we might therefore start by using CA, DCA or CCA. This initial analysis will provide a check on how unimodal the data are. If the lengths of the ordination axes are less than about 2 s.d., most of the response curves (or surfaces) will be monotonic, and we can consider using PCA or RDA. The advantage of using PCA and RDA is that in their biplot they provide more quantitative information than CA, DCA and (D)CCA in their joint plot, but this advantage would be outweighed by disadvantages when the data are strongly non-linear (ordination lengths greater than about 4 s.d.).

As illustrated by the Dune Meadow Data whose ordination lengths are about 3 s.d., DCA and PCA may result in similar configurations of site points (Figures 5.7 and 5.15). That they result in dissimilar configurations of species points, even if the ordination lengths are small, is simply due to the difference in meaning

154

of the species scores in DCA and PCA (Subsections 5.2.5 and 5.3.5).

Table 5.13 also shows the types of variables involved in regression, calibration, ordination and canonical ordination. We distinguish between response variables, explanatory variables and 'composite' variables, which in community ecology typically correspond to species presences or abundances, measured environmental variables and 'composite gradients', respectively. A composite gradient is either a linear combination of measured environmental variables or a theoretical variable. Which technique is the appropriate one to use largely depends on the research purpose and the type of data available. Ordination and cluster analysis (Chapter 6) are the only available techniques when one has no measured environmental data. Calibration must be considered if one wants to make inferences about values of a particular environmental variable from species data and existing knowledge of species–environment relations. Regression and canonical ordination are called for if one wants to build up and extend the knowledge of species–environment relations (Subsections 3.1.1 and 5.1.1).

Whether to use regression or to use canonical ordination depends on whether it is considered advantageous to analyse all species simultaneously or not. In a simultaneous analysis by canonical ordination, one implicitly assumes that all species are reacting to the same composite gradients of environmental variables according to a common response model. The assumption arises because canonical ordination constructs a few composite gradients for all species. By contrast in regression analysis, a separate composite gradient is constructed for each species. Regression may therefore result in more detailed descriptions and more accurate predictions of each particular species, at least if sufficient data are available. However ecological data that are collected over a large range of habitat variation require non-linear models; building good non-linear models by regression is not easy, because it requires construction of composite gradients that are non-linear combinations of environmental variables (Subsection 3.5.4). In CCA, the composite gradients are linear combinations of environmental variables, giving a much simpler analysis, and the non-linearity enters the model through a unimodal model for a few composite gradients, taken care of in CCA by weighted averaging. Canonical ordination is easier to apply and requires less data than regression. It provides a summary of the species–environment relations. The summary may lack the kind of detail that can in principle be provided by regression; on the other hand, the advantages of using regression, with its machinery of statistical tests, may be lost in practice, through the sheer complexity of non-linear model building and through lack of data. Because canonical ordination gives a more global picture than regression, it may be advantageous to apply canonical ordination in the early exploratory phase of the analysis of a particular data set and to apply regression in subsequent phases to selected species and environmental variables.

As already shown in the examples in Subsection 5.5.2, canonical ordination and ordination followed by environmental interpretation can be used fruitfully in combination. If the results do not differ much, then we know that no important environmental variables have been overlooked in the survey. But note that those included could merely be correlated with the functionally important ones. A further proviso is that the number of environmental variables (q) is small compared to the number of sites (n). If this proviso is not met, the species–environment correlation may yield values close to 1, even if none of the environmental variables affects the species. (Note the remarks about R^2 in Subsection 3.2.1.) In particular, canonical ordination and ordination give identical ordination axes if $q \geqslant n - 1$. If the results of ordination and canonical ordination do differ much, then we may have overlooked major environmental variables, or important non-linear combinations of environmental variables already included in the analysis. But note that the results will also differ if CA or DCA detect a few sites on their first axis that have an aberrant species composition and if these sites are not aberrant in the measured environmental variables. After deleting the aberrant sites, the ordinations provided by (D)CA and CCA may be much more alike.

The question whether we have overlooked major environmental variables can also be studied by combining ordination and canonical ordination in a single analysis. Suppose we believe that two environmental variables govern the species composition in a region. We may then choose two ordination axes as linear combinations of these variables by canonical ordination, and extract further (unrestricted) axes as in CA or PCA, i.e. by the usual iteration process, making the axes unrelated to the previous (canonical) axes in each cycle. The eigenvalues of the extra axes measure residual variation, i.e. variation that cannot be explained by linear combinations of the environmental variables already included in the analysis. Such combined analyses are called partial ordination. Partial PCA (Subsection 5.3.5) is a special case of this.

A further extension of the analytical power of ordination is partial canonical ordination. Suppose the effects of particular environmental variables are to be singled out from 'background' variation imposed by other variables. In an environmental impact study, for example, the effects of impact variables are to be separated from those of other sources of variation, represented by 'covariables'. One may then want to eliminate ('partial out') the effects of the covariables and to relate the residual variation to the impact variables. This is achieved in partial canonical ordination. Technically, partial canonical ordination can be carried out by any computer program for canonical ordination. The usual environmental variables are simply replaced by the residuals obtained by regressing each of the impact variables on the covariables. The theory of partial RDA and partial CCA is described by Davies & Tso (1982) and ter Braak (1988). Partial ordination and partial canonical ordination are available in the computer program CANOCO (ter Braak 1987b). The program also includes a Monte Carlo permutation procedure to investigate the statistical significance of the effects of the impact variables.

5.8 Bibliographic notes

A simple ordination technique of the early days was polar ordination (Bray & Curtis 1957; Gauch 1982), which has been recently reappraised by Beals (1985). PCA was developed early this century by K. Pearson and H. Hotelling (e.g. Mardia et al. 1979) and was introduced in ecology by Goodall (1954). PCA was popularized by Orlóci (1966). CA has been invented independently since 1935 by several authors working with different types of data and with different rationales. Mathematically

CA is the same as reciprocal averaging, canonical analysis of contingency tables, and optimal or dual scaling of nominal variables (Gifi 1981; Gittins 1985; Greenacre 1984; Nishisato 1980). Benzécri et al. (1973) developed CA in a geometric context. Neither of these different approaches to CA is particularly attractive in ecology. Hill (1973) developed an ecological rationale (Subsection 5.2.2). The dispersion of the species scores by which we introduced CA in Subsection 5.2.1 is formally identical to the 'squared correlation ratio' (η^2) used by Torgerson (1958, Section 12.7) and Nishisato (1980, p.23) and also follows from the reciprocal gravity problem in Heiser (1986). RDA is also known under several names (Israëls 1984): PCA of instrumental variables (Rao 1964), PCA of y with respect to x, reduced rank regression (Davies & Tso 1982). Ter Braak (1986a) proposed CCA. COR was derived by H. Hotelling in 1935 (Gittins 1985). Campbell & Atchley (1981) provide a good geometric and algebraic introduction to CVA and Williams (1983) discusses its use in ecology. Methods to obtain the maximum likelihood solutions for Gaussian ordination have been investigated, under the assumption of a normal distribution, a Poisson distribution and a Bernoulli distribution for the species data, by Gauch et al. (1974), Kooijman (1977) and Goodall & Johnson (1982), respectively. However the computational burden of these methods and, hence, the lack of reliable computer programs have so far prevented their use on a routine basis. Ihm & van Groenewoud (1984) and ter Braak (1985) compared Gaussian ordination and CA. Non-metric multidimensional scaling started with the work by Shepard (1962) and Kruskal (1964). Schiffman et al. (1981) provide a clear introduction. They refer to local non-metric scaling as (row) conditional scaling. Meulman & Heiser (1984) describe a canonical form of non-metric multidimensional scaling. Early applications of non-metric multidimensional scaling in ecology were Anderson (1971), Noy-Meir (1974), Austin (1976), Fasham (1977), Clymo (1980) and Prentice (1977; 1980). The simple unfolding model (response models with circular contours) can in principle be fitted by methods of multidimensional scaling (Kruskal & Carroll 1969; Dale 1975; de Sarbo & Rao 1984; Heiser 1987), but Schiffman et al. (1981) warn of practical numerical problems that may reduce the usefulness of this approach. Most of the problems have, however, been circumvented by Heiser (1987).

Many textbooks use matrix algebra to introduce multivariate analysis techniques, because it provides an elegant and concise notation (Gordon 1981; Mardia et al. 1979; Greenacre 1984; Rao 1973; Gittins 1985). For ecologists, the book of Pielou (1984) is particularly recommended. All techniques described in Chapter 5 can be derived from the singular-value decomposition of a matrix, leading to singular vectors and singular values (Section 5.9). The decomposition can be achieved by many numerical methods (e.g. Gourlay & Watson 1973), one of which is the power algorithm (Table 5.6). The power algorithm is used in Chapter 5 because it provides the insight that ordination is simultaneously regression and calibration, and because it does not require advanced mathematics. The power algorithm can easily be programmed on a computer, but is one of the slowest algorithms available to obtain a singular-value decomposition. Hill (1979a) and ter Braak (1987b) use the power algorithm with a device to accelerate the process. The iteration processes of Tables 5.2 and 5.6 are examples of alternating least-

157

squares methods (Gifi 1981) and are related to the EM algorithm (Everitt 1984). The power algorithm is also a major ingredient of partial least squares (Wold 1982).

Computer programs for PCA, COR and CVA are available in most statistical computer packages. CA and DCA are available in DECORANA (Hill 1979). The program CANOCO (ter Braak 1987b) is an extension of DECORANA and it also includes PCA, PCO, RDA, CCA, CVA and partial variants of these techniques. All these techniques can be specified in terms of matrix algebra (Section 5.9). With the facilities for matrix algebra operations in GENSTAT (Alvey et al. 1977) or SAS (SAS Institute Inc. 1982), one can therefore write one's own programs to analyse small to medium-sized data sets. Schiffman et al. (1981) describe various programs for multidimensional scaling.

Chapter 5 uses response models as a conceptual basis for ordination. Carroll (1972) defined a hierarchy of response models, from the linear model (Equation 5.11), through the model with circular contour lines (Equation 5.5) to the full quadratic model (Equation 3.24) with ellipsoidal contours of varying orientation. He terms these models the vector model, the (simple) unfolding model and the general unfolding model, respectively (also Davison 1983). By taking even more flexible response models, we can define even more general ordination techniques. However the more flexible the model, the greater the computational problems (Prentice 1980). Future research must point out how flexible the model can be to obtain useful practical solutions.

5.9 Ordination methods in terms of matrix algebra

What follows in this section is a short introduction to ordination methods in terms of matrix algebra:
- to facilitate communication between ecologists and the mathematicians they may happen to consult
- to bridge the gap between the approach followed in Chapter 5 and the mainstream of statistical literature on multivariate methods
- to suggest computational methods based on algorithms for singular-value decomposition of a matrix or to extract eigenvalues and eigenvectors from a symmetric matrix.

To start, please read Section 5.8 first.

5.9.1 Principal components analysis (PCA)

Let $Y = \{y_{ki}\}$ be an $m \times n$ matrix containing the data on m species (rows of the matrix) and n sites (columns of the matrix). In the most familiar form of PCA, species-centred PCA, the data are abundances with the species means already subtracted, so that $y_{k+} = 0$ as in Subsection 5.3.1. PCA is equivalent to the singular-value decomposition (SVD) of Y (e.g. Rao 1973; Mardia et al. 1979; Greenacre 1984)

$$Y = P \Lambda^{0.5} Q'$$

Equation 5.19

158

where $\mathbf{P}$ and $\mathbf{Q}$ are orthonormal matrices of dimensions $m \times r$ and $n \times r$, respectively, with $r = \min{(m,n)}$, i.e. $\mathbf{P'P} = \mathbf{I}$ and $\mathbf{Q'Q} = \mathbf{I}$, and Λ is a diagonal matrix with diagonal elements λ_s ($s = 1 \ldots r$), which are arranged in order of decreasing magnitude $\lambda_1 \geqslant \lambda_2 \geqslant \lambda_3 \geqslant \ldots \geqslant 0$.

The columns of $\mathbf{P}$ and $\mathbf{Q}$ contain the singular vectors of $\mathbf{Y}$, and $\lambda_s^{0.5}$ is the sth singular value of $\mathbf{Y}$. If the sth column of $\mathbf{P}$ is denoted by $\mathbf{p}_s$, an m vector, and the sth column of $\mathbf{Q}$ by $\mathbf{q}_s$, an n vector, Equation 5.19 can be written as

$$\mathbf{Y} = \Sigma_{s=1}^{r} \lambda_s^{0.5}\, \mathbf{p}_s\, \mathbf{q}_s' \qquad\qquad \text{Equation 5.20}$$

The least-squares approximation of $\mathbf{Y}$ in Equation 5.11 of Subsection 5.3.2 is obtained from Equation 5.20 by retaining only the first two terms of this summation, and by setting $\mathbf{b}_s = \lambda_s^{0.5}\, \mathbf{p}_s$ and $\mathbf{x}_s = \mathbf{q}_s$ ($s = 1, 2$).

The kth element of $\mathbf{b}_1$ then contains the species score b_{k1}, and the ith element of $\mathbf{x}_1$ contains the site score x_{i1} on the first axis of PCA. Similarly, $\mathbf{b}_2$, and $\mathbf{x}_2$ contain the species and sites scores on the second axis of PCA. The species and sites scores on both axes form the coordinates of the points for species and sites in the biplot (Subsection 5.3.4). The interpretation of the PCA biplot follows from Equation 5.11: inner products between species points and site points provide a least-squares approximation of the elements of the matrix $\mathbf{Y}$ (Gabriel 1971; 1978). Equation 5.20 shows that the total sum of squares $\Sigma_{ki}\, y_{ki}^2$ equals $\lambda_1 + \ldots + \lambda_r$, the sum of all eigenvalues, and that the total residual sum of squares

$$\Sigma_{k,i}\, [y_{ki} - (b_{k1}\, x_{i1} + b_{k2}\, x_{i2})]^2 = \lambda_3 + \lambda_4 + \ldots + \lambda_r.$$

An appropriate measure of goodness of fit is therefore $(\lambda_1 + \lambda_2)/(\text{sum of all eigenvalues})$. From $\mathbf{P'P} = \mathbf{I}$, $\mathbf{Q'Q} = \mathbf{I}$ and Equation 5.20, we obtain

$$\mathbf{b}_s = \mathbf{Y}\mathbf{x}_s \qquad\qquad \text{Equation 5.21}$$

and

$$\lambda_s\, \mathbf{x}_s = \mathbf{Y'}\mathbf{b}_s. \qquad\qquad \text{Equation 5.22}$$

Hence, the species scores are a weighted sum of the site scores and the site scores are proportional to a weighted sum of the species scores (Table 5.6 and Subsection 5.3.2). Equation 5.21 and Equation 5.22 show that $\mathbf{b}_s$ and $\mathbf{x}_s$ are eigenvectors of $\mathbf{YY'}$ and $\mathbf{Y'Y}$, respectively, and that λ_s is their common eigenvalue; whence, the R-mode and Q-mode algorithms of Subsection 5.3.6.

The SVD of the species-by-species cross-product matrix $\mathbf{YY'}$ is $\mathbf{P} \Lambda \mathbf{P'}$, as follows from Equation 5.19 by noting that $\mathbf{Q'Q} = \mathbf{I}$. A least-squares approximation of the matrix $\mathbf{YY'}$ in two dimensions is therefore given by the matrix $\mathbf{b}_1\mathbf{b}'_1 + \mathbf{b}_2\, \mathbf{b}'_2$. Since $\mathbf{YY'}/(n - 1)$ contains covariances between species, the biplot of $\mathbf{x}_s$ and $\mathbf{b}_s$ is termed the covariance biplot (Corsten & Gabriel 1976; ter Braak 1983).

159

The SVD of the site-by-site cross-product matrix $Y'Y$ is $Q \Lambda Q'$. A biplot of Y and $Y'Y$ is therefore obtained by redefining b_s and x_s as $b_s = p_s$ and $x_s = \lambda_s^{0.5} q_s$. The inter-site distances in this biplot approximate the Euclidean Distances between sites as defined by Equation 5.16; hence the name Euclidean Distance biplot. The approximation is, however, indirect namely through Equation 5.17 with c_{ij} the (i, j)th element of $Y'Y$. A consequence of this is that the inter-site distances are always smaller than the Euclidean Distances.

5.9.2 Correspondence analysis (CA)

In CA, the species-by-sites matrix Y contains the abundance values y_{ki} in which $y_{ki} \geqslant 0$. The data is not previously centred in CA. Let $M = \text{diag}(y_{k+})$, an $m \times m$ diagonal matrix containing the row totals of Y, $N = \text{diag}(y_{+i})$, an $n \times n$ diagonal matrix containing the column totals of Y.

As stated in Subsection 5.2.1, CA chooses standardized site scores x that maximize the dispersion of species scores, which are themselves weighted averages of the site scores (Equation 5.1). In matrix notation, the vector of species scores $u = (u_k)[k = 1, ..., m]$ is

$$u = M^{-1}Yx \qquad\qquad \text{Equation 5.23}$$

and the dispersion is

$$\delta = u'Mu / x'Nx = x'Y'M^{-1}Yx / x'Nx \qquad\qquad \text{Equation 5.24}$$

where the denominator takes account of the standardization of x (Table 5.2c), provided x is centred ($1'Nx = 0$).

The problem of maximizing δ with respect to x has as solution the second eigenvector of the eigenvalue equation

$$Y'M^{-1}Yx = \lambda N x \qquad\qquad \text{Equation 5.25}$$

with $\delta = \lambda$ (Rao 1973, Section 1f.2 and p.74; Mardia et al. 1979, Theorem A9.2).

This can be seen by noting that the first eigenvector is a trivial solution ($x = 1$; $\lambda = 1$); because the second eigenvector is orthogonal to the first eigenvector in the N metric, the second eigenvector maximizes δ subject to $1'Nx = 0$. What is called the first eigenvector of CA in Section 5.2 is thus the second eigenvector of Equation 5.25, i.e. its first non-trivial eigenvector. The second non-trivial eigenvector of Equation 5.25 is similarly seen to maximize δ, subject to being centred and to being orthogonal to the first non-trivial eigenvector, and so on for subsequent axes. Equation 5.25 can be rewritten as

$$\lambda x = N^{-1}Y'u \qquad\qquad \text{Equation 5.26}$$

Equations 5.23 and 5.26 form the 'transition equations' of CA. In words: the species scores are weighted averages of the site scores and the site scores are proportional to weighted averages of the species scores (Table 5.2 and Exercise 5.1.3).

The eigenvectors of CA can also be obtained from the SVD

$$\mathbf{M}^{-0.5} \mathbf{Y} \mathbf{N}^{-0.5} = \mathbf{P} \Lambda^{0.5}\mathbf{Q}' \qquad\qquad \text{Equation 5.27}$$

by setting $\mathbf{u}_s = \lambda_s^{0.5} \mathbf{M}^{-0.5} \mathbf{p}_s$ and $\mathbf{x}_s = \mathbf{N}^{-0.5} \mathbf{q}_s$, where $\mathbf{p}_s$ and $\mathbf{q}_s$ are the sth columns of $\mathbf{P}$ and $\mathbf{Q}$, respectively ($s = 1, ..., r$).

This can be seen by inserting the equations for $\mathbf{u}_s$ and $\mathbf{x}_s$ in Equations 5.23 and 5.26, and rearranging terms. It is argued in Subsection 5.2.7 that it is equally valid to distribute λ_s in other ways among $\mathbf{u}_s$ and $\mathbf{x}_s$, as is done, for example, in Hill's scaling (Subsection 5.2.2).

CA differs from PCA in the particular transformation of $\mathbf{Y}$ in Equation 5.27 and in the particular transformation of the singular vectors described just below that equation.

5.9.3 Canonical correlation analysis (COR)

As in species-centred PCA, let $\mathbf{Y}$ be an $m \times n$ matrix in which the kth row contains the centred abundance values of the kth species (i.e. $y_{k+} = 0$) and let $\mathbf{Z}$ be a $q \times n$ matrix in which the jth row contains the centred values of the jth environmental variable (i.e. $z_{j+} = 0$). Define

$$\mathbf{S}_{12} = \mathbf{YZ}', \mathbf{S}_{11} = \mathbf{YY}', \mathbf{S}_{22} = \mathbf{ZZ}' \text{ and } \mathbf{S}_{21} = \mathbf{S}'_{12}. \qquad \text{Equation 5.28}$$

The problem of COR is to determine coefficients for the species $\mathbf{b} = (b_k)[k = 1, ..., m]$ and for the environmental variables $\mathbf{c} = (c_j)[j = 1, ..., q]$ that maximize the correlation between $\mathbf{x}^* = \mathbf{Y}'\mathbf{b}$ and $\mathbf{x} = \mathbf{Z}'\mathbf{c}$. The solution for $\mathbf{b}$ and $\mathbf{c}$ is known to be the first eigenvector of the respective eigenvalue equations

$$\mathbf{S}_{12} \mathbf{S}_{22}^{-1} \mathbf{S}_{21} \mathbf{b} = \lambda \mathbf{S}_{11} \mathbf{b} \qquad\qquad \text{Equation 5.29}$$

$$\mathbf{S}_{21} \mathbf{S}_{11}^{-1} \mathbf{S}_{12} \mathbf{c} = \lambda \mathbf{S}_{22} \mathbf{c} \qquad\qquad \text{Equation 5.30}$$

The eigenvalue λ equals the squared canonical correlation (Rao 1973; Mardia et al. 1979; Gittins 1985).

Note that $\mathbf{b}$ can be derived from a multiple regression of $\mathbf{x}$ on the species, or from $\mathbf{c}$, by

$$\mathbf{b} = (\mathbf{YY}')^{-1} \mathbf{Yx} = \mathbf{S}_{11}^{-1} \mathbf{S}_{12} \mathbf{c} \qquad\qquad \text{Equation 5.31}$$

and, similarly, $\mathbf{c}$ can be derived from a multiple regression of $\mathbf{x}^*$ on the environmental variables, or from $\mathbf{b}$, by

$\lambda c = (ZZ')^{-1} Zx^* = S_{22}^{-1} S_{21} b$ Equation 5.32

It can be verified that b and c from Equations 5.31 and 5.32 satisfy Equations 5.29 and 5.30, by inserting b from Equation 5.31 into Equation 5.32 and by inserting c from Equation 5.32 into Equation 5.31; but note that we could have distributed λ in other ways among Equations 5.31 and 5.32. Equations 5.31 and 5.32 form the basis of the iteration algorithm of Table 5.12. Step 7 of Table 5.12 takes care of the eigenvalue: at convergence, x is divided by λ (Table 5.6c). Once convergence is attained, c should be divided by λ to ensure that the final site scores satisfy $x = Z'c$ (Step 5);hence the λ in Equation 5.32. The second and further axes obtained by Table 5.12 also maximize the correlation between x and x^*, but subject to being uncorrelated to the site scores of the axes already extracted.

COR can also be derived from the SVD of

$$S_{11}^{-0.5} S_{12} S_{22}^{-0.5} = P \Lambda^{0.5} Q'$$ Equation 5.33

The equivalence of Equation 5.31 with Equation 5.33 can be verified by pre-multiplying Equation 5.33 on both sides with $S_{11}^{-0.5}$ and post-multiplying Equation 5.33 on both sides with Q and by defining

$$B = S_{11}^{-0.5} P \Lambda^{0.5} \text{ and } C = S_{22}^{-0.5} Q.$$ Equation 5.34

The sth column of B and of C contain the canonical coefficients on the sth axis of the species and environmental variables, respectively. The equivalence of Equation 5.32 with Equation 5.33 can be shown similarly.

COR allows a biplot to be made in which the correlations between species and environmental variables are approximated. The problem to which the canonical correlation biplot is the solution can be formulated as follows: determine points for species and environmental variables in t-dimensional space in such a way that their inner products give a weighted least-squares approximation to the elements of the covariance matrix S_{12}. In the approximation, the species and the environmental variables are weighted inversely with their covariance matrices S_{11} and S_{22}, respectively. Let the coordinates of the points for the species be collected in the $m \times t$ matrix G and those for the environmental variables in the $q \times t$ matrix H. The problem is then to minimize

$$\|S_{11}^{-0.5} (S_{12} - GH') S_{22}^{-0.5}\|^2 = \|S_{11}^{-0.5} S_{12} S_{22}^{-0.5} - (S_{11}^{-0.5} G)(S_{22}^{-0.5} H)'\|^2$$
Equation 5.35

with respect to the matrices G and H, where $\|\bullet\|$ is the Euclidean matrix norm, e.g. $\|Y\|^2 = \Sigma_{k,i} y_{ki}^2$.

From the properties of an SVD (Subsection 5.9.1), it follows that the minimum is attained when $S_{11}^{-0.5} G$ and $S_{22}^{-0.5} H$ correspond to the first t columns of the matrices $P\Lambda^{0.5}$ and Q of Equation 5.33, respectively. The required least-squares approximation is thus obtained by setting $\mathbf{G}$ and $\mathbf{H}$ equal to the first t columns of $\mathbf{S}_{11}{}^{0.5}\,\mathbf{P}\Lambda^{0.5}$ and $\mathbf{S}_{22}{}^{0.5}\,\mathbf{Q}$, respectively. Again, Λ can be distributed in other ways among $\mathbf{P}$ and $\mathbf{Q}$. For computational purposes, note that

$$\mathbf{S}_{11}{}^{0.5}\,\mathbf{P}\,\Lambda^{0.5} = \mathbf{S}_{11}\,\mathbf{S}_{11}{}^{-0.5}\,\mathbf{P}\,\Lambda^{0.5} = \mathbf{S}_{11}\,\mathbf{B} = \mathbf{YY'B} = \mathbf{YX} \qquad \text{Equation 5.36}$$

and

$$\mathbf{S}_{22}{}^{0.5}\,\mathbf{Q} = \mathbf{S}_{22}\,\mathbf{S}_{22}{}^{-0.5}\,\mathbf{Q} = \mathbf{S}_{22}\,\mathbf{C} = \mathbf{ZX} \qquad \text{Equation 5.37}$$

where $\mathbf{X} = \mathbf{Z'C}$. Because $\mathbf{X'X} = \mathbf{I}$, the biplot can thus be constructed from the inter-set correlations of the species and the intra-set correlations of the environmental variables (which are the correlations of the site scores $\mathbf{x}$ with the species variables and environmental variables, respectively). This construction rule requires the assumption that the species and environmental variables are standardized to unit variance, so that $\mathbf{S}_{12}$ is actually a correlation matrix. The angles between arrows in the biplot are, however, not affected by whether either covariances or correlations between species and environment are approximated in the canonical correlation biplot.

5.9.4 Redundancy analysis (RDA)

RDA is obtained by redefining $\mathbf{S}_{11}$ in subsection 5.9.3 to be the identity matrix (Rao 1973, p.594-595). In the RDA biplot, as described in Subsection 5.5.3, the coordinates of the point for the species and the variables are given in the matrices $\mathbf{P}$ and $\mathbf{S}_{22}{}^{0.5}\,\mathbf{Q}\,\Lambda^{0.5}$, respectively.

5.9.5 Canonical correspondence analysis (CCA)

CCA maximizes Equation 5.24 subject to Equation 5.14, provided $\mathbf{x}$ is centred. If the matrix $\mathbf{Z}$ is extended with a row of ones, Equation 5.14 becomes $\mathbf{x} = \mathbf{Z'c}$, with $\mathbf{c} = (c_j)[j = 0, 1, ..., q]$. By inserting $\mathbf{x} = \mathbf{Z'c}$ in Equation 5.24 and (re)defining, with $\mathbf{Y}$ non-centred,

$$\mathbf{S}_{12} = \mathbf{YZ'},\;\; \mathbf{S}_{11} = \mathbf{M} = \text{diag}\,(y_{k+}) \text{ and } \mathbf{S}_{22} = \mathbf{ZNZ'} \qquad \text{Equation 5.38}$$

we obtain

$$\delta = \mathbf{c'S}_{21}\,\mathbf{S}_{11}{}^{-1}\,\mathbf{S}_{12}\,\mathbf{c}\,/\,\mathbf{c'}\,\mathbf{S}_{22}\,\mathbf{c} \qquad \text{Equation 5.39}$$

The solutions of CCA can therefore be derived from the eigenvalue Equation 5.30 with $\mathbf{S}_{12}$, $\mathbf{S}_{11}$ and $\mathbf{S}_{22}$ defined as in Equation 5.38. If defined in this way, CCA has a trivial solution $\mathbf{c'} = (1, 0, 0, ..., 0)$, $\lambda = 1$, $\mathbf{x} = \mathbf{1}$ and the first non-trivial eigenvector maximizes δ subject to $\mathbf{1'Nx} = \mathbf{1'NZ'c} = 0$ and the maximum δ equals the eigenvalue. A convenient way to exclude the trivial solution is to subtract from each environmental variable its weighted mean $\bar{z}_j = \Sigma_i\, y_{+i}\, z_{ji}/\, y_{++}$

163

(and to remove the added row of ones in the matrix $\mathbf{Z}$). Then, the matrix $\mathbf{Z}$ has weighted row means equal to 0: $\Sigma_i \, y_{+i} \, z_{ji} = 0$. The species scores and the canonical coefficients of the environmental variables can be obtained from Equation 5.33 and Equation 5.34, by using the definitions of Equation 5.38.

As described in Subsection 5.5.2, the solution of CCA can also be obtained by extending the iteration algorithm of Table 5.2. Steps 1, 4, 5 and 6 remain the same as in Table 5.2. In matrix notation, the other steps are

Step 2 $\mathbf{b} = \mathbf{M}^{-1} \mathbf{Y} \mathbf{x}$ Equation 5.40

Step 3a $\mathbf{x}^* = \mathbf{N}^{-1} \mathbf{Y}'\mathbf{b}$ Equation 5.41

Step 3b $\mathbf{c} = (\mathbf{Z}\mathbf{N}\mathbf{Z}')^{-1} \mathbf{Z}\mathbf{N} \mathbf{x}^*$ Equation 5.42

Step 3c $\mathbf{x} = \mathbf{Z}'\mathbf{c}$ Equation 5.43

with $\mathbf{b} = \mathbf{u}$, the m vector containing the species scores u_k ($k = 1, ..., m$).

Once convergence has been attained, to ensure that the final site scores satisfy $\mathbf{x} = \mathbf{Z}'\mathbf{c}$, $\mathbf{c}$ should be divided by λ, as in COR (below Equation 5.32). This amounts to replacing $\mathbf{c}$ in Equation 5.42 by $\lambda\mathbf{c}$ (as in Equation 5.32). To show that the algorithm gives a solution of Equation 5.30, we start with Equation 5.42, modified in this way, insert $\mathbf{x}^*$ of Equation 5.41 in Equation 5.42, next insert $\mathbf{b}$ by using Equation 5.40, next insert $\mathbf{x}$ by using Equation 5.43 and finally use the definitions of $\mathbf{S}_{11}$, $\mathbf{S}_{12}$ and $\mathbf{S}_{22}$ in CCA.

CCA allows a biplot to be made, in which the inner products between points for species and points for environmental variables give a weighted least-squares approximation of the elements of the $m \times q$ matrix

$$\mathbf{W} = \mathbf{M}^{-1}\mathbf{Y}\mathbf{Z}',$$

the (k,j)th element of which is the weighted average of species k with respect to the (centred) environmental variable j. In the approximation, the species are given weight proportional to their total abundance (y_{k+}) and the environmental variables are weighted inversely with their covariance matrix $\mathbf{S}_{22}$. The possibility for such a biplot arises because

$$\mathbf{M}^{0.5} \, \mathbf{W} \, \mathbf{S}_{22}^{-0.5} = \mathbf{S}_{11}^{-0.5} \, \mathbf{S}_{12} \, \mathbf{S}_{22}^{-0.5} \qquad \text{Equation 5.44}$$

so that, from Equations 5.44 and 5.33, after rearranging terms,

$$\mathbf{W} = (\mathbf{S}_{11}^{-0.5} \, \mathbf{P}) \, \Lambda^{0.5} \, (\mathbf{S}_{22}^{0.5} \, \mathbf{Q})' \qquad \text{Equation 5.45}$$

Apart from particular considerations of scale (Subsection 5.2.2), the coordinates of the points for species and environmental variables in the CCA biplot are thus

164

given by the first t columns of $S_{11}^{-0.5} P \Lambda^{0.5}$ and $S_{22}^{0.5} Q$, respectively. The matrix $S_{11}^{-0.5} P \Lambda^{0.5}$ actually contains the species scores, as follows from Equation 5.34. The other matrix required for the biplot can be obtained by

$$S_{22}^{0.5}Q = S_{22} S_{22}^{-0.5} Q = S_{22} C = ZNZ'C = ZNX \qquad \text{Equation 5.46}$$

5.10 Exercises

Exercise 5.1 Correspondence analysis: the algorithm

This exercise illustrates the two-way weighted-averaging algorithm of CA (Table 5.2) with the small table of artificial data given below.

Species	Sites				
	1	2	3	4	5
A	1	0	0	1	0
B	0	0	1	0	1
C	0	2	0	1	0
D	3	0	0	1	1

The data appear rather chaotic now, but they will show a clear structure after having extracted the first CA ordination axis. The first axis is dealt with in Exercises 5.1.1-3, and the second axis in Exercises 5.1.4-6.

Exercise 5.1.1 Take as site scores the values 1, 2, ..., 5 as shown above the data table. Now, standardize the site scores by using the standardization procedure described in Table 5.2c.

Exercise 5.1.2 Use the site scores so standardized as initial site scores in the iteration process (Table 5.2a). Carry out at least five iteration cycles and in each cycle calculate the dispersion of the species scores. (Use an accuracy of three decimal places in the calculations for the site and species scores and of four decimal places for s.) Note that the scores keep changing from iteration to iteration, but that the rank order of the site scores and of the species scores remains the same from Iteration 4 onwards. Rearrange the species and sites of the table according to their rank order. Note also that the dispersion increases during the iterations.

Exercise 5.1.3 After 19 iterations, the site scores obtained are 0.101, –1.527, 1.998, –0.524, 1.113. Verify these scores for the first CA axis (within an accuracy of two decimal places) by carrying out one extra iteration cycle. What is the eigenvalue of this axis? Verify that Equation 5.1 holds true for the species scores and site scores finally obtained, but that Equation 5.2 does not hold true. Modify Equation 5.2 so that it does hold true.

Exercise 5.1.4 We now derive the second CA axis by using the same initial site scores as in Exercise 5.1.2. Orthogonalize these scores first with respect to the first axis by using the orthogonalization procedure described in Table 5.2b, and next standardize them (round the site scores of the first axis to two decimals and use four decimals for *v* and *s* and three for the new scores).

Exercise 5.1.5 Use the site scores so obtained as initial site scores to derive the second axis. The scores stabilize in four iterations (within an accuracy of two decimal places).

Exercise 5.1.6 Construct an ordination diagram of the first two CA axes. The diagram shows one of the major 'faults' in CA. What is this fault?

Exercise 5.2 Adding extra sites and species to a CA ordination

Exercise 5.2.1 We may want to add extra species to an existing CA ordination. In the Dune Meadow Data, *Hippophae rhamnoides* is such a species, occurring at Sites 9, 18 and 19 with abundances 1, 2 and 1, respectively. Calculate from the site scores in Table 5.1c the score for this species on the first CA axis in the way this is done with CA. Plot the abundance of the species against the site score. What does the species score mean in this plot? At which place does the species appear in Table 5.1c? Answer the same questions for *Poa annua*, which occurs at Sites 1, 2, 3, 4, 7, 9, 10, 11, 13 and 18 with abundances 3, 3, 6, 4, 2, 2, 3, 2, 3 and 4, respectively, and for *Ranunculus acris*, which occurs at Sites 5, 6, 7, 9, 14 and 15 with abundances 2, 3, 2, 2, 1 and 1, respectively.

Exercise 5.2.2 Similarly, we may want to add an extra site to an existing CA ordination. Calculate the score of the site where the species *Bellis perennis*, *Poa pratensis* and *Rumex acetosa* are present with abundances 5, 4 and 3, respectively (imaginary data). (Hint: recall how the site scores were obtained from the species scores in Exercise 5.1.3.) Species and sites so added to an ordination are called passive, to distinguish them from the active species and sites of Table 5.1. The scores on higher-order axes are obtained in the same way.

Exercise 5.2.3 Rescale the scores of Table 5.1c to Hill's scaling and verify that the resulting scores were used in Figure 5.4.

Exercise 5.3 Principal components analysis

Add the extra species and the extra site of Exercise 5.2 to the PCA ordination of Table 5.5c. Plot the abundance of the extra species against the site scores. What does the species score mean in this plot? At which places do the species appear in Table 5.5c?

Exercise 5.4 Length of gradient in DCA

Suppose DCA is applied to a table of abundances of species at sites and that the length of the first axis is 1.5 s.d. If, for each species, we made a plot of its abundance against the site scores of the first axis, would most plots suggest monotonic curves or unimodal curves? And what would the plots suggest if the length of the axis was 10 s.d.?

Exercise 5.5 Interpretation of joint plot and biplot

Exercise 5.5.1 Rank the sites in order of abundance of *Juncus bufonius* as inferred from Figure 5.7, as inferred from Figure 5.15 and as observed in Table 5.1a. Do the same for *Eleocharis palustris*.

Exercise 5.5.2 If Figure 5.15 is interpreted erroneously as a joint plot of DCA, one gets different inferred rank orders and, when Figure 5.7 is interpreted erroneously as a biplot, one also gets different rank orders. Is the difference in interpretation greatest for species that lie near the centre of an ordination diagram or for species that lie on the edge of an ordination diagram?

Exercise 5.6 Detrended canonical correspondence analysis

Cramer (1986) studied vegetational succession on the rising sea-shore of an island in the Stockholm Archipelago. In 1978 and 1984, the field layer was sampled on 135 plots of 1 m^2 along 4 transects. The transects ran from water level into mature forest. One of the questions was whether the vegetational succession keeps track with the land uplift (about 0.5 cm per year) or whether it lags behind. In both cases, the vegetation zones 'run down the shore', but in the latter case too slowly. Because succession in the forest plots was not expected to be due to land uplift, only the 63 plots up to the forest edge were used. These plots contained 68 species with a total of about 1000 occurrences on the two sampling occasions. An attempt was made to answer the question by using detrended canonical correspondence analysis (DCCA) with two explanatory variables, namely altitude above water level in 1984 (not corrected for land uplift; so each plot received the same value in 1978 as in 1984) and time (0 for 1978, 6 for 1984). The altitude ranged from –14 to 56 cm. The first two axes gave eigenvalues 0.56 and 0.10, lengths 4.4 and 0.9 s.d. and species–environment correlations 0.95 and 0.74, respectively. Table 5.14 shows that the first axis is strongly correlated with altitude and almost uncorrelated with time, whereas the second axis is strongly correlated with time and almost uncorrelated with altitude. However the canonical coefficients tell a more interesting story.

Exercise 5.6.1 With Table 5.14, show that the linear combination of altitude and time best separating the species in the sense of Section 5.5.2 is

$$x = 0.054 z_1 + 0.041 z_2 \qquad \text{Equation 5.47}$$

167

Table 5.14 Detrended canonical correspondence analysis of rising shore vegetation data: canonical coefficients ($100 \times c$) and intra-set correlations ($100 \times r$) for standardized environmental variables. In brackets, the approximate standard errors of the canonical coefficients. Also given are the mean and standard deviation (s.d.) of the variables.

Variable	Coefficients		Correlations		mean	s.d.
	Axis 1	Axis 2	Axis 1	Axis 2		
Altitude (cm)	100 (3)	4 (4)	99	19	22	18.5
Time (years)	12 (3)	-34 (3)	7	-99	3	2.9

where z_1 is the numeric value of altitude (cm) and z_2 is the numeric value of time (years) and where the intercept is, arbitrary, set to zero.

Hint: note that Table 5.14 shows standardized canonical coefficients, i.e. canonical coefficients corresponding to the standardized variables $z_1^* = (z_1 - 22)/18.5$ and $z_2^* = (z_2 - 3)/2.9$.
Similarly, show that the standard errors of estimate of $c_1 = 0.054$ is 0.0016 and of $c_2 = 0.041$ is 0.010.

Exercise 5.6.2 Each value of x in Equation 5.47 stands for a particular species composition (Figures 5.8 and 5.18) and changes in the value of x express species turnover along the altitude gradient in multiples of s.d. With Equation 5.47, calculate the species turnover between two plots that were 15 cm and 25 cm above water level in 1984, respectively. Does the answer depend on the particular altitudes of these plots or only on the difference in altitude? What is, according to Equation 5.47, the species turnover between these plots in 1978?

Exercise 5.6.3 With Equation 5.47, calculate the species turnover between 1978 and 1984 for a plot with an altitude of 15 cm in 1984? Does the answer depend on altitude?

Exercise 5.6.4 With Equation 5.47, calculate the altitude that gives the same species turnover as one year of succession.

Exercise 5.6.5 Is there evidence that the vegetational succession lags behind uplift?

Exercise 5.6.6 Roughly how long would it take to turn the species composition of the plot closest to the sea into that of the plot that is on the edge of the forest? Hint: use the length of the first axis. Is there evidence from the analysis that there might also be changes in species composition that are unrelated to land uplift? Hint: consider the length of the second axis.

168

5.11 Solutions to exercises

Exercise 5.1   Correspondence analysis: the algorithm

Exercise 5.1.1   The centroid of the site scores is $z = (4 \times 1 + 2 \times 2 + 1 \times 3 + 3 \times 4 + 2 \times 5)/12 = 2.750$ and their dispersion is $s^2 = [4 \times (1 - 2.750)^2 + \dots + 2 \times (5 - 2.750)^2]/12 = 2.353$, thus $s = 1.5343$. The standardized initial score for the first site is thus $x_1 = (1 - 2.750)/1.5343 = -1.141$. The other scores are listed on the second line of Table 5.15.

Exercise 5.1.2   In the first iteration cycle at Step 2, we obtain for Species C, for example, the score $[2 \times (-0.489) + 1 \times 0.815)]/(2 + 1) = -0.054$, and for Site 5 at Step 3, the score $(0.815 - 0.228)/(1 + 1) = 0.294$. The dispersion of the species scores in the first iteration cycle is $\delta = (2 \times 0.163^2 + 2 \times 0.815^2 + 3 \times 0.054^2 + 5 \times 0.228^2)/12 = 0.138$. See further Table 5.15. In the iterations shown $z = 0.000$, apart from Iteration 3, where $z = -0.001$ (Step 5). The rearranged data table shows a Petrie matrix (Subsection 5.2.3).

Exercise 5.1.3   The standardized site scores obtained in the 19th and 20th iteration are equal within the accuracy of two decimal places; so the iteration process has converged (Table 5.15). The eigenvalue of the first axis is $\lambda_1 = 0.7799$, the value of s calculated last. Equation 5.2 does not hold true for the final site and species scores. But the site scores calculated in Step 3 are weighted averages of the species scores and are divided in the 20th iteration by $s = 0.7799$ to obtain the final site scores. On convergence, s equals the eigenvalue λ; thus the final site and species scores satisfy the relation $\lambda\, x_i = \Sigma_{k=1}^{m}\, y_{ki}\, u_k / \Sigma_{k=1}^{m} y_{ki}$. Applying Steps 3, (4) and 5 to the eigenvector (the scores x_i) thus transforms the eigenvector into a multiple of itself. The multiple is the 'eigenvalue' of the eigenvector. Note that δ equals λ within arithmetic accuracy.

Exercise 5.1.4   In Step 4.2, we obtain $v = [4 \times (-1.141) \times 0.10 + 2 \times (-0.489) \times (-0.53) + 1 \times 0.163 \times 2.00 + 3 \times 0.815 \times (-0.53) + 2 \times 1.466 \times 1.11]/12 = 0.2771$ and for Site 1 at Step 4.3, the score $-1.141 - 0.277\ 1 \times 0.10 = -1.169$. See further the first four lines of Table 5.16.

Exercise 5.1.5   See Table 5.16.

Exercise 5.1.6   The configuration of the site points looks like the letter V, with Site 1 at the bottom and Sites 2 and 3 at the two extremities. This is the arch effect of CA (Section 5.2.3).

Exercise 5.2   Adding extra sites and species to a CA ordination

Exercise 5.2.1   In CA, Equation 5.1 is used to obtain species scores from site scores. Thus the score for *Hippophae rhamnoides* is $[1 \times 0.09 + 2 \times (-0.31)$

Table 5.15 Two-way weighted averaging algorithm applied to the data of Exercise 5.1 to obtain the first ordination axis of CA. The initial site scores (Line 1) are first standardized (Line 2). The values in brackets are rank numbers of the scores of the line above. Column 1, iteration number; Column 2, step number in Table 5.2 ; Column 3, x is site score and u is species score; Column 4, dispersion of the species scores (δ) when preceded by u, or otherwise the square root of the dispersion of the site scores of the line above (s).

Column				Sites					Species			
1	2	3	4	1	2	3	4	5	A	B	C	D
0	1	x		1.000	2.000	3.000	4.000	5.000				
0	5	x	1.5343	-1.141	-0.489	0.163	0.815	1.466				
1	2	u	0.1375						-0.163	0.815	-0.054	-0.228
1	3	x		-0.212	-0.054	0.815	-0.148	0.294	(2)	(4)	(3)	(1)
1	5	x	0.3012	-0.704	-0.179	2.706	-0.491	0.976				
2	2	u	0.6885						-0.598	1.841	-0.283	-0.325
2	3	x		-0.393	-0.283	1.841	-0.402	0.758	(1)	(4)	(3)	(2)
2	5	x	0.6953	-0.567	-0.408	2.646	-0.580	1.089				
3	2	u	0.7171						-0.574	1.868	-0.465	-0.238
3	3	x		-0.322	-0.465	1.868	-0.426	0.815	(1)	(4)	(2)	(3)
3	5	x	0.7193	-0.448	-0.646	2.597	-0.592	1.133				
4	2	u	0.7342						-0.520	1.865	-0.628	-0.161
4	3	x		-0.251	-0.628	1.865	-0.436	0.852	(2)	(4)	(1)	(3)
4	5	x	0.7383	-0.340	-0.851	2.526	-0.591	1.154				
5	2	u	0.7498						-0.466	1.840	-0.764	-0.091
5	3	x		-0.185	-0.764	1.840	-0.440	0.875	(2)	(4)	(1)	(3)
5	5	x	0.7529	-0.246	-1.015	2.444	-0.584	1.162				
.	.	.		0.7606	(3)	(1)	(5)	(2)	(4)			
20	2	u	0.7800						-0.211	1.556	-1.193	0.178
20	3	x		0.081	-1.193	1.556	-0.409	0.867	(2)	(4)	(1)	(3)
20	5	x	0.7799	0.104	-1.530	1.995	-0.524	1.112				
				(3)	(1)	(5)	(2)	(4)				

+ 1 × (-0.68)]/(1 + 2 + 1) = -0.30, for *Poa annua* -0.33 and for *Ranunculus acris* -0.19. All three species come in Table 5.1c between *Elymus repens* and *Leontodon autumnalis*. The plots asked for suggest unimodal response curves for *Hippophae rhamnoides* and *Poa annua*, but a bimodal curve for *Ranunculus acris*. The species score is the centroid (centre of gravity) of the site scores in which they occur. The score gives an indication of the optimum of the response curve for the former two species, but has no clear meaning for the latter species. In general, species with a score close to the centre of the ordination may either be unimodal, bimodal or unrelated to the axes (Subsection 5.2.5).

Exercise 5.2.2 The weighted average for the site is [3 × (-0.65) + 5 × (-0.50) + 4 × (-0.39)]/(3 + 5 + 4) = -0.50, which must be divided as in Exercise 5.1.3 by λ (= 0.536) to obtain the site score -0.93. If we calculated the score for the

170

Table 5.16 Two-way weighted averaging algorithm applied to the data of Exercise 5.1 to obtain the second ordination axis of CA. The first line shows the site scores of the first ordination axis (*f*). The scores on the second line are used as the initial scores after orthogonalizing them with respect to the first axis (Line 3) and standardizing them (Line 4). Column 5 is *v*, defined in Table 5.2; the other columns are defined in Table 5.15.

Column					Sites					Species			
1	2	3	4	5	1	2	3	4	5	A	B	C	D
0	4.1	f			0.10	-1.53	2.00	-0.53	1.11				
0	4.1	x			-1.141	-0.489	0.163	0.815	1.466				
0	4.3	x		0.2771	-1.169	-0.065	-0.391	0.962	1.158				
0	5.3	x	0.9612		-1.216	-0.068	-0.407	1.001	1.205				
1	2	u	0.0837							-0.107	0.399	0.288	-0.288
1	3	x			-0.243	0.288	0.399	-0.036	0.056				
1	4	x		0.0001	-0.243	0.288	0.399	-0.036	0.056				
1	5	x	0.2182		-1.114	1.320	1.829	-0.165	0.257				
2	2	u	0.5956							-0.639	1.043	0.825	-0.650
2	3	x			-0.647	0.825	1.043	-0.155	0.197				
2	4	x		-0.0011	-0.647	0.823	1.045	-0.156	0.198				
2	5	x	0.5967		-1.084	1.379	1.751	-0.261	0.332				
3	2	u	0.5980							-0.673	1.042	0.832	-0.636
3	3	x			-0.645	0.832	1.042	-0.159	0.203				
3	4	x		-0.0014	-0.645	0.830	1.045	-0.160	0.205				
3	5	x	0.5982		-1.078	1.387	1.747	-0.267	0.343				
4	2	u	0.5984							-0.672	1.045	0.836	-0.632
4	3	x			-0.642	0.836	1.045	-0.156	0.206				
4	4	x		-0.0016	-0.642	0.834	1.048	-0.157	0.208				
4	5	x	0.5985		-1.073	1.393	1.751	-0.262	0.348				

second axis by the same method, the extra site would come somewhat below Site 5 in the ordination diagram (Figure 5.4).

Exercise 5.2.3 The site scores of Table 5.1c must be divided by $\sqrt{(1 - \lambda)/\lambda}$ = $\sqrt{(0.464/0.536)}$ = 0.93 and the species scores by $\sqrt{\lambda(1 - \lambda)}$ = $\sqrt{(0.536 \times 0.464)}$ = 0.50 (Subsection 5.2.2). For Site 20, for example, we obtain the score $1.95/0.93 = 2.10$ and for *Juncus articulatus* $1.28/0.50 = 2.56$. In Hill's scaling, the scores satisfy Equation 5.2 whereas Equation 5.1 must be modified analogously to the modification of Equation 5.2 in Exercise 5.1.3

Exercise 5.3   Principal components analysis

The mean abundance of *Hippophae rhamnoides* is 0.2. With Equation 5.8, we obtain the score $(0-0.2) \times (-0.31) + (0 -0.2) \times (-0.30) + ... + (2-0.2) \times (-0.04) + (1-0.2) \times 0.00 + ... + (0-0.2) \times 0.45 = -0.03$. Similarly we obtain the scores –3.22 for *Poa annua* and –1.48 for *Ranunculus acris*. The plots suggest monotonic decreasing relations for the latter two species, and a unimodal relation (if any) for the first species. If straight lines are fitted in these plots, the slope of regression turns out to equal the species score (Subsection 5.3.1). The species come at different places in Table 5.5c. For example, *Poa annua* comes just after *Bromus hordaceus*. The score for the extra site is calculated by dividing the weighted sum (Equation 5.9) by the eigenvalue: $3.90/471 = 0.008$.

Exercise 5.4 Length of gradient in DCA

In DCA, axes are scaled such that the standard deviation (tolerance) of the response curve of each species is close to one and is on average equal to one. Each response curve will therefore rise and decline over an interval of about 4 s.d. (Figure 3.6; Figure 5.3b). If the length of the first axis equals 1.5 s.d., the length of the axis covers only a small part of the response curve of each species. Most plots will therefore suggest monotonic curves, although the true response curves may be unimodal (Figure 3.3). If the length of the first axis is 10 s.d., the response curves of many species are contained within the length of the axis, so that many of the plots will suggest unimodal response curves.

5.5 Interpretation of joint plot and biplot

Exercise 5.5.1   Inferred rank orders of abundance are for *Juncus bufonius*

from Figure 5.7 (DCA) sites $12 > 8 > 13 > 9 > 4 \approx 18$
from Figure 5.15 (PCA) sites $13 \approx 3 > 4 > 9 \approx 12$
from Table 5.1a (data) sites $9 = 12 > 13 > 7 - -$

and for *Eleocharis palustris*

from Figure 5.7 (DCA) sites $16 > 14 \approx 15 > 20 > 8$
from Figure 5.15 (PCA) sites $16 > 20 > 15 > 14 > 19$
from Table 5.1a (data) Sites $16 > 15 > 8 = 14 = 20$.

Exercise 5.5.2   The difference in interpretation is greatest for species that lie at the centre of the ordination diagram. In a DCA diagram, the inferred abundance drops with distance from the species point in any direction, whereas in a PCA diagram the inferred abundance decreases or increases with distance from the species point, depending on the direction. This difference is rather unimportant for species that lie on the edge of the diagram, because the site points all lie on one side of the species point. One comes to the same conclusion by noting that a species point in a DCA diagram is its inferred optimum; if the optimum lies far outside the region of the sites the inferred abundance changes monotonically across the region of site points (*Eleocharis palustris* in Figure 5.7).

Exercise 5.6 Detrended canonical correspondence analysis

Exercise 5.6.1   From Table 5.14, we see that the best linear combination is

172

$x = 1.00 \ z_1^* + 0.12 \ z_2^*$. In terms of unstandardized variables, we obtain $x = 1.00 \times (z_1 - 22)/18.5 + 0.12 \times (z_2 - 3)/2.9 = (1.00/18.5)z_1 + (0.12/2.9)z_2 - 22/18.5 - 0.12 \times 3/2.9 = 0.054 \ z_1 + 0.041 \ z_2 - 1.31$. The standard error of c_1 is $0.03/18.5 = 0.00162$ and of c_2 is $0.03/2.9 = 0.010$.

Exercise 5.6.2 The value of x for the plot that was 15 cm above water level in 1984 is $x = 0.054 \times 15 + 0.041 \times 6 = 1.056$ s.d. For the plot 25 cm above water level, we obtain $x = 0.054 \times 25 + 0.041 \times 6 = 1.596$ s.d. Hence, the species turnover is $1.596 - 1.056 = 0.54$ s.d. According to Equation 5.47, turnover depends only on the difference in altitude between the plots: $0.054 \times (25 - 15) = 0.54$, and does not depend on the particular altitudes of the plots nor on the year of sampling. The species turnover between plots differing 10 cm in altitude is therefore 0.54 s.d. on both occasions of sampling.

Exercise 5.6.3 The value of x for a plot with an altitude of 15 cm in 1984 was 1.056 s.d. in 1984 (Exercise 5.6.2) and was $0.054 \times 15 + 0.041 \times 0 = 0.81$ s.d. in 1978. (Note that in the model altitude was not corrected for uplift; hence $z_1 = 15$ in 1984 and in 1978.) The species turnover is $1.056 - 0.81 = 0.246$ s.d., which equals 0.041×6 s.d. and which is independent of altitude. Hence, each plot changes about a quarter standard deviation in 6 years.

Exercise 5.6.4 The species turnover rate is 0.041 s.d. per year, whereas the species turnover due to altitude is 0.054 s.d. per centimetre. The change in altitude that results in 0.041 s.d. species turnover is therefore $0.041/0.054 = 0.76$ cm. An approximate 95% confidence interval can be obtained for this ratio from the standard error of c_1 and c_2 and their covariance by using Fieller's theorem (Finney 1964). Here the covariance is about zero. In this way, we so obtained the interval (0.4 cm, 1.1 cm).

Exercise 5.6.5 From Exercise 5.6.4, we would expect each particular species composition to occur next year 0.76 cm lower than its present position. Uplift (about 0.5 cm per year) is less; hence, there is no evidence that the vegetational succession lags behind the land uplift. The known uplift falls within the confidence interval given above. Further, because the value 0 cm lies outside the confidence interval, the effect of uplift on species composition is demonstrated. Uplift apparently drives the vegetational succession without lag.

Exercise 5.6.6 The length of the first axis is 4.4 s.d. From Exercise 5.6.3, we know that each plot changes about 0.25 s.d. in 6 years. The change from vegetation near the sea to vegetation at the edge of the forest therefore takes roughly $(4.4/0.25) \times 6$ years ≈ 100 years. The second axis is 0.9 s.d. and mainly represents the differences in species composition between the two sampling occasions that are unrelated to altitude and land uplift. More precisely, the canonical coefficient of time on the second axis is $-0.34/2.9 = -0.117$. It therefore accounts for 0.117×6 s.d. $= 0.70$ s.d. of the length of the second axis, whereas time accounted for 0.25 s.d. of the length of the first axis. There are apparently more changes going on than can be accounted for by uplift.

173

6 Cluster analysis

O.F.R. van Tongeren

6.1 Introduction

6.1.1 Aims and use

For ecological data, cluster analysis is a type of analysis that classifies sites, species or variables. Classification is intrinsic in observation: people observe objects or phenomena, compare them with other, earlier, observations and then assign them a name. Therefore one of the major methods used since the start of the study of ecology is the rearrangement of data tables of species by sites, followed by the definition of community types, each characterized by its characteristic species combination (Westhoff & van der Maarel 1978; Becking 1957). Scientists of different schools have different ideas about the characterization of community types and the borders between these types. In vegetation science, for instance, the Scandinavian school and the Zürich–Montpellier school differ markedly, the Scandinavians emphasizing the dominants and the Zürich–Montpellier school giving more weight to characteristic and differential species, which are supposed to have a narrower ecological amplitude and are therefore better indicators for the environment. Cluster analysis is an explicit way of identifying groups in raw data and helps us to find structure in the data. However even if there is a continuous structure in the data, cluster analysis may impose a group structure: a continuum is then arbitrarily partitioned into a discontinuous system of types or classes.

Aims of classification are:
- to give information on the concurrence of species (internal data structure)
- to establish community types for descriptive studies (syntaxonomy and mapping)
- to detect relations between communities and the environment by analysis of the groups formed by the cluster analysis with respect to the environmental variables (external analysis).

In Chapter 6, an introduction will be given to several types of cluster analysis. This chapter aims at a better understanding of the properties of several methods to facilitate the choice of a method, without pretending to show you how to find the one and only best structure in your data. It is impossible to choose a 'best' method because of the heuristic nature of the methods. If there is a markedly discontinuous structure, it will be detected by almost any method, a continuous structure will almost always be obscured by cluster analysis.

6.1.2   Types of cluster analysis

There are several types of cluster analysis, based on different ideas about the cluster concept. Reviews are found mainly in the taxonomic literature (Lance & Williams 1967; Dunn & Everitt 1982). Here a brief summary will be given of the main groups.

A major distinction can be made between divisive and agglomerative methods. Divisive methods start with all objects (in ecology mostly samples; in taxonomy operational taxonomic units, OTUs) as a group. First this group is divided into two smaller groups, which is repeated subsequently for all previously formed groups, until some kind of 'stopping rule' is satisfied. The idea of this way of clustering is that large differences should prevail over the less important smaller differences: the global structure of a group should determine the subgroups. Alternatively agglomerative methods start with individual objects, which are combined into groups by collection of objects or groups into larger groups. Here 'local' similarity prevails over the larger differences. Divisive methods will be described in Section 6.3, and agglomerative methods in Section 6.2. Most agglomerative methods require a similarity or dissimilarity matrix (site by site) to start from. Several indices of (dis)similarity will be introduced in Subsection 6.2.3.

A second way of distinguishing methods is to classify them by hierarchical and non-hierarchical methods. Hierarchical methods start from the idea that the groups can be arranged in a hierarchical system. In ecology, one could say that a certain difference is more important than another one and therefore should prevail: be expressed at a higher hierarchical level. Non-hierarchical methods do not impose such a hierarchical structure on the data. For data reduction, non-hierarchical methods are usually used.

Non-hierarchical classification handles
- redundancy: sites that are much like many other sites are grouped without considering the relations to other less similar sites
- noise: before subsequent hierarchical clustering, a 'composite sample' may be constructed
- outliers, which can be identified because they appear in small clusters or as single samples.

6.2   Agglomerative methods

6.2.1   Introduction

Agglomerative cluster analysis starts from single objects, which are agglomerated into larger clusters. In many sciences, agglomerative techniques are employed much more frequently than divisive techniques. The historical reason for this is the inefficient way early polythetic divisive techniques used computer resources, while the agglomerative ones were more efficient. Now, the opposite seems true. Nevertheless, there is a very large range of agglomerative techniques, each emphasizing other aspects of the data and therefore very useful.

All agglomerative methods share the idea that some kind of (dis)similarity

175

function between (groups of) objects (usually sites) is decisive for the fusions. Different methods, however, are based on different ideas on distance between clusters. Within most methods, there is also a choice between different 'indices of similarity or dissimilarity' (distance functions). Most of this section is devoted to similarity and dissimilarity indices.

6.2.2 Similarity and dissimilarity

Grouping of sites and species in many ecological studies is a matter of personal judgment on the part of the investigator: different investigators may have different opinions or different aims; they therefore obtain different results. There are, however, many different objective functions available with which to express similarity.

Ideally, similarity of two sites or species should express their ecological relation or resemblance; dissimilarity of two sites or species is the complement of their similarity. Since this idea of similarity includes an ecological relation, it is important which ecological relation is focused upon – so the objectives of a study may help to determine the applicability of certain indices. Most of the indices used in ecology do not have a firm theoretical basis. My attitude towards this problem is that practical experience, as well as some general characteristics of the indices, can help us choose the right one. Numerous indices of similarity or dissimilarity have been published, some of them are widely used, others are highly specific.

The aim of this section is to make the concepts of similarity and dissimilarity familiar and to examine some of the popular indices. Although most indices can be used to compute (dis)similarities between sites as well as between species, they are demonstrated here as if the site is the statistical 'sampling unit'. Computations of similarity can be made directly from the species-abundance values of sites or indirectly, after using some ordination technique from the site scores on the ordination axes. With indirect computation, dissimilarities refer to distances between sites in the ordination space.

Comparison of sites on the basis of presence–absence data

If detailed information on species abundance is irrelevant for our problem or if our data are qualitative (e.g. species lists), we use an index of similarity for qualitative characters. The basis of all similarity indices for qualitative characters is that two sites are more similar if they share more species and that they are more dissimilar if there are more species unique for one of both (two species are more similar if their distribution over the sites is more similar). One of the earliest indices is the index according to Jaccard (1912). This Jaccard index is the proportion of species out of the total species list of two sites, which is common to both sites:

$$SJ = c/(a + b + c)$$

Equation 6.1

where
SJ is the similarity index of Jaccard
c is the number of species shared by the two sites
a and b are the numbers of species unique to each of the sites.

Often the equation is written in a different way:

$$SJ = c/(A + B - c) \qquad\qquad \text{Equation} \quad 6.2$$

where c is the number of species shared and A and B are the total numbers of species for the samples: $A = a + c$ and $B = b + c$).

Sørensen (1948) proposed another similarity index, often referred to as coefficient of community *(CC)*.

$$CC = 2c/(A + B) \text{ or } 2c/(a + b + 2c). \qquad\qquad \text{Equation} \quad 6.3$$

Instead of dividing the number of species shared by the total number of species in both samples, the number of species shared is divided by the average number of species. Faith (1983) discusses the asymmetry of these indices with respect to presence or absence.

Comparison of samples on the basis of quantitative data

Quantitative data on species abundances always have many zeros (i.e. species are absent in many sites); the problems arising from this fact have been mentioned in Section 3.4. Therefore an index of similarity for quantitative characters should also consider the qualitative aspects of the data. The similarity indices in this subsection are different with respect to the weight that is given to presence or absence (the qualitative aspect) with regard to differences in abundance when the species is present. Some of them emphasize quantitative components more than others. Two of them are very much related to the Jaccard index and the coefficient of community, respectively: similarity ratio (Ball 1966) and percentage similarity (e.g. Gauch 1982). The other indices can easily be interpreted geometrically.

The similarity ratio is:

$$SR_{ij} = \Sigma_k y_{ki}\, y_{kj}/(\Sigma_k y_{ki}^2 + \Sigma_k y_{kj}^2 - \Sigma_k y_{ki}\, y_{kj}) \qquad \text{Equation} \quad 6.4$$

where y_{ki} is the abundance of the k-th species at site i, so sites i and j are compared. For presence–absence data (0 indicating absence and 1 presence), this can be easily reduced to Equation 6.1, indicating that the Jaccard index is a special case of the similarity ratio. For Sørensen's index, Equation 6.3, the same can be said in respect to percentage similarity:

$$PS_{ij} = 200\, \Sigma_k \min\,(y_{ki}, y_{kj})/(\Sigma_k y_{ki} + \Sigma_k y_{kj}) \qquad \text{Equation} \quad 6.5$$

177

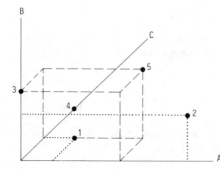

Figure 6.1 Five sites (1-5) in a three-dimensional space of which the axes are the thre species A, B and C. Site 1 is characterized by low abundance of species A and specie C, and absence of species B. In site 2, species A is dominant, species B is less importan and species C is absent. Sites 3 and 4 are monocultures of species B and C, respectively Site 5 has a mixture of all three species.

where min (y_{ki}, y_{kj}) is the minimum of y_{ki} and y_{kj}.

Some indices can be explained geometrically. To explain these indices, it i necessary to represent the sites by a set of points in a multi-dimensional spac (with as many dimensions as there are species). One can imagine such a spac with a maximum of three species (Figure 6.1) but conceptually there is no differenc if we use more species (see Subsection 5.3.3).

The position of a site is given by using the abundances of the species as coordinate (Figure 6.1), and therefore sites with similar species composition occupy nearb positions in species space. The Euclidean Distance, *ED*, between two sites is a obvious measure of dissimilarity:

$$ED = \sqrt{\Sigma_k (y_{ki} - y_{kj})^2}$$

Equation 6.

Figure 6.1 shows that quantitative aspects play a major role in Euclidean Distance the distance between Sites 1 and 2, which share one species, is much larger tha the distance between Sites 1 and 3, not sharing a species.

More emphasis is given to qualitative aspects by not considering a site as point but as a vector (Figure 6.2). Understandably, the direction of this vecto tells us something about the relative abundances of species. The similarity o two sites can be expressed as some function of the angle between the vector of these sites. Quite common is the use of the cosine (or Ochiai coefficient):

$$cos = OS = \Sigma_k y_{ki} y_{kj} / \sqrt{\Sigma_k y_{ki}^2 \, \Sigma_k y_{kj}^2}$$

Equation 6.

A dissimilarity index that is more sensitive to qualitative aspects than the Euclidea Distance is the chord distance:

178

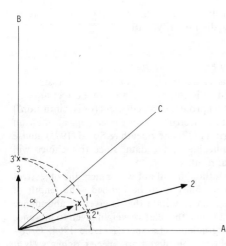

Figure 6.2 The same space as in Figure 6.1. Samples are now indicated by vectors. Crosses indicate where the sample vectors intersect the unit sphere (broken lines). Note that the distance between 1′ and 2′ is much lower than the distance between either of them and 3′. The angle between sample vectors 1 and 3 is indicated by α.

$$CD = \sqrt{\Sigma_k \left[y_{ki} / (\Sigma_k y_{ki}^2)^{\frac{1}{2}} - y_{kj} / (\Sigma_k y_{kj}^2)^{\frac{1}{2}} \right]^2}$$ Equation 6.8

This chord distance is geometrically represented by the distance between the points where the sample vectors intersect the unit sphere (Figure 6.2).

Conversion of similarity to dissimilarity and vice versa

For some applications, one may have to convert a similarity index into a dissimilarity index. This conversion must be made if, for instance, no dissimilarity index with the desired properties is available, but the cluster algorithm needs an index of dissimilarity. For cluster algorithms merely using the rank order of dissimilarities, any conversion reversing the rank order is reasonable, but care must be taken for cluster algorithms that use the dissimilarities in a quantitative way (as forming an interval or ratio scale (Subsection 2.4.2). We mention two ways of making the conversion:
- subtracting each similarity value from a certain value: in this way the intervals between the values are preserved. Bray & Curtiss (1957), for instance, subtract similarity values from the expected similarity among replicate samples, the so-called internal association. In practice, the best estimate of internal association *(IA)* is the maximum similarity observed. Thus percentage similarity is converted to percentage distance, *PD*, using this subtraction:

$$PD = IA - PS$$ Equation 6.9

179

– taking the reciprocal of each similarity value. In this way, the ratios between similarity values are preserved in the dissimilarity matrix.

6.2.3 Properties of the indices

Despite many studies (e.g. Williams et al. 1966; Hajdu 1982) addressing the problem of which index should be used, it is still difficult to give a direct answer. The choice of index must be guided by best professional judgment (or is it intuition?) of the investigator, by the type of data collected and by the ecological question that should be answered. Dunn & Everitt (1982) and Sneath & Sokal (1973) advise to choose the simplest coefficient applicable to the data, since this choice will generally ease the interpretation of final results.

However one can say a little bit more, though still not very much: the objectives of a study may help in deciding which index is to be applied. The length of the sampled gradient is important: the relative weight that is given to abundance (quantity) should be larger for short gradients, the relative weight given to presence or absence should be larger for long gradients (Lambert & Dale 1964; Greig-Smith 1971). Other criteria that should be considered are species richness (Is it very different at different sites?) and dominance or diversity of the sites (Are there substantial differences between sites?). The easiest way of getting some feeling for these aspects is to construct hypothetical matrices of species abundances and see how the various indices respond to changes in different aspects of the data. However this gives only an indication and one must be aware of complications whenever more characteristics of the data are different between samples.

To demonstrate the major responses to dominance/diversity, species richness and length of gradient a set of artificial species-by-site data, each referring to one major aspect of ecological samples, is given, together with graphs, to indicate the responses of the indices (Tables 6.1-6.4). To obtain comparable graphs (Figures

Table 6.1 Artificial species-by-sites table. Total abundance for each sample is 10, the number of species (α-diversity) is lower on the right side and the 'evenness' is constant (equal scores for all species in each sample).

Site	1	2	3	4	5	6	7	8	9	10
Species										
A	1.00	1.11	1.25	1.43	1.67	2.00	2.50	3.33	5.00	10.00
B	1.00	1.11	1.25	1.43	1.67	2.00	2.50	3.33	5.00	
C	1.00	1.11	1.25	1.43	1.67	2.00	2.50	3.33		
D	1.00	1.11	1.25	1.43	1.67	2.00	2.50			
E	1.00	1.11	1.25	1.43	1.67	2.00				
F	1.00	1.11	1.25	1.43	1.67					
G	1.00	1.11	1.25	1.43						
H	1.00	1.11	1.25							
I	1.00	1.11								
J	1.00									

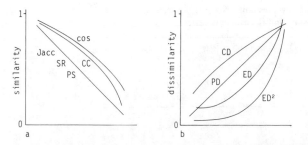

Figure 6.3 Standardized (dis)similarity (ordinate) between the first site and each of the other sites in Table 6.1 in corresponding order on the abscissa. Note that squared Euclidean Distance (ED²) is strongly affected by higher abundances. a: similarity indices. b: dissimilarity indices.

Table 6.2 Artificial species-by-sites table. Evenness and species number are constant, the sample totals varying largely.

Site	1	2	3	4	5	6	7	8	9	10
Species										
A	1	2	3	4	5	6	7	8	9	10
B	1	2	3	4	5	6	7	8	9	10
C	1	2	3	4	5	6	7	8	9	10
D	1	2	3	4	5	6	7	8	9	10
E	1	2	3	4	5	6	7	8	9	10
F	1	2	3	4	5	6	7	8	9	10
G	1	2	3	4	5	6	7	8	9	10
H	1	2	3	4	5	6	7	8	9	10
I	1	2	3	4	5	6	7	8	9	10
J	1	2	3	4	5	6	7	8	9	10

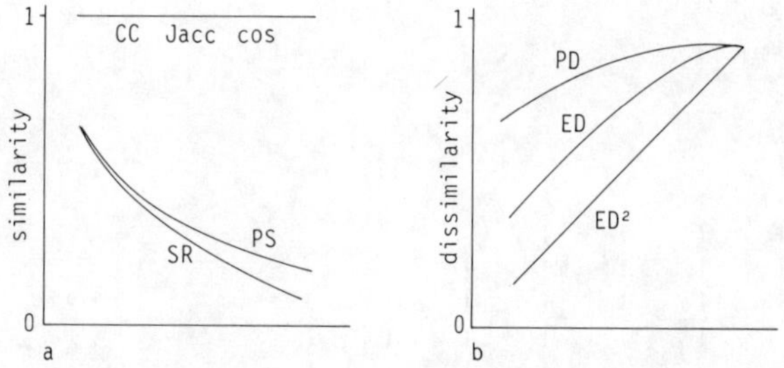

Figure 6.4 Standardized (dis)similarity (ordinate) between the first site and each of the other sites in Table 6.2 in corresponding order on the abscissa. Note that coefficient of community (CC), Jaccard index (Jacc) and cosine are at their maximum for all sites compared with the first site. a: similarity indices. b: dissimilarity indices.

181

Table 6.3 Artificial species-by-sites table. Number of species (2) and total
abundance (10) are constant but the evenness varies.

Site	1	2	3	4	5	6	7	8	9
Species									
A	1	2	3	4	5	6	7	8	9
B	9	8	7	6	5	4	3	2	1

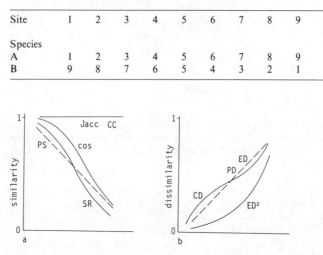

Figure 6.5 Standardized (dis)similarity (ordinate) between the first site and each of the
other sites in Table 6.3 in corresponding order on the abscissa. Note the difference with
Figure 6.4: the cosine is not at its maximum value for other sites as compared to the
first site. a: similarity indices. b: dissimilarity indices.

Table 6.4 Artificial species-by-sites table. A regular gradient with equal
species numbers in the samples, equal scores for all species: at each 'step'
along the gradient one species is replaced by a new one.

Site	1	2	3	4	5	6	7	8	9
Species									
A	1								
B	1	1							
C	1	1	1						
D	1	1	1	1					
E	1	1	1	1	1				
F	1	1	1	1	1	1			
G		1	1	1	1	1	1		
H			1	1	1	1	1	1	
I				1	1	1	1	1	1
J					1	1	1	1	1
K						1	1	1	1
L							1	1	1
M								1	1
N									1

6.3-6.6), all indices are scaled from 0 to 1. Comparisons are always made between the first site of the artificial data and the other sites within that table. The captions to Tables 6.1-6.4 and Figures 6.3-6.6 give more information on the properties of the artificial data. Table 6.5 summarizes the major characteristics of the indices but it is only indicative.

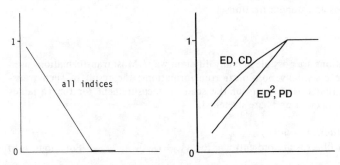

Figure 6.6 Standardized (dis)similarity (ordinate) between the first site and each of the other sites in Table 6.4 in corresponding order on the abscissa. Except for Euclidean Distance (ED) and Chord Distance (CD), all indices are linear until a certain maximum (or minimum) is reached, a: similarity indices. b: dissimilarity indices.

Table 6.5 Characteristics of the (dis)similarity indices. The asterisk (*) indicates qualitative characteristics. Sensitivity for certain properties of the data is indicated by: − not sensitive; + sensitive; ++ and +++ strongly sensitive.

	abbreviation	qualitative	quantitative	dissimilarity	similarity	sensitivity to species richness	sensitivity to dominant species	sensitivity to sample total
Similarity Ratio	SR	*	*		*		++	++	++
Percentage Similarity	PS	*	*		*		++	+	+
Cosine	Cos	*	*		*		+	+	
Jaccard Index	SJ	*			*		++	−	−
Coefficient of Community	CC	*			*		+	−	−
Cord Distance	CD	*	*	*			+	+	−
Percentage Dissimilarity	PD	*	*	*			++	+	+
Euclidean Distance	ED	*	*	*			++	++	++
Squared Euclidean Distance	ED²	*	*	*			+++	+++	+++

183

6.2.4 Transformation, standardization and weighting

Transformation, standardization and weighting of data are other ways of letting certain characteristics of the data express themselves more or less strongly. This paragraph is meant to give you some idea of how certain manipulations can be made with the data and what are the reasons for and the consequences of transformations and standardizations.

Transformation

Transformations are possible in many different ways. Most transformations used in ecology are essentially non-linear transformations: the result of such transformations is that certain parts of the scale of measurement for the variables are shrunk, while other parts are stretched.

Logarithmic transformation.
$$y_{ij}^* = \log_e y_{ij} \text{ or (if zeros are present) } y_{ij}^* = \log_e (y_{ij} + 1) \qquad \text{Equation 6.10}$$

This transformation is often used for three essentially different purposes:
- to obtain the statistically attractive property of normal distribution for log-normally distributed variables (as in Subsection 2.4.4)
- to give less weight to dominant species, in other words to give more weight to qualitative aspects of the data
- in environmental variables, to reflect the linear response of many species to the logarithm of toxic agents or (in a limited range) to the logarithm of nutrient concentrations. Instead of '+1', take the minimum non-zero value.

Square-root transformation.
$$y_{ij}^* = y_{ij}^{1/2} \qquad \text{Equation 6.11}$$

This transformation is used
- before analysis of Poisson-distributed variables (e.g. number of individuals of certain species caught in a trap over time)
- to give less weight to dominant species.

Exponential transformation.
$$y_{ij}^* = a^{y_{ij}}. \qquad \text{Equation 6.12}$$

If a is a real number greater than 1, the dominants are emphasized.

Transformation to an ordinal scale. The species abundances are combined into classes. The higher the class number, the higher the abundance. A higher class number always means a higher abundance, but an equal class number does not always mean an equal abundance: intervals between classes are almost meaningless. Dependent on the class limits, one can influence the results of a classification in all possible ways. An extreme is the transformation to presence–absence scale

184

(1/0). A transformation to ordinal scale always includes loss of information: if continuous data are available any other transformation is to be preferred. However it can be very useful to collect data on an ordinal scale (as is done in the Zürich–Montpellier school of vegetation science) for reduction of the work in the field.

Standardization

Several aspects of standardization have been treated in Subsection 2.4.4. Here we discuss some other types of standardization that are used in cluster analysis. Standardization can here be defined as the application of a certain standard to all variables (species) or objects (sites) before the computation of the (dis)similarities or before the application of cluster analysis. Possible ways of standardizing are as follows.

Standardization to site total. The abundances for each species in a site are summed and each abundance is divided by the total: in this way relative abundances for the species are computed, a correction is made for 'size' of the site (total number of individuals collected at the site or total biomass). Care should be taken if these sizes are very different, because rare species tend to appear in large sites: (dis)similarity measures that are sensitive to qualitative aspects of the data might still be inappropriate.

Standardization to species total. For each species the abundances are summed over all sites and then divided by the total. This standardization strongly over-weights the rare species and down-weights the common species. It is therefore recommended to use this standardization only if the species frequencies in the table do not differ too much. This type of standardization can be applied when different trophic levels are represented in the species list, because the higher trophic levels are less abundant (pyramids of biomass and numbers).

Standardization to site maximum. All species abundances are divided by the maximum abundance reached by any species in the site. This standardization is applied for the same reason as standardization to site total. It is less sensitive to species richness, but care should be taken if there are large differences in the 'evenness' of sites. If an index is used with a large weighting for abundance, sites with many equal scores will become extremely different from sites with a large range in their scores.

Standardization to species maximum. The reason for this standardization is that, in the opinion of many ecologists, less abundant species (in terms of biomass or numbers) should be equally weighted. As the standardization to species total, this type of standardization is recommended when different trophic levels are present in the species list. This standardization also makes data less dependent on the kind of data (biomass or numbers or cover) collected.

Standardization to unit site vector length. By dividing the species abundance in a site by the square root of their summed squared abundances, all end-points of the site vectors are situated on the unit sphere in species-space. Euclidean Distance then reduces to chord distance.

185

Weighting

There are several reasons for weighting species or sites. Depending on the reason for down-weighting several kinds of down-weighting can be applied.

Down-weighting of rare species. A lower weight dependent on the species frequency, is assigned to rare species to let them influence the final result to a lesser extent. This should be done if the occurrence of these species is merely by chance and if the (dis)similarity index or cluster technique is sensitive to rare species.

Down-weighting of species indicated by the ecologist. A lower weight is assigned to species (or sites) that are less reliable (determination of a species is difficult; a sample is taken by an inexperienced field-worker) or to species that are ecologically less relevant (planted trees; the crop species in a field). This kind of down-weighting is ad hoc and therefore arbitrary.

6.2.5 Agglomerative cluster algorithms

All agglomerative methods are based on fusion of single entities (sites) or clusters (groups of sites) into larger groups. The two groups that closest resemble each other are always fused, but the definition of (dis)similarity between groups differs between methods.

Often the results of hierarchical clustering are presented in the form of a dendrogram (tree-diagram, e.g. Figure 6.8). Such a dendrogram shows the relations between sites and groups of sites. The hierarchical structure is indicated by the branching pattern.

Single-linkage or nearest-neighbour clustering

The distance between two clusters is given by the minimum distance that can be measured between any two members of the clusters (Figure 6.7). A dendrogram of the classification of the Dune Meadow Data with single-linkage clustering,

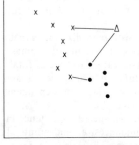

Figure 6.7 Distances (solid lines) between clusters in single linkage: samples within the same cluster are indicated with the same symbol.

186

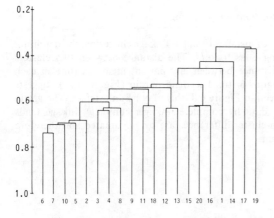

Figure 6.8 Dendrogram of single linkage, using the Dune Meadow Data and the similarity ratio.

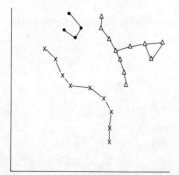

Figure 6.9 Hypothetical example of 'chaining', a problem occurring in single-linkage clustering.

using similarity ratio, is given in Figure 6.8. The dendrogram shows us that there are not very well defined clusters: our data are more or less continuous. Single-linkage clustering can be used very well to detect discontinuities in our data. For other research in community ecology, it is not appropriate because of its tendency to produce straggly clusters, distant sites being connected by chains of sites between them (Figure 6.9).

187

Complete-linkage or furthest-neighbour clustering

In contrast to the definition of distance in single-linkage clustering, the definition in complete-linkage clustering is as follows. The distance between two clusters is given by the maximum distance between any pair of members (one in each cluster) of both clusters (Figure 6.10). The dendrogram (Figure 6.11) suggests clear groups but, as can be seen in Figure 6.8, this may be an artefact. The group structure is imposed on the data by complete linkage: complete linkage tends to tight clusters, but between-cluster differences are over-estimated and therefore exaggerated in the dendrogram.

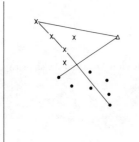

Figure 6.10 Distances (solid lines) between clusters in complete linkage: samples within the same cluster are indicated with the same symbol.

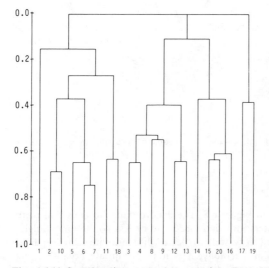

Figure 6.11 Complete-linkage dendrogram of the Dune Meadow Data using the similarity ratio.

188

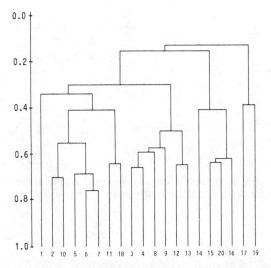

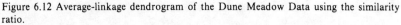

Figure 6.12 Average-linkage dendrogram of the Dune Meadow Data using the similarity ratio.

Average-linkage clustering

In average-linkage clustering, the between-group (dis)similarity is defined as the average (dis)similarity between all possible pairs of members (one of each group). This method is most widely used in ecology and in systematics (taxonomy). The algorithm maximizes the 'cophenetic correlation', the correlation between the original (dis)similarities and the (dis)similarities between samples, as can be derived from the dendrogram. For any sample pair, it is the lowest dissimilarity (or highest similarity) required to join them in the dendrogram (Sneath & Sokal 1973). As can be seen in the dendrogram of average linkage (Figure 6.12), this method is intermediate between complete and single linkage. The preceding explanation refers to UPGMA, the unweighted-pair groups method (Sokal & Michener 1958). There are variants of this technique in which a weighted average is computed (e.g. Lance & Williams 1967).

Centroid clustering

In centroid clustering, between-cluster distance is computed as the distance between the centroids of the clusters. These centroids are the points in species space defined by the average abundance value of each species over all sites in a cluster (Figure 6.13). Figure 6.14 shows subsequent steps in centroid clustering. For the Dune Meadow Data, the dendrogram (which is not presented) closely resembles the average-linkage dendrogram (Figure 6.12).

189

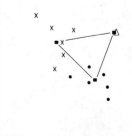

Figure 6.13 Between-cluster distances (solid lines) in centroid clustering: samples within the same cluster are indicated with the same symbol; cluster centroids are indicated by squares.

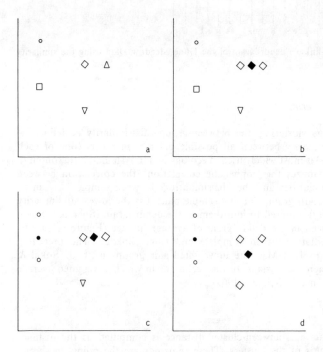

Figure 6.14 Subsequent steps in centroid clustering. Sites belonging to the same cluster are indicated with the same open symbol. Cluster centroids are indicated by the corresponding filled symbols. Upon fusion of two clusters, the symbols of sites change to indicate the new cluster to which they belong.

190

Ward's method, also known as Orlóci's (1967) error sum of squares clustering, is in some respects similar to average-linkage clustering and centroid clustering. Between-cluster distance can either be computed as a squared Euclidean distance between all pairs of sites in a cluster weighted by cluster size (resembling average-linkage clustering) or as an increment in squared distances towards the cluster centroid when two clusters are fused (resembling centroid clustering). Penalty by squared distance and cluster size makes the clusters tighter than those in centroid clustering and average linkage, and more like those obtained in complete linkage. The algorithm proceeds as follows: with all samples in a separate cluster, the sum of squared distances is zero, since each sample coincides with the centroid of its cluster. In each step, the pair of clusters is fused, which minimizes the total within-group sum of squares (Subsection 3.2.1, residual sum of squares), which is equal to minimizing the increment (dE) in the total sum of squares:

$$dE = E_{p+q} - E_p - E_q$$

where
E is the total error sum of squares
E_{p+q} is the within-group sums of squares for the cluster in which p and q are fused together
E_p and E_q the sums of squares for the individual clusters p and q.

The within-group sum of squares for a cluster is:

$$E_p = 1/N \, \Sigma_{i \varepsilon p} \, \Sigma_k \, (y_{ki} - \bar{y}_k)^2$$

where the first summation is over all members of cluster p and the second summation is over all species.

The dendrograms of Ward's clustering, average linkage and complete linkage using squared Euclidean Distance are given in Figure 6.15.

6.3 Divisive methods

6.3.1 Introduction

Divisive methods have long been neglected. The reason for this is that they were developed in the early years of numerical data analysis. At that time they failed either because of inefficiency (too many computational requirements) or because the classification obtained was inappropriate. Williams & Lambert (1960) developed the first efficient method for divisive clustering: association analysis. This method is monothetic: divisions are made on the basis of one attribute (e.g. character or species). Although it is not used very often now, some authors still use association analysis or variations of association analysis (e.g. Kirkpatrick et

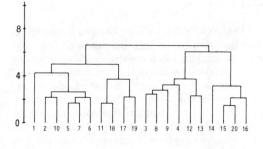

a

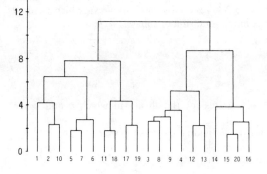

b

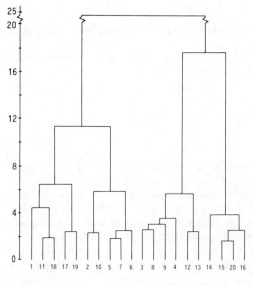

c

192

al. 1985). Subsection 6.3.2 briefly describes association analysis. Efficient polythetic methods for divisive clustering appeared after Roux & Roux (1967) introduced the partitioning of ordination space for purposes of classification. Lambert et al. (Williams 1976b) wrote a program to perform a division based on partitioning of the first axis of principal component analysis. A divisive clustering is obtained by repeating this method of partitioning on the groups obtained in the previous step. More recently, Hill (Hill et al. 1975; Hill 1979b) developed a method based on the partitioning of the first axis of CA. Since this method has some remarkable features and in most cases leads to very interpretable solutions it will be treated in detail in Subsection 6.3.3.

6.3.2 Association analysis and related methods

Association analysis (Williams & Lambert 1959 1960 1961) starts selecting the species that is maximally associated to the other species: association between species is estimated as the qualitative correlation coefficient for presence–absence data, regardless of its sign. For each species, the sum of all associations is computed. The species having the highest summed association value is chosen to define the division. One group is the group of sites in which the species is absent, the other group is the group of sites in which the species is present. Because it is sensitive to the presence of rare species and to the absence of more common ones this method is not often used in its original form. Other functions defining association, chi-square and information statistics have been proposed. These functions produce better solutions. Groups obtained in monothetic methods are less homogeneous than groups resulting from polythetic methods, because in the latter case more than one character determines the division. Therefore if a polythetic method is available it should always be preferred over a monothetic one (Coetzee & Werger 1975; Hill et al. 1975).

6.3.3 Two Way INdicator SPecies ANalysis

This section deals with the method of Two Way INdicator SPecies ANalysis (TWINSPAN). The TWINSPAN program by Hill (1979b) not only classifies the sites, but also constructs an ordered two-way table from a sites-by-species matrix. The process of clustering sites and species and the construction of the two-way table are explained step by step to illustrate TWINSPAN's many features, some of which are available in other programs too. However the combination of these features in TWINSPAN has made it one of the most widely used programs in community ecology.

Figure 6.15 Comparison of average linkage, complete linkage and Ward's method using squared Euclidean Distance. a: average linkage. b: complete linkage. c: Ward's method. The dendrograms for average linkage and complete linkage are similar. By the use of squared Euclidean Distance, the larger distances have a higher weighting in average linkage. The result of Ward's method is different from both other methods, even at the four-cluster level.

Pseudo-species

One of the basic ideas in TWINSPAN stems from the original idea in phytosociology that each group of sites can be characterized by a group of differential species, species that appear to prevail in one side of a dichotomy. The interpretation of TWINSPAN results is, in this respect, similar to the interpretation of a table rearranged by hand. Since the idea of a differential species is essentially qualitative, but quantitative data must be handled effectively also, Hill et al. (1975) developed a qualitative equivalent of species abundance, the so-called pseudo-species (see Section 3.4). Each species abundance is replaced by the presence of one or more pseudo-species. The more abundant a species is, the more pseudo-species are defined. Each pseudo-species is defined by a minimum abundance of the corresponding species, the 'cut level'. This way of substituting a quantitative variable by several qualitative variables is called conjoint coding (Heiser 1981). An advantage of this conjoint coding is that if a species' abundance shows a unimodal response curve along a gradient, each pseudo-species also shows a unimodal response curve (see Section 3.4), and if the response curve for abundance is skewed, then the pseudo-species response curves differ in their optimum.

Making a dichotomy; iterative character weighting

A crude dichotomy is made by ordinating the samples. In TWINSPAN, this is done by the method of correspondence analysis (Hill 1973; Section 5.2) and division of the first ordination axis at its centre of gravity (the centroid). The groups formed are called the negative (left-hand) and positive (right-hand) side of the dichotomy. After this division the arrangement is improved by a process that is comparable to iterative character weighting (Hogeweg 1976) or to the application of a transfer algorithm (Gower 1974) that uses a simple discriminant function (Hill 1977). What follows is an account of this process of iterative character weighting in some more details; the reader may skip the rest of this passage at first reading.

A new dichotomy is constructed by using the frequencies of the species on the positive and negative sides of the first, crude dichotomy: differential species (species preferential for one of the sides of the dichotomy) are identified by computing a preference score. Positive scores are assigned to the species with preference for the positive side of the dichotomy, negative scores for those preferential for the negative side. An absolute preference score of 1 is assigned to each pseudo-species that is at least three times more frequent on one side of the dichotomy as on the other side. Rare pseudo-species and pseudo-species that are less markedly preferential are down-weighted. A first ordering of the sites is obtained by adding the species preference scores to each other as in PCA (Chapter 5, Equation 5.9). This weighted sum is standardized so that the maximum absolute value is 1. A second ordering is constructed by computing for each site the average preference scores (similar to the computation of weighted averages in correspondence analysis (Chapter 5, Equation 5.2)) without down-weighting

194

of the rare species. In comparison to the first ordering, this one polarizes less strongly when there are many common (non-preferential) species, which is to be expected at the lower levels of the hierarchy. At the higher levels of the hierarchy it polarizes more strongly than the first ordination because more rare species can be expected at the higher levels. Hill's preference scores have a maximum absolute value of 1, so the scores for the sites in this second ordering range from −1 to 1. The scores in both orderings are added to each other and this so-called refined ordination is divided at an appropriate point near its centre (see Hill 1979b). The refined ordination is repeated using the refined classification. With the exception of a few 'borderline' cases, this refined ordination determines the dichotomy. For borderline cases (sites that are close to the point where the refined ordination is divided), the final decision is made by a third ordination: the indicator ordination. The main aim of this indicator ordination is, however, not to assign these borderline cases to one of the sides of the dichotomy, but to reproduce the dichotomy suggested by the refined ordination by using simple discriminant functions based on a few of the most highly preferential species.

Hill (1979b) warns of confusion arising from the terms 'Indicator Species Analysis' in TWINSPAN's name, because indicator ordination is an appendage, not the real basis, of the method. He suggests the name 'dichotomized ordination analysis' as a generic term to describe a wide variety of similar methods (e.g. the program POLYDIV of Williams (1976b)). The indicator species (the set of most highly preferential species that reproduce as good a refined ordination as possible) can be used afterwards in the field to assign a not-previously-sampled stand to one of the types indicated by TWINSPAN.

The construction of a species-by-sites table

For the construction of a species-by-sites table two additional features are necessary. First, the dichotomies must be ordered and, second, the species must be classified. The order of the site groups is determined by comparison of the two site groups formed at any level with site groups at two higher hierarchical levels. Consider the hierarchy in Figure 6.16. Assume that the groups 4, 5, 6 and 7 have already been ordered. The ordering of subsequent groups is now decided upon. Without ordering we are free to swivel each of the dichotomies, and therefore

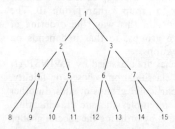

Figure 6.16 TWINSPAN dichotomy; cluster numbers are the numbers used by TWINSPAN.

195

Table 6.6 TWINSPAN table of the demonstration
samples. Options are default TWINSPAN options,
except the cut levels, which are 1, 2, 3, ... 9. Zeros
and ones on the right-hand side and at the bottom
of the table indicate the dichotomies.

```
                1111   i        111112
                17895670123489234560

 3 Air pra .2.3................ 00000
12 Emp nig ...2................ 00000
13 Hyp rad 22.5................ 00000
28 Vic lat 2.1....1............ 00000
 5 Ant odo .4.44324............ 00001
18 Pla lan 323.5553............ 00010
 1 Ach mil .2..222413.......... 000110
26 Tri pra ....252............. 000110
 6 Bel per ..2.2..2.322........ 000111
 7 Bro hor ....2.24.4.3........ 000111
 9 Cir arv ..........2......... 000111
11 Ely rep ....4...4444.6...... 001
17 Lol per 7.2.2666756542...... 001
19 Poa pra 413.2344445444.2.... 001
23 Rum ace ....563......22..... 001
16 Leo aut 52563333.522322222.2 01
20 Poa tri ....645427654549..2. 01
27 Tri rep 3.222526.521233261.. 01
29 Bra rut 4.632622..22224..444 01
 4 Alo gen ........2725385..4.  10
24 Sag pro 2..3.......52242.... 10
25 Sal rep ..33..............5  10
 2 Agr sto ........4843454475   110
10 Ele pal ............4...4584 11100
21 Pot pal ................22.. 11100
22 Ran fla ...........2..22224  11100
30 Cal cus ...............4.33  11100
14 Jun art ............44...334 11101
 8 Che alb ................1.... 1111
15 Jun buf ......2......443.... 1111

                00000000000011111111
                00001111111100001111
                    00001111
```

this hierarchical structure only indicates that 8 should be next to 9, 10 next to
11, etc. The groups (e.g. 10 and 11) are ordered 11, 10 if group 11 is more similar
to group 4 than group 10 and also less similar to group 3 than group 10. The
ordering 10, 11 is better when the reverse holds. In this way, the ordering of
the dichotomy is determined by relatively large groups, so that it depends on
general relations more than on accidental observations.

After completing the site classification the species are classified by TWINSPAN
in the light of the site classification. The species classification is based on fidelity,
i.e. the degree to which species are confined to particular groups of sites. In other
aspects the classification of the species closely resembles the site classification.
A structured table is made from both classifications by ordering the species groups
in such a way that an approximate 'positive diagonal' (from upper left to lower

196

right) is formed. A TWINSPAN table of the Dune Meadow example is given in Table 6.6.

6.4 Non-hierarchical clustering

A non-hierarchical clustering can be constructed by selecting sites to act as an initial point for a cluster and then assigning the other sites to the clusters. The methods vary in details. Gauch (1979) starts picking up a random site and clusters all sites within a specified radius from that site. COMPCLUS, as his program is called (composite clustering), repeats this process until all sites are accounted for. In a second phase, sites from small clusters are reassigned to larger clusters by specifying a larger radius. Janssen (1975) essentially proposes the same approach but picks up the first site from the data as initiating point for the first cluster. This method is applied in CLUSLA (Louppen & van der Maarel 1979). As soon as a site lies further away from the first site than specified by the radius, this site is the initiating point for the next cluster. Subsequent sites are compared to all previously formed clusters. In this way there is a strong dependence on the sequence in which the sites enter the classification. A second step in CLUSLA is introduced to reallocate the sites to the 'best' cluster. This is done by comparing all sites with all clusters: if they are more similar to another cluster then to their parent cluster at that moment, they are placed in the cluster to which they are most similar. In contrast to COMPCLUS, not only within-cluster homogeneity, but also between cluster distances are used by CLUSLA. A method combining the benefits of both methods is used in FLEXCLUS (van Tongeren 1986). From the total set of sites a certain number is selected at random or indicated by the user. All other sites are assigned to the nearest of the set. By relocation until stability is reached, a better clustering is achieved. Outliers are removed by reduction of the radius of the clusters afterwards. Variations of these methods are numerous; others have been presented by, for example, Benzécri (1973), Salton & Wong (1978) and Swain (1978).

Although hierarchical clustering has the advantage over non-hierarchical clustering that between-group relations are expressed in the classification, there is no guarantee that each level of the hierarchy is optimal. A combination of hierarchical and non-hierarchical methods can be made by allowing sites to be relocated, to improve clustering. Since clusters change by relocations, this can be repeated in an iterative process until no further changes occur.

If there is a clear group structure at any level of the hierarchy, no relocations will be made. An example of such a method is relocative centroid sorting. This method is demonstrated in Figure 6.17. Because of the possibility of relocations, a dendrogram cannot be constructed. By using relocative centroid sorting in a slightly different way – assigning each site to a cluster at random or by a sub-optimal, quick method – as shown in Figure 6.17, computing time can be saved because computation of a site-by-site (dis)similarity matrix can be replaced by computation of a site-by-cluster matrix. This is used in the table rearrangement program TABORD (van der Maarel et al. 1978), and also in CLUSLA (Louppen & van der Maarel 1979) and FLEXCLUS (van Tongeren 1986).

197

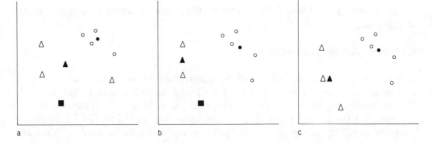

Figure 6.17 Three steps in relocative centroid sorting. a: arbitrary initial clustering. b: after relocation. c: after fusion of the most similar clusters. Samples in the same cluster are indicated by the same open symbol.The corresponding closed symbols indicate the cluster centroids.

6.5 Optimality of a clustering

It is difficult to decide which solution of the cluster analysis to choose. There are very different criteria used to do so: one can distinguish between external and internal criteria.

External criteria are not dependent on the method of clustering. Other data are used to test whether the clustering result is useful.

- In syntaxonomy (Westhoff & van der Maarel 1978), we look for sufficient differences in floristic composition to be able to interpret the results, for instance within the meaning of syntaxa, characteristic and differential species.
- In synecology, if we only used the information on the species composition for our clustering, we have the possibility to test for differences in other variables between the clusters (e.g. analysis of variance for continuous data or chi-square test for nominal variables, cf. Subsection 3.3.1).
- In survey mapping, we can have restrictions on the number of legend units, dependent on scale and technical facilities.

Internal criteria are dependent on the data used for obtaining the clustering and usually also the method of clustering. There are almost as many methods to decide which cluster method or which hierarchical level is best as there are methods for clustering. Some of these methods use statistical tests, but usually it would be better to use the word pseudo-statistics: the conditions for application of the tests are never satisfied because the same characters that are used to group the data are used to test for differences. Most other methods (a review can be found in Popma et al. 1983) use two criteria for the determination of the optimum solution:

- Homogeneity of the clusters (average (dis)similarity of the members of a cluster or some analogue).
- Separation of the clusters (average (dis)similarity of each cluster to its nearest neighbour, or some analogous criterion).

198

There are many possible definitions of homogeneity and separation of clusters, and each definition might indicate another clustering as the best one. The use of methods to determine the optimum clustering should therefore be restricted to evaluation of subsequent steps in one type of analysis (Hogeweg 1976; Popma et al. 1983).

6.6 Presentation of the results

Results of a classification can be presented in different ways. We have already mentioned:
- the species-by-sites table, giving as much information as possible on all separate sites and species abundances. In vegetation science, additional environmental information and information on the number of species in a site is usually provided in the head of the table.
- The dendrogram, a representation in which the hierarchical structure of the site groups is expressed.

When there are many sites, a species-by-sites table becomes quite large and it is not very easy to interpret the table. This is the reason for the construction of a so-called synoptical table. A synoptical table summarizes the results for each cluster. Classical synoptical tables in the Braun–Blanquet school of vegetation science present a presence class and minimum and maximum values for cover/abundance in each vegetation type for all species. Table 8.3 is an example of such a table. In Table 8.3 the presence classes I to V represent within cluster species frequencies (0-20, 20-40, 40-60, 60-80, 80-100%, respectively). Many other ways of presenting the summarized data in a synoptical table are possible. For example, one can form cross-tabulations of species groups by clusters, or instead of presence class and minimum and maximum scores, average values and standard deviations can be entered into the table.

A dendrogram and a species-by-sites table cannot be used for presentation in more than one dimension. Therefore it can be very useful to present the results of the classification in an ordination diagram of the same set of sites. In such a diagram, more complex relations with the environment can be clearly elucidated (cf. Figure 8.7).

6.7 Relation between community types and environment

The relation between community types as they are defined by cluster analysis on the basis of species data and their environment can be explored in several ways. A hierarchical classification can be evaluated at each level: at each dichotomy, a test can be done for differences between the two groups or all clusters at a certain level are evaluated simultaneously. In a non-hierarchical classification we are restricted to simultaneous comparison of all resulting clusters.The range of methods goes from exploratory analysis (Subsection 6.7.1) to statistical tests of significance (Subsection 6.7.2).

Table 6.7 FLEXCLUS table, similarity ratio, centroid sorting with relocations. Sample 1, which formed a separate cluster, is added to the second cluster by hand. Environmental data are summarized and added to the table.

```
Sites:
          :11 1 11 :        : 11:1112:
          :79617085: 123489: 23:4560:

Leo aut   :26353353: 52232: 45:4475:
Bra rut   : 3942262:  2222: 4 : 444:
Tri rep   : 2532622: 52123: 32:61 :
Agr sto   :        :  4843: 45:4475:
Ach mil   :2 2 24 2:13    :   :    :
Ant odo   :443 24 4:      :   :    :
Pla lan   :2 535335:      :   :    :
Poa pra   :1 344432:44544 : 2 :    :
Lol per   : 676622:75642 :   :    :
Bel per   :    222: 322  :   :    :
Ely rep   :      4:4444 6:   :    :
Alo gen   :        : 27253: 85: 4 :
Poa tri   : 4 54 6:276545: 49: 2 :
Sag pro   : 3 2    :  522: 42:    :
Jun buf   :    2   :    4: 43:    :
Jun art   :        :   44:   :34 :
Cal cus   :        :      :4 33:
Ele pal   :        :  4 :  :4584:
Ran fla   :        :  2 : 2:2224:
Air pra   :23      :      :   :    :
Bro hor   :   24 2: 4 3  :   :    :
Hyp rad   :25 2    :      :   :    :
Pot pal   :        :      :22 :
Rum ace   : 6 3  5:   2 : 2 :    :
Sal rep   : 3   3  :      :   :  5:
Tri pra   : 5 2  2:      :   :    :
Vic lat   :   2 11 :      :   :    :
Che alb   :        :      : 1:    :
Cir arv   :        :   2  :   :    :
Emp nig   : 2      :      :   :    :
--------------------------------------
Environmental parameters:
--------------------------------------
Dept A1 :        :      :   :    :
mean    :   4.0  :  3.8:5.9: 7.5:
s.d.    :   1.1  :  0.6:0.1: 3.6:
% HF    :   38   :  33 :0  : 0  :
% NM    :   38   :  0  :0  :75  :
--------:--------:------:---:----:
Moisture:        :      :   :    :
 cl 1   :*****   : *    :   :    :
 cl 2   :**      : **   :   :    :
 cl 3   :        :      :   :    :
 cl 4   :        : *    : * :    :
 cl 5   :*       : *    : * :****:
--------:--------:------:---:----:
Manure  :        :      :   :    :
 cl 1   :*****   : *    :   :    :
 cl 2   :**      : *    :   :    :
 cl 3   :*       : *    : * :*   :
 cl 4   :        : ***  :   :    :
```

6.7.1 The use of simple descriptive statistics

For a continuous variable, mean and standard deviation can be computed for each cluster, if there is no reason to expect outliers or skewed distributions. For skewed distributions the alternative is to inspect the median and the mid-range, or range for each cluster. Table 6.7 gives means and standard deviations for the depth of the A1 soil horizon. There seems to be a weak relation between the clusters and the depth of the A1 horizon. A good alternative for continuous variables is the construction of stem and leaf diagrams for each cluster separately.

For ordinal and nominal variables, such as moisture class and fertilization class in the Dune Meadow Data, the construction of histograms give us quick insight into the relations. Table 6.7 clearly shows that moisture might be the most important environmental variable affecting the species composition. For nominal variables, frequencies or within cluster proportions of occurrence might also give insight in the relations. The proportion of plots managed by the Department of Forests, the Dutch governmental institution administering the country's nature reserves is, for instance, high in cluster 4, which are the nutrient-poor meadows.

In the preceding subsections, all variables have been evaluated separately and on one hierarchical level. A different, quick way to evaluate all levels of a hierarchical classification in the light of all appropriate environmental variables is the use of simple discriminant functions for each dichotomy as performed by DISCRIM (ter Braak 1982 1986). In the DISCRIM method, simple discriminant functions are constructed in which those environmental variables are selected that optimally predict the classification. Figure 6.18 shows the TWINSPAN classification of the sites (Table 6.6) in the form of a dendrogram and it shows at each branch the most discriminating variables selected by DISCRIM.

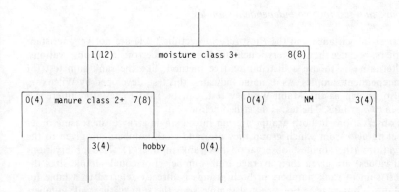

Figure 6.18 This is the same TWINSPAN site classification as in Table 6.6, but now presented as a dendrogram. At each branch the most discriminating variables, selected by DISCRIM, are shown. Numbers at the branches indicate the number of sites for which the conditions are true. Numbers in brackets indicate the number of sites in the clusters.

6.7.2 The use of tests of significance

Tests of significance are introduced in Chapter 3, Section 3.2. To test whether any environmental variable might be controlling the species composition (or might be controlled by the species composition, or simply related to the species composition) in the communities, we take as null hypothesis that the species composition is independent from the environmental variable. Rejecting the null hypothesis indicates that the environmental variable is related to the species composition of our community types in some way or another.

Analysis of variance

Analysis of variance is explained in Subsection 3.2.1. It can be used to detect relations between community types and continuous environmental variables. The systematic part consists of the expected values of the environmental variable, one for each community type and the error part is the variation in the values within each community type. Analysis of variance is based on the normal distribution. Therefore environmental variables must often be transformed, e.g. by using the logarithm of their values (see Subsection 2.4.4).

Chi-square test

Subsection 3.3.1 describes the chi-square test for $r \times k$ contingency tables. The chi-square test is used to test the null hypothesis: that a nominal environmental variable is not related to the community types. Care should be taken if the numbers of data are small or if the frequency of occurrence of the nominal variable is low (see Subsection 3.3.1).

The rank sum test for two independent samples

Analysis of variance and the t test (cf. Subsection 3.2.1) are not very resistant to outliers, because they are very much affected by gross errors in the observations. An alternative is to use a distribution-free method, like the rank sum test for two independent samples. As its name indicates, this test, developed by Wilcoxon, but also known as the Mann-Whitney test, can be used to test for differences between two groups. The test is described below.

All observations in both groups are put into a single array in increasing order (indicating also from which group they are) and rank numbers are given to the observations (the smallest observation has rank number 1). Tied observations (equal values) are given their average rank number. For equal sample sizes the smallest sum of rank numbers of both groups is directly referred to a table for the Mann-Whitney test. For unequal sample sizes the sum of the rank numbers is computed for the smallest sample (T_1). A second T (T_2) is computed:

$$T_2 = n_1 (n_1+n_2+1) - T_1$$

where n_1 and n_2 are the sizes of the smaller and the larger sample, respectively.

The test criterion T is the smaller of T_1 and T_2. For sample sizes outside the limits of the table, an approximate normal deviate Z is referred to the tables of the normal distribution to obtain the significance probability P:

$$Z = (|\mu - T| - 0.5)/\sigma$$

where $\mu = (n_1 + n_2 + 1)/2$ and $\sigma = \sqrt{(n_2\,\mu/6)}$.

With this test two small groups (minimum sizes of 4 and 4, 3 and 5 or 2 and 8) can be compared without any assumption on the distribution of the environmental variable. For further details on this test and other tests of significance refer to a statistical handbook (e.g. Snedecor & Cochran 1980).

6.8 Bibliography

Scientific classification of communities can be traced back in history to von Humboldt (1807): he used associations of plants to define community types. Jaccard (1912) took the first step in the direction of multivariate analysis by the development of his index of similarity. Many years later he was followed by Sørensen (1948), who developed his 'method of establishing groups of equal amplitude in plant sociology' based on similarity of species content. The increasing access scientists have had to computers over the last thirty years has led to rapid developments in multivariate methods. An early work that is devoted to the use of multivariate methods in taxonomy is a book written by Sokal & Sneath (1963). In the late sixties, and 1970s there was a rapid increase in the use of cluster analysis (and ordination) by ecologists. Pielou (1969), Goodall (1970) and Williams (1976a) give a theoretical background to these methods.

Numerical classification in the phytosociological context is elucidated by Goodall (1973) and Mueller-Dombois & Ellenberg(1974). Everitt (1980) and Dunn & Everitt (1982) are more recent introductions to numerical taxonomy. Gauch (1982) gives an introduction to classification of communities and mentions many applications of cluster analysis.

6.9 Exercises

Exercise 6.1   Single-linkage clustering with Jaccard similarity

Exercise 6.1a   Compute the Jaccard similarities for the sites in the artificial species-by-site table given below. Since the similarities are symmetrical and the diagonal elements all equal 1, you should only compute the elements below the diagonal of the site-by-site similarity matrix (cf. Exercise 6.1c for the species).

Site		1	2	3	4	5	6
Species	A		4	1		2	2
	B				1		
	C	1					
	D	2	1	1		1	
	E		1		4		5
	F		3	1	1		3
	G	1	1	3		1	

Exercise 6.1b   Perform single-linkage clustering for the sites.

Exercise 6.1c   The species similarities are:

	A	B	C	D	E	F
B	0					
C	0	0				
D	0.60	0	0.25			
E	0.40	0.33	0	0.17		
F	0.60	0.25	0	0.33	0.75	
G	0.60	0	0.25	1.0	0.17	0.33

Perform single-linkage clustering for the species.

Exercise 6.1d   Rearrange the sites and the species to represent the hierarchical structure. Try also to obtain the best possible ordering.

Exercise 6.2   Complete-linkage clustering with percentage similarity

Exercise 6.2a Compute the missing elements (denoted by *) of the site-by-site similarity matrix (percentage similarity) for the table of Exercise 6.1a.

	1	2	3	4	5
2	29				
3	40	*			
4	0	25	17		
5	50	58	60	*	
6	0	60	*	63	29

Exercise 6.2b Perform complete-linkage clustering for all sites.

Exercise 6.3   Single-linkage clustering with Euclidean Distance

Exercise 6.3a Compute the missing elements (denoted by *) of the site-by-site Euclidean Distance matrix for the table of Exercise 6.1a.

	1	2	3	4	5
2	5.3				
3	*	4.2			
4	4.9	5.7	5.3		
5	2.5	3.8	2.5	4.9	
6	6.3	4.7	6.3	2.5	*

Exercise 6.3b Perform single linkage and try to find out why the result is so different from Exercise 6.1.

Exercise 6.4   Divisive clustering

This exercise is a demonstration of a simple classification procedure using a divisive strategy with iterative character weighting (this procedure is different from the procedures used in TWINSPAN and by Hogeweg (1976)). The species-by-sites table of Exercise 6.1 is used here too.

Step a Divide the sites in two groups, a positive and a negative one. You may choose a random division or monothetic division based on the presence of one species. In the solution we intially place Sites 1,3 and 6 in the negative group and Sites 5,2 and 4 in the positive group.

Step b Compute the sum of abundances for each species for both sides of the dichotomy (SPOS for the positive scores, SNEG for the negative scores).

Step c Compute a preference score for each species: pref = (SPOS−SNEG)/ (SPOS+SNEG).

Step d Compute a weighted sum (or weighted average) of the species abundances for each site.

Step e Find maximum and minimum site score (MAX and MIN) and the midpoint of the site scores (MID = (MAX+MIN)/2).

Step f Assign all sites with a score less than the midpoint to the 'negative' group and all the other sites to the 'positive' group.

Step g Repeat steps b–f until the division is stable.

Step h Repeat steps a–g for each subgroup.

Exercice 6.5   *Cluster interpretation with nominal environmental data*

Do the chi-square test for moisture classes 1 and 2 combined, and 3, 4, and 5 combined, for:

Exercise 6.5a   The first division of TWINSPAN (Table 6.6)

Exercise 6.5b   The first three clusters of Table 6.7 combined and the last cluster (highest hierarchical level). What is your null-hypothesis? Has it to be rejected or not? Is it correct to use the chi-square test in this case?

Exercise 6.6.   *Cluster interpretation with ordinal environmental data*

Perform Wilcoxon's test for 'Depth of A1' for the first division of TWINSPAN (Table 6.6).

6.10 Solutions to exercises

Exercise 6.1 Single-linkage clustering with Jaccard similarity

Exercise 6.1a The number of species per site (A) and the values of c and $(A+B-c)$ in the notation of Subsection 6.2.2 are given below:

Site number	1	2	3	4	5	6
A (or B)	3	5	4	3	3	3
c	2: 2					
	3: 2	4				
	4: 0	2	1			
	5: 2	3	3	0		
	6: 0	3	2	2	1	
$(A+B-c)$	2: 6					
	3: 5	5				
	4: 6	6	6			
	5: 4	5	4	6		
	6: 6	5	6	4	5	

Jaccard similarity is obtained by dividing corresponding elements of the tables: c and $(A+B-c)$

		1	2	3	4	5
Jaccard	2	0.33				
	3	0.40	0.80			
	4	0.00	0.33	0.17		
	5	0.50	0.06	0.75	0.00	
	6	0.00	0.60	0.40	0.50	0.20

Exercise 6.1b In order to obtain the single-linkage clustering we only have to find the highest remaining similarity as demonstrated in the following table:

Fusion	Highest similarity	Between sites	Clusters fused
1	0.80	2,3	2,3
2	0.75	3,5	(2,3),5
3	0.60	2,6	(2,3,5),6
4	0.50	1,5	1,(2,3,5,6)
5	0.50	4,6	(1,2,3,5,6),4

Exercise 6.1c

Fusion	Highest similarity	Between species	Clusters fused
1	1.0	D,G	D,G
2	0.75	E,F	E,F
3	0.60	A,D	A,(D,G)
4	0.60	A,F	(A,D,G),(E,F)
5	0.33	B,E	(A,D,E,F,G),B
6	0.25	C,D	(A,B,D,E,F,G),C

Exercise 6.1d The hierarchy can be represented in a dendrogram in which each dichotomy can be swivelled. In order to obtain the best possible ordering each site is placed next to its nearest neighbour.

2 next to 3:	2	3				
5 next to 3:	2	3	5			
6 next to 2:	6	2	3	5		
1 next to 5:	6	2	3	5	1	
4 next to 6:	4	6	2	3	5	1
or the reverse	1	5	3	2	6	4

If we use the same procedure for the species, there is a problem for species C.

D next to	G: D G	or	G D
E next to	F: E F	or	F E
A next to	D: A D G	or	G D A
A next to	F: E F A D G	or	G D A F E
B next to	E: B E F A D G	or	G D A F E B

C between A and D is not a good solution because A resembles D much more than C does.

C next to G is better, but still there are two possibilities, C between D and G or C at the end. Because C resembles A less than D and G do, C is placed at the end of the ordering:

B E A D G C or C G A F E B

The rearranged table:

	1	5	3	2	6	4
C	1					
G	1	1	3	1		
D	2	1	1	1		
A		2	1	4	2	
F		1	3	3	1	
E			1	5	4	
B				1		

Exercise 6.2   Complete linkage clustering with percentage similarity

Exercise 6.2a   Computation of the similarities:

$PS_{2,3}$: Sum of scores site 2: 10
        Sum of scores site 3: 6
        Minimum scores: A = 1, B = 0, C = 0, D = 1, E = 0, F = 1, G = 1
        Sum of minimum scores: $c = 1+0+0+1+0+1+1 = 4$
        $PS_{2,3} = 200 \times 4/(10+6) = 50$
$PS_{3,6}$: $c = 1+0+0+0+0+1 = 2$
        $PS_{3,6} = 200 \times 2/(6+10) = 25$
$PS_{4,5}$: no common species: $c = 0$ $PS_{4,5} = 0$

The complete similarity matrix is now:

	1	2	3	4	5
2	29				
3	40	50			
4	0	25	17		
5	50	58	60	0	
6	0	60	25	63	29

Exercise 6.2b   The first step is the same as with single linkage: the two most similar samples are fused. Then we construct a new similarity matrix:

fusion 1:4 and 6, similarity 63

	1	2	3	(4,6)
2	29			
3	40	50		
(4,6)	0	25')	17")	
5	50	58	60	0"')

Note: ')min (25,60) = 25
        ")min (17,25) = 17
        "')min (0,29) = 0

Fusion 2: 5 and 3, similarity 60

new similarity matrix:

	1	2	(3,5)
2	29		
(3,5)	40')	50")	
(4,6)	0	25	

Note: ') min (40,50) = 40
        ") min (50,58) = 50

Fusion 3: 2 and (3,5), similarity 50

new similarity matrix:

	1	(2,3,5)
(2,3,5)	29	
(4,6)	0	0

Fusion 4: (2,3,5) and 1, similarity 29

last fusion: (1,2,3,5) and (4,6), similarity 0

Exercise 6.3   Single-linkage clustering with Euclidean Distance

Exercise 6.3a

$$ED_{1,3} = [(0-1)^2+(0-0)^2+(1-0)^2+(2-1)^2+ (0-0)^2 +(0-1)^2+(1-3)^2]^{1/2}$$
$$= (1+0+1+1+0+1+4)^{1/2} = 8^{1/2} = 2.8$$

$$ED_{5,6} = [(2-0)^2+(0-0)^2+(0-5)^2+(0-3)^2+ (1-0)^2]^{1/2} = (4+1+25+9+1)^{1/2}$$
$$= 40^{1/2} = 6.3$$

Exercise 6.3b   Instead of looking for the highest similarity values, we look for the lowest dissimilarity values.

Fusion	Dissimilarity	Between sites	Clusters fused
1	2.5	1,5	1,5
2	2.5	3,5	(1,5),3
3	2.5	4,6	4,6
4	4.2	2,3	(1,3,5),2
5	4.7	2,6	(1,2,3,5),(4,6)

Now Sites 4 and 6 group together because of the dominance of Species E. Sites 2 and 3 are more different and fuse therefore later, because of the different abundances for Species A, F and G. Sites 1 and 5 are fused first, because of their low species abundances, and so on.

Exercise 6.4   Divisive clustering

Steps a–g   The species-by-sites table (Table 6.8) is rearranged according to the solution from Exercise 6.1.

Table 6.8   Species-by-sites table rearranged according to the solution for Exercise 6.1, with SPOS, SNEG and PREF computed in Steps b and c of the iteration algorithm of Exercise 6.4.

					step 1 b			step 1 c step 2 b			step 2 c step 3 b			step 3 c
					SPOS	SNEG	PREF	SPOS	SNEG	PREF	SPOS	SNEG	PREF
1					0	1	−1	0	1	−1	0	1	−1
1	1	3	1		2	4	−0.33	2	4	−0.33	1	5	−0.67
2	1	1	1		2	3	−0.20	2	3	−0.20	1	4	−0.75
	2	1	4	2	6	3	0.33	8	1	0.78	6	3	0.33
	1	3	3	1	4	4	0	7	1	0.75	7	1	0.75
	1	5	4	5	5	5	0	10	0	1	10	0	1
		1	1		0	1	0	1	0	1	1	0	1

Step 1a   Initial choice: Sites 1, 3 and 6 in the negative group; Sites 2, 4 and 5 in the positive group.

Step 1b–1c   See Table 6.8.

Step 1d   For Sites 1–6 the weighted sums of the preference scores are −1.73, 0.79, −0.86, 1, 0.13 and 0.66, respectively.

Step 1e   MID = (−1.73 + 1)/2 = −0.36.

Step 1f   Sites 1 and 3 in the negative group; the other sites in the positive group.

Step 2b–2c   See Table 6.8.

Step 2d   For Sites 1–6 the weighted sums are −1.73, 5.84, 0.34, 5.75, 1.03 and 8.81, respectively.

Step 2e   MID = (−1.73 + 8.81)/2 = 3.56.

Step 2f   Sites 1, 3 and 5 in the negative group; the other sites in the positive group.

Step 3b–3c   See Table 6.8.

Step 3d   For Sites 1–6 the weighted sums are −3.17, 2.15, −1.68, 5.75, −0.76 and 7.91, respectively.

Step 3e   MID = (−3.17 + 7.91)/2 = 2.37.

Step 3f   Same as Step 2f, so the classification is stable now.

Step h   This is solved in essentially the same way for further subdivisions.

Exercise 6.5   Cluster interpretation with nominal environmental data
We start by making two-way cross-tabulations of the observed frequencies (*o*):

	TWINSPAN	(Table 6.6)		FLEXCLUS	(Table 6.7)	
Cluster number:	0	1	total	1+2+3	4	total
Moisture class: 1+2	11	0	11	11	0	11
3+4+5	1	8	9	5	4	9
Total:	12	8	20	16	4	20

The expected cell frequencies can be obtained by dividing the product of the corresponding row and column totals by the overall total, e.g. $(11 \times 12)/20 = 6.6$ (first cell of the 'TWINSPAN' table) The two-way cross-tabulation of the expected cell frequencies (e) becomes:

Cluster number:	0	1	total	1+2+3	4	total
Moisture class: 1+2	6.6	4.4	11	8.8	2.2	11
3+4+5	5.4	3.6	9	7.2	1.8	9
Total:	12	8	20	16	4	20

For a two-by-two table all deviations from the expected values are equal (check this for yourself): $o-e = 4.4$ and $(o-e)^2 = 19.36$ (TWINSPAN), $o-e = 2.2$ and $(o-e)^2 = 4.84$ (FLEXCLUS, highest level).

$$\chi^2 = 19.36(1/6.6+1/4.4+1/5.4+1/3.6) = 16.30 \text{ (TWINSPAN)}$$

$$\chi^2 = 4.84(1/8.8+1/2.2+1/7.2+1/1.8) = 6.11 \text{ (FLEXCLUS, highest level).}$$

Our null hypothesis is that the classification is not related to moisture class. We have 1 degree of freedom since the number of rows and the number of columns both are equal to 2 : $v =$ (number of rows – 1) $\times$ (number of columns – 1). Referring to a table of the chi-square distribution we see that the null hypothesis should be rejected ($P < 0.005$ for the TWINSPAN classification and $0.01 < P < 0.025$ for the FLEXCLUS classification), which means that the types are related to moisture level for both classifications.

We should not use the chi-square test because the expected cell frequencies are too low.

Exercise 6.6   Cluster interpretation with ordinal environmental data

The following table shows the sites ordered by increasing depth of the A1 horizon; the sites belonging to the right-hand side of the TWINSPAN dichotomy are indicated with an asterisk. Rank numbers are assigned, in case of ties the average rank number is used. Site 18 is not included in the list since its value for this variable is missing.

Site	7	1	10	2	11	20*	9*	19	17	4	8*	6	3	16*	12*	13*	5	14*	15*
A1	2.8	2.8	3.3	3.5	3.5	3.5	3.7	3.7	4.0	4.2	4.2	4.3	4.3	5.7	5.8	6.0	6.3	9.3	12.
Rank	1.5	1.5	3	5	5	5	7.5	7.5	9	10.	10.	12.	12.	14	15	16	17	18	19

Rank numbers 10. and 12. indicate 10.5 and 12.5 respectively. Value of T_1 (for the smaller, right-hand-side group) is the sum of the rank numbers: 105. Value of $T_2 = 8(8+11+1)-105 = 55$. T, which is referred to a table of Wilcoxon's test, is the smaller of these two: 55 Looking up this value in the table, we conclude that we cannot reject our null hypothesis. There is no evidence that the types are related to the depth of the A1 horizon.

7 Spatial aspects of ecological data

P.A. Burrough

7.1 Introduction

Many kinds of environmental mapping, such as those practised by physical geographers, geologists, soil scientists, biogeographers, ecologists and land-use specialists, use the choropleth map as the major means of recording and displaying the spatial variation of the phenomenon under study. Choropleth mapping assumes that uniform areas can be delineated by sharp boundaries (i.e. that all important change occurs at the boundaries of the delineated spatial units, Figure 7.1.). Sometimes the boundaries may be those of administrative areas or land-use types, but very frequently they will have been interpreted from external aspects of the landscape such as can be seen from stereo aerial photographs, remotely sensed imagery, or in the field.

The major, and frequently forgotten, principle of choropleth mapping is that the values of the quantitative variables under study are necessarily assumed to be constant *within* the delineated areas (or the values have a constant mean with a small, residual error variation). Because all important changes are assumed to take place at boundaries between delineated units, choropleth maps model variation by a stepped surface. In statistical terms, the choropleth map can be represented

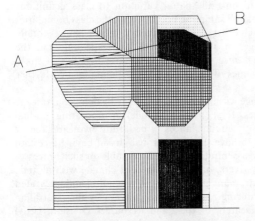

Figure 7.1 A simple choropleth map and its cross-section through the line A–B. All important variation is assumed to occur at the boundaries.

by the technique of one-way analysis of variance (see Subsections 3.2.1 and 3.5.5), in which the value of a quantitative variable Z at a point x is given by

$$Z = b_0 + b_j + \varepsilon \qquad\qquad\qquad\qquad \text{Equation 7.1}$$

where

b_0 is the overall mean of the attribute for the study area as a whole

b_j is the difference between the mean of delineated area j, in which point x falls, and the overall mean ($\Sigma_j b_j = 0$)

ε is an independent, normally distributed random error term having zero mean and variance σ^2.

This simple model of the variation of environmental phenomena over the landscape is frequently inadequate because spatial change does not always take place at well-defined, abrupt boundaries. Boundaries that have been interpreted from one set of criteria may be inadequate for resolving the variation of other attributes (Beckett & Burrough 1971). Reclassification of delineated areas, (e.g. Dent 1985) and drawing boundaries in different places (e.g. Bie & Beckett 1973) may give maps that have a completely different visual appearance, but which may still inadequately describe the variation over the landscape of the phenomenon of interest.

A common and well-established approach to the problem of spatial variation *within* mapping unit boundaries is to remap the area at larger scales (e.g. Vink 1963), thereby repeating the choropleth principle, but with smaller spatial subdivisions. As Burrough (1983a) has pointed out, this process of nested delineation can go on indefinitely – it is a simple approach to dealing with complex problems (see Mandelbrot 1982).

The basic weakness of the choropleth approach, as expressed by Equation 7.1, is that it requires the scientist to link all spatial variation to class definitions and boundaries, and to assume that all sampling points within the class boundaries are statistically independent of each other. The choropleth principle ignores totally the possibility of *gradual* change, or *trend*, within a landscape unit and the phenomenon of spatial *autocovariance*, i.e. that points close together in space are more likely to have similar values of an attribute than points located further apart.

The realization of gradual change within major geographically defined landscape units has important consequences. First, a spatial model incorporating gradual spatial change would seem in many circumstances to be intuitively nearer reality than one that requires all points to have the same probability of having a certain value. For example, gradual change is intuitively more acceptable for such processes as the deposition of alluvium by rivers, or the variation of climate with altitude or distance from a coast. To be fair, many published soil, vegetation and integrated resource maps (e.g. those made during reconnaissance surveys by the Australian C.S.I.R.O. Division of Land Research, or the Land Resources Division, U.K.) do contain information about within-unit spatial variation. Because of the nature of the surveys, however, this spatial variation is usually only given as a qualitative description in the memoir that accompanies the printed map.

One major reason for using choropleth mapping techniques is that they allow the surveyor to make generalizations from a large number of point observations (soil borings, quadrat counts, etc.). Before computers became generally available for processing large amounts of complex spatial data, classification and choropleth mapping was the only approach to the problem of mapping the spatial variation of complex ecological phenomena. More recently, the availability of the computer and of new mathematical techniques for modelling gradual spatial change, have allowed us to approach the problem of within-landscape-unit variation in ways that do not suffer from the restrictions of the choropleth technique. This chapter describes two of these ways, namely trend-surface analysis and the study of spatial autocovariance, and it explains how they can be used to help discover the spatial structure of a phenomenon or be applied for extending the data sampled at points to map larger areas.

7.2 Trend-surface analysis

7.2.1 Introduction

The simplest way to model gradual, long-range spatial variation is to fit a regression line, surface or volume to the observed point data. This technique, known as trend-surface analysis, because of its frequent application to data gathered from two-dimensional space, is a special case of the usual multiple regression techniques in which observed values of the variable of interest, Z, are regressed against the independent position variables, which act as the explanatory variables (Section 3.2 and Subsection 3.5.2). The resulting line, surface or volume represents the long-range, gradually changing aspects of the spatial variation of Z (the systematic part of the model). The short-range, random component is then represented by the deviations from regression (the error part of the model).

7.2.2 One-dimensional trends

Consider the value of an environmental variable Z that has been measured along a transect. Assume that Z increases monotonically with X, so that a plot of Z against X has the form shown in Figure 7.2. The long-range variation of Z over X can be expressed by a regression line. The regression line can be fitted in several ways. If the line is to be a 'best fit' it means that the squared deviations of the data points from the line must be a minimum. Clearly, deviations from the regression line can be calculated in three ways. These are:
- the deviation of Z from the fitted line
- the joint deviation of Z and X from the fitted line
- the deviation of X from the fitted line.

Because Z is the variable to be explained by X (location), the first case is appropriate, so that we have the regression model

$$Z = b_0 + b_1 X + \varepsilon \qquad\qquad \text{Equation 7.2}$$

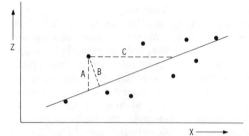

Figure 7.2 The deviations from a linear trend. A: minimum deviation in Z. B: minimum joint deviation in X and Z. C: minimum deviation in X. For trend surfaces, A is required.

where, in addition, the error ε is assumed to be independent and normally distributed. The parameters b_0 and b_1 of this model can be estimated by the least-squares methods of Subsection 3.2.2. The t value of b_1 (see Subsection 3.2.2.) or the correlation coefficient between Z and X show whether there is a long-range trend or that short-range erratic components dominate. The long-range variation or 'trend' is expressed by the slope of the regression line b_1.

7.2.3 Extensions to non-normally distributed variables

So far, the simplest case has been considered, i.e. when Z is normally distributed and linearly dependent on X. Very often environmental variables are non-normally distributed. The causes are diverse, but physical properties of soil, for example, are often log-normally distributed. In these situations it is usual to first examine the distribution of Z to check for normality, and then if the distribution is markedly skewed, to apply a transformation to bring the distribution back to normality. For example, a positively skewed distribution can often be transformed to a normal distribution by taking the logarithm of the variable (see Section 2.4). The transformed data are then used for the regression. Note that the actual requirement for using the least-squares method to fit Equation 7.2 is that the errors – and not the observed values – have a normal distribution. Because the errors cannot be examined in advance, and because the distribution of variable and errors are often of the same kind, in practice the distribution of Z is used to choose the transformation required. Care must be taken when interpreting the results to realize that the slope and intercept parameter refer to the transformed values.

7.2.4 Non-linear dependence of Z and X

In many circumstances Z is not a linearly increasing or decreasing function of X, but one that may vary in another way with location (Figure 7.3). In these situations higher-order regression lines can be used to describe the long-range variations. An example is:

216

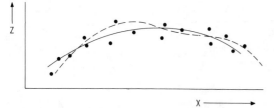

Figure 7.3 Quadratic (——) and cubic (------) trend polynomials in one dimension.

$$Z = b_0 + b_1X + b_2X^2 + \varepsilon \qquad \text{Equation 7.3}$$

which is known as a second-order equation. As before, the problem is to minimize the sums of squares and the goodness of fit as indicated by the multiple correlation or the fraction of variance accounted for (see Subsection 3.2.1). Higher-order equations can also be used. By increasing the number of terms it is possible to fit any complicated curve exactly. Against this is the real problem of matching up the mathematical model with physical reality, i.e. finding a plausible interpretation. Subsection 3.2.3 explains how to find a compromise between a small number of terms and a large correlation.

7.2.5 Two-dimensional surfaces

For two dimensions, represented by orthogonal axes X_1 and X_2, the value of the variable Z might be expressed as a polynomial function:

$$Z = \Sigma_{r+s \leqslant p} \, b_{rs} X_1^r X_2^s + \varepsilon \qquad \text{Equation 7.4}$$

which is fitted by least squares.

The first few functions are:

b_0	flat
$b_0 + b_1X_1 + b_2X_2$	linear (1st order)
$b_0 + b_1X_1 + b_2X_2 + b_3X_1^2 + b_4X_1 X_2 + b_5X_2^2$	quadratic (2nd order)

These are illustrated in Figure 7.4. The integer p is the *order* of the trend surface. There are $(p+1)(p+2)/2$ terms in the function for the trend surface.

The problem of finding the b_i coefficients is a standard problem in multiple regression (see Subsection 3.5.2). Ripley (1981) warns that polynomial regression is 'an ill-conditioned least squares problem that needs careful numerical analysis'.

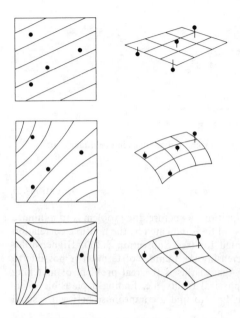

Figure 7.4 1st, 2nd and 3rd order trend surfaces showing some sample points and their deviations from the fitted surfaces.

He recommends that all distances be re-scaled in the range -1 to $+1$. Ripley also points out that if the number of parameters in the polynomial equals the number of data points the surface will be an exact fit; when this is so the value of R^2 (the coefficient of determination) and the distinction between long-range and short-range variation is meaningless.

7.2.6 Trend surfaces as interpolators

Just as in ordinary regression, the regression equation of the trend surface can be used to predict, or interpolate, values of the property at unsampled sites. Trend surfaces are not exact interpolators, however, because the surface usually does not go through the data points. The deviations between observed values and the interpolated values are known as residuals. The better the fit of the surface to the data, the lower the sums of squares of the deviations of the residuals. This least-squares criterion is often used to judge the goodness of fit of the trend surface to the data. As in the case of simple two-variable regression, the fraction of the variance accounted for estimates how much of the variation in the data has been taken up, or 'explained', by the regression surface.

The overall significance of the fit of a multiple regression trend surface can be tested by the usual variance ratio, F (see Section 3.2):

218

$F =$ Mean square regression / Mean square residual.

This F test assumes that the residuals are independently and normally distributed with zero mean and constant variance σ^2 (i.e. the residuals have no spatial correlation). This is unrealistic because there is much evidence to show that deviations from trend surfaces tend to be correlated over short distances. Plotting the deviations from a low-order trend surface may reveal more about the spatial behaviour of the variable in question than the trend itself, which may only be describing obvious, or easily understood, variation. Strongly positively autocorrelated residuals may lead to attempts to fit surfaces of too high an order.

7.2.7 Use of trend surfaces

Trend surfaces are mostly used to describe gradual, long-range variation. Although they are easy to compute, they suffer from a number of problems, both procedural and conceptual, so that they should only be used with great care. These problems can be summarized as follows:

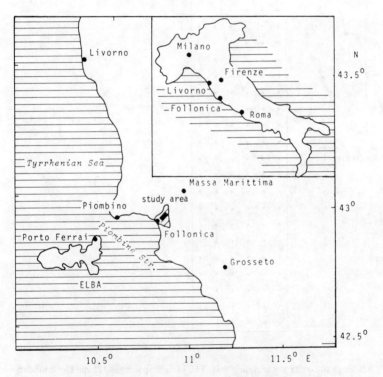

Figure 7.5 Location of the sites in northern Italy from which the test data were collected.

Conceptual problems
- Fitting a trend surface has little point unless the trend can be shown to have a physical explanation.
- The regression models assume that all deviations from regression (the residuals) are normally distributed and spatially independent. That this is not the case can be often seen by mapping the residuals, which may often reveal spatial clustering. This clustering can reveal important characteristics of the data.

Practical problems
- When data points are few, extreme values can seriously distort the surface.

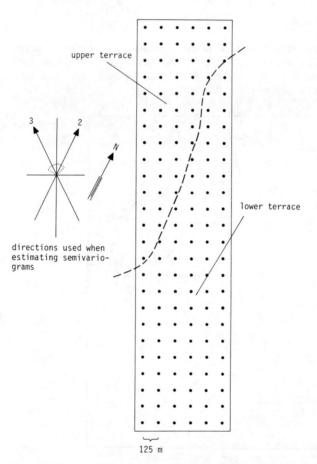

Figure 7.6 Design of the 25 x 6 sampling grid. The sample spacing is 125 m. The boundary between the Pleistocene terrace and the Holocene river valley is shown as (------).

- The surfaces are extremely susceptible to edge effects. Higher-order polynomials can turn abruptly near area edges leading to unrealistic values.
- Trend surfaces are inexact interpolators. Because they are long-range models, extreme values of distant data points can exert an unduly large influence and result in poor, local estimates of the value of the property of interest.

The following example uses soil data from a study of soil variation made in northern Italy (Kilic 1979). The study area is located near Follonica, Italy, about 200 km north of Rome. Figure 7.5 shows the general location. Figure 7.6 shows the detailed layout of the 25 x 6 sampling grid (spacing 125 m) and its location over the boundary of an upper alluvial terrace with a lower, recent alluvial terrace. Both terraces have a general north–south trend in altitude, from hilly land in the north down to a coastal alluvial plain in the south.

The soil was examined at each grid point by auger to a depth of 120 cm. Although a full profile description was made at each point, the following discussion will be restricted to a single soil property that clearly illustrates the anisotropy and variation at several scales. The property chosen is the percentage clay content of the top-soil (0-20cm).

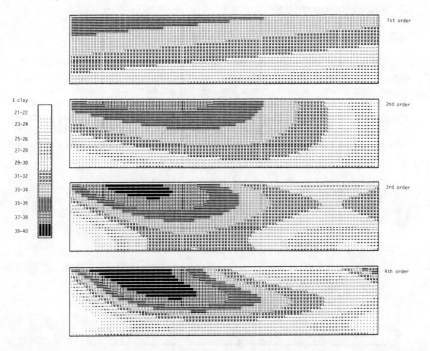

Figure 7.7 1st, 2nd and 3rd order trend surfaces fitted to all 150 data points for percentage clay of the top-soil.

221

Figure 7.7 shows the 1st, 2nd, 3rd and 4th order trend surfaces that were fitted to the original data.

7.2.8 Local trends

An alternative to fitting a global trend surface is to fit a series of local trend surfaces. A regular grid of points at which interpolations will be made is laid over the study area. A regular square area, or 'window', is chosen and is laid over the first grid point at the lower left-hand corner of the area to be mapped. A trend surface is fitted to those data points falling within the window, and it is then used to estimate the value of the property of interest at a point on a regular grid located at the centre of the window. The window is then moved up to lie over the next grid point and the process repeated. The advantage of the method is that major outliers do not affect results except within their own locality. The disadvantage is that there is no easy way to choose the optimum size of the window, nor the order of the local trend. Figure 7.8 shows 1st order and 2nd order local trends computed for the data from the study in Italy.

7.3 Spatial autocovariation

7.3.1 Introduction

Very frequently, the spatial variation of ecological attributes varies continuously within a spatial unit in a way that cannot easily be described by a simple regression polynomial. Often the short-range variation of an attribute, as seen at a set of observation points, varies essentially in a correlated, but random manner, at least at the scale at which the observations have been made. That is to say, sample points that are close together tend to be more similar than points further apart, but there is no easy, direct relation between sample site location and the value

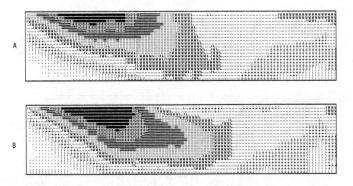

Figure 7.8 Local trend surfaces fitted within a search radius of 450 m. A: 1st order local trend. B: 2nd order local trend. For more details see Figure 7.7.

222

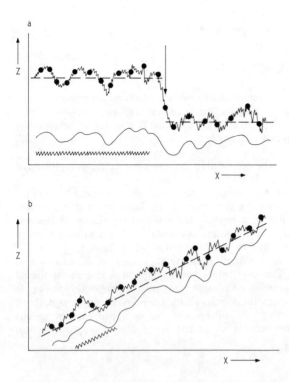

Figure 7.9 The mean components of spatial variation. a: variation composed of a structural difference in mean values (————), random, but spatially correlated variation (wavy line) and uncorrelated variation or 'noise' (saw-tooth line). Samples taken at observation sites (●) reflect the sum of all sources of variation. b: as a, but the structural variation is displayed here as a linear trend. The random but spatially correlated variation and the noise component are as with a.

of the attribute, as would be implied by a trend surface.

This line of thought brings us to a model of spatial variation that contains at least three components. The first is a major structural component that represents the average value of the variable within a physiographically or ecologically defined area or mapping unit (Figure 7.9a). Alternatively, this major structure might be better described by a trend (Figure 7.9b). Superimposed on the major structure is a second kind of structural variation that, at the scale of observation, cannot be identified with a deterministic component of the landscape. This second kind of variation is spatially correlated, gradual variation. Finally, there may also be a third component that consists of essentially uncorrelated random variation, such as would be caused by observational or analytical error and by spatial variations at scales too small to be resolved by the sampling network.

If the vector x describes a position in one, two or three dimensions, the value

223

of a variable Z at point x can be represented by

$$Z(x) = m(x) + \varepsilon'(x) + \varepsilon'' \qquad\qquad \text{Equation 7.5}$$

where
$m(x)$ is the term describing the major 'deterministic', structural component
$\varepsilon'(x)$ is the term describing the spatially correlated, but random, variation
ε'' is a residual, spatially independent noise term that is assumed to be normally distributed with zero mean and variance σ^2.

Variables whose variation can be described in this way are often called regionalized variables (Journel & Huibregts 1978).

Very often, we can assume that the major 'structural' component can be specified by reference to a constant mean or a given trend for the landscape unit in question, so the problem becomes one of describing the remaining variation in the best way possible. In particular, it is important to know the relative balance between the spatially correlated $\varepsilon'(x)$ and the uncorrelated ε'' terms. In order to do this, it is necessary to make some assumptions about the nature of the variation of the variable under study. The most important assumption is that the statistical properties of the spatially correlated variation, $\varepsilon'(x)$, are the same within the whole of the major landscape or 'structural' unit. This is often known as the hypothesis of statistical stationarity, which, as will be seen, can be approached in various ways. Without such assumptions it would not be possible to apply statistical methods to spatial analysis.

7.3.2 Stationarity

The idea of statistical stationarity or uniformity, as stated above, is a necessary assumption. Consider a series of observation points laid out at equal intervals along a linear transect. Suppose that a soil property Z is estimated at each point x. Formally we can say that if the joint distribution of the n random variables $Z(x_1) \dots Z(x_n)$ is the same as the joint distribution of $Z(x_1 + h) \dots Z(x_n + h)$ for all $x_1 \dots x_n$ and h, then the series is said to be strictly *stationary*. Put simply, the statistical properties of the series are not affected by moving the sample points a distance h from x_i to $x_i + h$. It is usual to replace the above definition with that of 'second-order stationarity' or 'weak stationarity', in which the mean is constant, the autocovariance depends only on sampling interval, and the variance is finite and constant. The major difference with conventional, non-spatial statistics is the recognition that the variance includes a covariance term that depends on sample spacing.

7.3.3 Statistical moments of spatial series

The statistical properties of a spatial series are described by the following moments.

First-order moment: the *mean*, which is given by

$$E[Z(x)] = \mu \qquad\qquad \text{Equation 7.6}$$

where E means the expectation of the value of attribute Z at point x. Put simply, the expectation is the most likely value. The mean is estimated by

$$\bar{z} = \Sigma^n_{i=1} \; Z(x_i)/n \qquad\qquad \text{Equation 7.7}$$

where n is the number of observations in the sample.

Second-order moments: there are three second-order moments, the variance, the (auto)covariance, and the semivariance. The *variance* is defined by

$$\text{var } Z(x) = E\,[Z(x) - \mu]^2 \qquad\qquad \text{Equation 7.8}$$

It is estimated by

$$\hat{C}\,(0) = \Sigma^n_{i=1}\,[Z(x_i) - \bar{z}\,]^2/n \qquad\qquad \text{Equation 7.9}$$

The *covariance* between the value of Z and two points x_1 and x_2, spaced distance h apart, is defined by

$$C(x_1, x_2) = E\,[Z(x_1) - \mu]\,[Z(x_2) - \mu] \qquad\qquad \text{Equation 7.10}$$

For a one-dimensional transect in which $x_i = x_i + h_{(i-1)}$, h being the lag, the covariance at lag h is estimated by

$$\hat{C}(h) = \Sigma^{n-h}_{i=1}\,[Z(x_i + h) - \bar{z}]\,[Z(x_i) - \bar{z}]/(n-h) \qquad\qquad \text{Equation 7.11}$$

When the lag is equal to 0, then the covariance $C(h) = C(0)$, i.e. the variance.

It is sometimes necessary to compare the spatial behaviour of different attributes that have been measured on different scales, and thus have different variances. This can be done by scaling the covariances to *autocorrelations* by dividing by the variance, i.e.

$$r(h) = \hat{C}(h)/\hat{C}(0) \qquad\qquad \text{Equation 7.12}$$

The autocorrelation varies between -1 and $+1$.

The *semivariance* is defined as one-half the variance of the increment $Z(x_1) - Z(x_2)$ and is written as

$$\gamma(h) = \tfrac{1}{2}\,\text{var}\,[Z(x_1) - Z(x_2)] \qquad\qquad \text{Equation 7.13}$$

For points separated by distance h, this can be written

$$\gamma(h) = \tfrac{1}{2}\,E[Z(x+h) - Z(x)]^2 \qquad\qquad \text{Equation 7.14}$$

The semivariance is estimated by

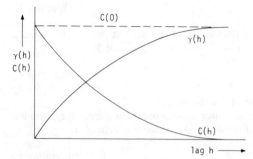

Figure 7.10 The variation of semivariance $\gamma(h)$ and autocovariance $C(h)$ with lag h for a second-order stationary process.

$$\hat{\gamma}(h) = \Sigma_{i=1}^{n-h} [Z(x_i+h) - Z(x_i)]^2/[2(n-h)] \qquad \text{Equation 7.15}$$

It can be seen that for a second-order stationary series $\gamma(h) = C(0)-C(h)$

and that

$$r(h) = 1 - \hat{\gamma}(h)/\hat{C}(0) \qquad \text{Equation 7.16}$$

When the autocovariance is plotted against sampling interval or lag, the resulting graph is known as an autocovariogram. The equivalent graph for semivariance is called a semivariogram. The autocovariogram and semivariogram of a second-order stationary series are mirror equivalents (Figure 7.10).

7.3.4 Autocovariograms and semivariograms

The following simple example shows how autocovariances and semivariances are calculated for various lags for a one-dimensional series. Consider the series shown in Figure 7.11.

(figuur 7.11)

1. The mean is estimated by $24/8 = 3$.

2. The variance is estimated by
$\hat{C}(0) = [(1–3)^2 + (3–3)^2 + (4–3)^2 + (6–3)^2 + (4–3)^2 + (3–3)^2 + (3–1)^2 + (2–3)^2]/8 = 20/8 = 2.5$.

3. The covariance at lag 1 is estimated by
$\hat{C}(1) = [(3–3)(1–3) + (4–3)(3–3) + (6–3)(4–3) + (4–3)(6–3) + (3–3)(4–3) + (1–3)(3–3) + (2–3)(1–3)]/7 = 8/7 = 1.143$.

226

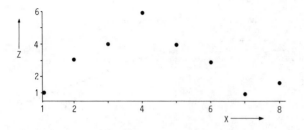

Figure 7.11 A simple example of a one-dimensional spatial series.

The autocorrelation at lag 1, $r(1) = 1.143/2.5 = 0.46$. The covariance at lag 2 is estimated by $\hat{C}(2) = [(4-3)(1-3) + (6-3)(3-3) + (4-3)(4-3) + (3-3)(6-3) + (1-3)(4-3) + (2-3)(3-3)]/6 = -3/6 = -0.5$. The autocorrelation at lag 2, $r(2) = -0.5/2.5 = -0.2$.

4. The semivariance at lag 1 is estimated by
$\hat{\gamma}(1) = [(1-3)^2 + (3-4)^2 + (4-6)^2 + (6-4)^2 + (4-3)^2 + (3-1)^2 + (1-2)^2]/(2\times7) = 19/14 = 1.357$.

The semivariance at lag 2 is estimated by
$\hat{\gamma}(2) = [(1-4)^2 + (3-6)^2 + (4-4)^2 + (6-3)^2 + (4-1)^2 + (3-2)^2]/(2\times6) = 37/12 = 3.08$.

The calculation proceeds in similar manner for larger lags.

7.3.5 Example of the use of autocorrelation analysis

The example chosen here is taken from a study of gradual variation of thermal emittance from the ploughed surface of experimental fields at the Wageningen Agricultural University's Experimental Farm at Swifterbant, in the Flevopolders, the Netherlands (ten Berge et al. 1983). The thermal emittance was recorded by a Daedalus line scanner carried in an aeroplane at 1500 ft on 26 March 1982 at 1300 hours. The resolution of the scanner on the ground was an area of about 1.5 x 1.5 m. Figure 7.12a shows the equivalent temperatures recorded for a set of 50 pixels lying along a 200 m transect, spaced 4 m apart. Figure 7.12b shows the autocorrelogram of these data; it shows that sites close together appeared to vary similarly, as is shown by Figure 7.12a. Indeed, the data could be modelled by a polynomial regression equation or 'trend'.

Figure 7.12c is a plot of data collected from a hand held scanner from the same sites and at the same time. Unlike the aeroplane data, these data are more scattered. The autocorrelogram (Figure 7.12d) shows that even for points only a few metres apart there is no significant correlation.

These results seem strange until one remembers that the aperture of the hand-held scanner was only 15 cm x 15 cm; i.e. it sampled an area only 1/100 of

227

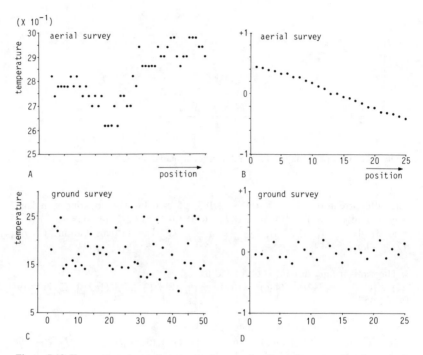

Figure 7.12 Transects and sample autocorrelograms for thermal scanner data from the Ir. A.P. Minderhoudhoeve Experimental Research Farm, Swifterbant, the Netherlands. A & B: estimated equivalent temperature as recorded by airborne DAEDALUS scanner (transect and autocorrelogram, respectively). C & D: estimated ground temperature recorded by hand-held thermal scanner (transect and autocorrelogram, respectively).

that sampled by the airborne scanner. The hand-held scanner was reacting to temperature variations caused by the low sun on the sides of the ploughed furrows – it was sampling a different phenomenon from that sampled by the scanner in the aeroplane. There was little point in calibrating the aeroplane data against the 'ground truth' because the two measurements refer to different spatial scales.

7.3.6 Considerations of non-stationarity

Whether spatial variation can be regarded as meeting stationarity assumptions is often a matter of scale. If we consider Figure 7.13, it will immediately become obvious that the concept of stationarity must be related to the scale of the observations. If we consider the trace as a whole, then the assumption of second-order stationarity appears plausible. But if we were to sample intensively between

228

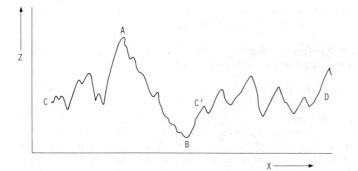

Figure 7.13 Stationarity is a question of scale. While the total transect, C–D, could reasonably be said to be 'stationary', the section A–B, if analysed separately, would have a clear trend.

points A and B we would be dealing with a trend. If we look at the section C'–D in detail, we could regard the variation as second-order stationary at the same scale as that used for the trend of section A–B.

Four important aspects of non-stationarity can be recognized:
 – non-normal distribution of the variable being studied
 – non-stationarity of the mean
 – non-stationarity of the variance
 – anisotropy (the variations are dependent on direction).

Non-normal distributions are common for many environmental properties, e.g. soil properties such as moisture tension. Transforming the variable by taking square roots, logarithms, logit or arc-tangents may help to transform non-normal distributions to approximations of the normal distribution (see Section 2.4).

Non-stationarity of the mean can be removed by 'pre-whitening'. This can mean subtraction of the original values from the fitted trend to leave the residuals, or by subtracting class means. All these techniques essentially attempt to filter out long-range variation. By implication, it is hoped that these components of long-range variation can easily be 'explained' in terms of easily recognizable aspects of the environment. Another common method to remove trends is 'differencing'. The initial series $Z(x_i)$ is replaced by $Z(x_i)-Z(x_i+h)$, where h is the sampling interval. It is usually tacitly assumed that these first-order differences are normally distributed with mean 0 and variance σ^2. Short-range variations can be removed by 'smoothing' (i.e. by using moving averages).

Non-stationarity of the mean leads to problems when estimating the covariance and the autocorrelation, because both estimates depend on a constant mean. Non-stationarity of the variance leads to a meaningless expression for the autocorrelation, $r(h) = C(h)/C(0)$, because the denominator is not constant.

The semivariance does not have either of these problems because it relies only on the differences between successive points. Consequently, when non-stationarity of the variance is expected, it is advisable to work with the semivariance instead of the covariance. If the mean can be assumed to be stationary, or the series

229

can be transformed to give a stationary mean, the assumption of second-order stationarity can be replaced by a weaker hypothesis known as the intrinsic hypothesis.

The intrinsic hypothesis assumes that the increment $Z(x+h)-Z(x)$ has mean zero and a finite variance that does not depend on x, i.e.

$$\text{var}[Z(x+h) - Z(x)] = E[Z(x+h) - Z(x)]^2 = 2\,\gamma(h) \qquad \text{Equation 7.17}$$

Note that the second-order stationarity implies the intrinsic hypothesis but the reverse is not so. Here, the second-order stationarity is limited to the increments of the random variable $Z(x)$. If data follow the intrinsic hypothesis, the semivariogram does not necessarily have an asymptote as in Figure 7.10.

7.3.7 The semivariogram, its properties and uses

Because of its ease of calculation and lack of problems when the variance is non-stationary (non-stationarity of the mean can be adjusted for), the semivariance is a better tool for describing the spatial variation of regionalized variables than the autocorrelation. The semivariogram has two main applications:
- optimum interpolation
- structure recognition.

The ideal semivariogram of a second-order stationary process has the form shown in Figure 7.14a. The semivariance rises gradually to a point, called the range, at which it levels out. This level is called the *sill*, and is theoretically equal to the variance of the series. The range is the distance within which sample points are spatially dependent. The presence of the sill, and a constant variance at lags greater than the range means that observations separated by distances greater than the range can be treated as being statistically independent. The physical implication is that the sample spacing is too large to resolve any pattern or structure in the attribute being studied. Consequently, sampling intervals that are greater than the range should *not* be used for mapping, unless there is very clear evidence from the external appearance of the landscape that sensible delineations can be made. For other applications, such as estimating the mean of an area, however, samples should be spaced *more widely* than the range, to avoid the effects of spatial correlation. If the data are not second-order stationary, but conform only to the intrinsic hypothesis, then the semivariogram will have no clear sill (Figure 7.14d).

Theoretically, the semivariogram should pass through the (0,0) point, because differences between points and themselves are always zero. Very often, the semivariogram appears to cut the $\gamma(h)$ axis at a positive value of $\gamma(h)$. This means that at the shortest sampling interval (lag $= 1$) there is a residual variation that is random and not spatially correlated. This nugget variance (the term comes from gold mining), shown in Figure 7.14a, is the sum of two components, the measurement errors associated with estimating $Z(x)$, and all sources of unseen variation that occur between the sampled points. The nugget variance is equivalent to the term ε'' in Equation 7.5.

Many experimentally determined semivariograms do not show a simple monotonic linear increase of semivariance with lag, but the experimental estimates of semivariance are scattered along a definite region. It is necessary both for the analysis of structure and for the use of the semivariogram for interpolation mapping (see Section 7.4) to fit a mathematical model to the experimentally observed data. The model can be fitted to the data using either least squares or maximum likelihood criteria; Webster & Burgess (1984) explain why they prefer to use maximum likelihood criteria.

Commonly used theoretical semivariogram models have been classified by Journel & Huijbregts (1978) into three classes.

Models with a sill (also called transition models)
 – Spherical model (Figure 7.14b)

$$\gamma(h) = c_0 + c_1 [(3h/2a) - 0.5 (h/a)^3] \text{ for } 0 < h < a$$

$$= c_0 + c_1 \qquad\qquad \text{for } h \geqslant a \qquad\qquad \text{Equation 7.18}$$

where
a is the range
h is the lag
c_0 is the nugget variance
$c_0 + c_1$ equals the sill.

 – Linear model with sill (one-dimensional series only, Figure 7.14a)

$$\gamma(h) = c_0 + bh \qquad\qquad \text{for } 0 < h < a$$

$$= c_0 + c_1 \qquad\qquad \text{for } h \geqslant a \qquad\qquad \text{Equation 7.19}$$

 – Exponential model (Figure 7.14c)

$$\gamma(h) = c_0 + c_1 [1 - \exp(-h/a)] \qquad\qquad\qquad \text{Equation 7.20}$$

 – Gaussian model

$$\gamma(h) = c_0 + c_1[1 - \exp(-h^2/a^2)] \qquad\qquad\qquad \text{Equation 7.21}$$

Models without a sill
 – Linear model

$$\gamma(h) = c_0 + bh \qquad\qquad\qquad\qquad\qquad \text{Equation 7.22}$$

 – Logarithmic model

$$\gamma(h) = c_0 + c_1 \log_e h \qquad\qquad\qquad\qquad \text{Equation 7.23}$$

231

Model with zero range or 100% nugget effect
This model (Figure 7.14f) is all sill, any structure having been unresolved at the scale of sampling.

$$\gamma(h) = c_0 \qquad\qquad\qquad \text{Equation 7.24}$$

The spherical model and the exponential model have been found to be most useful for fitting the experimentally observed semivariograms of soil data (Webster & Burgess 1984), followed by the linear model without sill for situations in which the variance is not stable (intrinsic hypothesis). Data that vary smoothly, such as ground-water levels and land forms, often have semivariograms with an inflection that can best be modelled by the Gaussian model.

In some situations the sampling may have detected two separate scales of

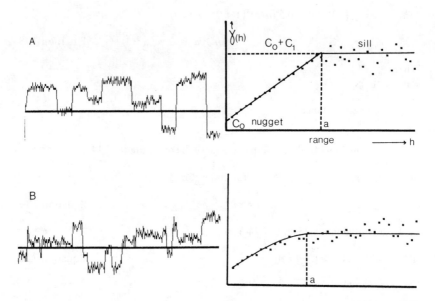

Figure 7.14 Some forms of semivariograms (right) and the possible kinds of spatial variation they describe (left). A: abrupt boundaries at discrete, regular spacings, or range a. The typical semivariogram model in one dimension is linear with a sill. Then the nugget variance describes the residual between boundary variation. B: the same as A, but with no single clearly defined distance between abrupt changes. The range, a, is the distance at which the semivariance ceases to rise further. The spherical model is best for this kind of variation (also for A type in two dimensions). C: abrupt changes occur at all distances, with spacings between changes distributed according to the Poisson distribution. The exponential model is best for this kind of variation. D: a linear trend gives a semivariogram that increases in steepness with increasing lag. E: a periodic signal results in a cyclic semivariogram. F: structureless variation or 'noise' results in a semivariogram that is 100% nugget variance.

232

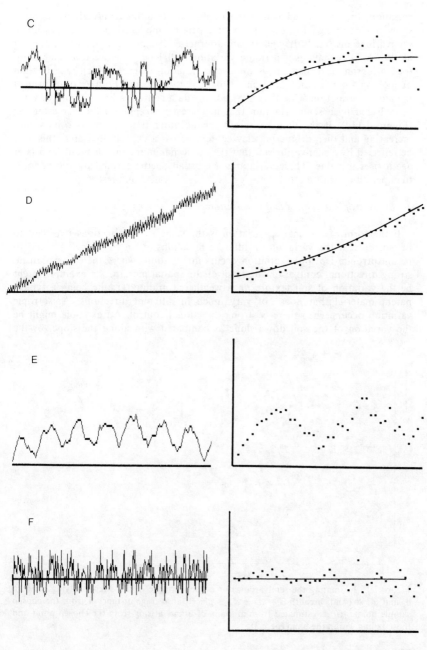

variation, the one nested within the other. The resulting semivariograms often show a clear range for each scale and can be modelled by a set of simple semivariograms (cf. Burrough 1983b; Journel & Huijbregts 1978; McBratney et al. 1982). Data having a periodic variation will have a semivariogram or autocovariogram that also shows the cyclic variation (Figure 7.14e). One method of looking for periodic variations in data is through the Fourier transform of the autocovariogram (Chatfield 1981). If the data contain a linear trend, this will be reflected by a semivariogram that increases in steepness with increasing lag (Figure 7.14d). If the data contain a second-order trend, such as occurs with increasing and then decreasing clay content across a depositional area, this may be reflected by a semivariogram that rises to a maximum and then falls and later again rises in value. These variations are usually better termed pseudo-periodic than periodic because often there is no true cyclic variation present.

7.3.8 Isotropic and anisotropic variation

If the form of the spatial variation is the same in all directions it is said to be isotropic. If the variation in different directions is not the same it is said to be anisotropic. Isotropic variation occurs in situations where there has been no strong directional control on the form of the spatial pattern. An example might be the variation of soil texture on a weathered granite, granite being a typical parent material that does not vary much in different directions. Anisotropic variation occurs when there is strong directional control. An example might be the variation of the soil down-slope as compared with along the slope, or the

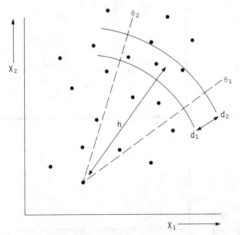

Figure 7.15 Estimating the semivariance or autocovariance in the direction lying between θ_1 and θ_2. A band, breadth $d = d_2 - d_1$, is placed at average distance h from the chosen sample point. $\gamma(h)$ is estimated from all pairs of points arising from the chosen point and those falling within the band (here 3).

variation across the strike of dipping sediments compared with those parallel with the outcrops. If variation is isotropic, the estimates of semivariance should yield similar values for sill, range and nugget when computed for all directions. If the variation is anisotropic, the semivariograms will reflect this by returning different ranges, and possibly also different sills and nuggets for each direction considered.

Anisotropy can be looked for in two ways. The simplest method is to lay out several linear transects over the landscape in different directions (two perpendicular to each other or three at 120° are common practice) and compute the semivariance in each direction. The resulting semivariograms can then be compared. If they show different ranges, then they may reflect anisotropy and should be interpreted separately. If they return the same results then they can be combined. Care should be taken when interpreting anisotropy because true anisotropy should not be confused with the normal variation in the form of the semivariogram that comes from sampling. Semivariograms can also be computed from arrays of points on grids or other irregular two-dimensional arrays. The semivariance is calculated for point-pairs falling in given distance classes (instead of a fixed distance as with linear transects) and for given directions (see Figure 7.15).

7.3.9 The semivariogram as a structural tool

Because the semivariogram can reveal information about the range, sill and nugget semivariances, it is a useful tool for analysing the spatial pattern of soil properties, particularly when their patterns are not clearly revealed in the landscape. The semivariogram is thus a tool that complements the existing landscape tools of aerial-photo and landscape analysis.

7.3.10 An example of using semivariograms to analyse spatial variation in soil

This example uses the data from northern Italy described in Subsection 7.2.7 to illustrate the effects of anisotropy and variation at several scales on the form of the semivariograms.

Semivariances were calculated for two directions, Direction 2 (Figure 7.6), parallel with the boundary (interpreted from aerial photos) between the upper and lower terrace, and Direction 3, almost perpendicular to this boundary. These semivariograms are given for the total area and for the upper and lower terrace separately in Figures 7.16a, b and c, respectively.

The semivariogram for the total area (Figure 7.16a) computed in Direction 2, parallel with the terrace boundary, keeps increasing, at least to a lag exceeding 1800 m. The semivariogram computed perpendicular to the boundary shows a rise to a lag of 600 m, remains constant for another 350 m and then declines. Clearly, the variation in percentage clay is markedly different in the two directions, as can be seen from the maps already shown in Figures 7.7 and 7.8. The semivariogram for north–south variation (Direction 2) continues to increase, suggesting a north–south trend, which perhaps can be explained in terms of differential deposition as one goes from the higher (north) to lower (south) parts

235

of the area. The variation in Direction 3 appears to be affected by two components, i.e. variation within the lower terrace and variation between the terraces.

To see what the effects of between-terraces differences are on the form of the semivariograms, the semivariances were estimated for each terrace separately. The results are illuminating. Variation of percentage clay on both directions on the

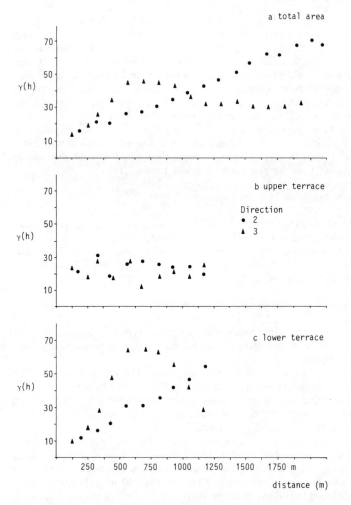

Figure 7.16 Sample semivariograms of the percentage clay of the top-soil of the data from northern Italy estimated for a: the total area. b: the upper (Pleistocene) terrace only. c: the lower (Holocene) valley area only. Semivariograms are estimated for two directions (see Figure 7.6).

upper terrace results in semivariograms that show 100% nugget variance (Figure 7.16b). The implication is that the variation of top-soil clay on the upper terrace cannot be distinguished from that of a non-correlated random variable – i.e. at this sample spacing (125 m) one cannot distinguish between long-range variation ('signal' variation) and unrecognizable short-range variation, or noise. As everything looks the same – equally noisy – the upper terrace area sampled is said to be homogeneous with respect to a sampling interval of 125 m. In contrast, the variation of percentage clay on the lower terrace (Figure 7.16c) shows even more clearly than for the total area the effects of anisotropy. Note, too, that the absolute values of the semivariance on the lower terrace are estimated to be higher than those on the upper terrace, or for the area as a whole. The semivariograms on the lower terrace show the trend (Direction 2) that occurs because of deposition parallel to the Pecora River and to the terrace boundary, and a pseudo-cyclical variation (Direction 3) perpendicular to the stream. Both patterns can clearly be seen in the contour plot.

7.4 Spatial interpolation

7.4.1 Introduction

As was shown in Section 7.2, trend surfaces can be used to interpolate the value of an attribute at unvisited sites. This was seen to be not entirely satisfactory, however, largely because of the problems of linking the polynomials to an understanding of physical process, the long-range nature of the surfaces, and the effect that a few extreme values could have on local estimates.

7.4.2 Weighted moving averages

As we have already noted, observations located close together tend to be more alike than observations spaced further apart, even if they happen to be within the same delineation of the same mapping unit. It is natural to feel that the contribution that a given sample point makes to an average interpolated value at an unvisited site should be weighted by a function of the distance between that observation and the site. So we can compute a weighted moving average:

$$\hat{Z}(x_0) = \sum_{i=1}^{n} w_i Z(x_i) / \sum_{i=1}^{n} w_i \qquad \text{Equation 7.25}$$

where the weights w_i are given by a function $\varphi(d(x, x_i))$ of the distance between x_0 and x_i. A requirement for this function is that $\varphi(d) \rightarrow 0$ as $d \rightarrow \infty$, which is given by the commonly used reciprocal or negative exponential functions, $\exp(-\alpha d)$ and $\exp(-\alpha d^2)$ for some constant $\alpha > 0$. Perhaps the most common form of $\varphi(d)$ is the inverse squared distance weighting:

$$\hat{Z}(x_j) = \sum_{i=1}^{n} [Z(x_i) / d_{ij}^2] / \sum_{i=1}^{n} (1 / d_{ij}^2) \qquad \text{Equation 7.26}$$

where the x_j are the points at which the surface is to be interpolated (usually

237

these points lie on a regular grid and d_{ij} is the distance between x_i and x_j).

The form of the interpolated surface can depend on the function, or on the parameters of the function used (Figure 7.17), and on the size of the domain or window, i.e. the area from which points are used for calculating the weighted average (Ripley 1981). Ripley also points out that the values estimated by moving averages are susceptible to clustering in the data points, and also to whether the observations are affected by a planar trend. He describes some ways of avoiding these drawbacks using methods of distance-weighted least-squares.

The size of the domain not only affects the value of the average value estimated at a point, but also controls the amount of computer time required for interpolation. Usually the size of the domain or window is set to include a certain minimum and maximum number of data points in an effort to balance computational efficiency against precision. The number of points used, n, may vary between 4 and 22, but is usually in the range 8-12, particularly if the original data lie on a regular grid. Alternatively, one may be able use a fixed number of data points to compute the average. This will make no difference to estimates from data on a regular grid, but when data are irregularly distributed, each interpolation will be made using a window of different size, shape and orientation.

The clustering problem in irregularly spaced data points has partly been solved by Shepard (1968), Giloi (1978) and Pavlidis (1982).

The interpolated values at the grid points can be displayed directly as grey-

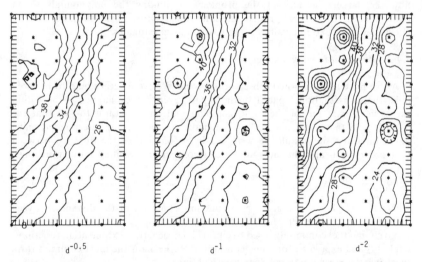

$$d^{-0.5} \qquad d^{-1} \qquad d^{-2}$$

Figure 7.17 These plots show how the value of the weighting parameters in inverse-distance interpolation affects the appearance of the resulting maps. The more quickly the value of the inverse-distance function declines with distance, the greater the likelihood that the map will show 'spotty' peaks and hollows centred on data-points (asterisks). There is no a-priori way of knowing which value of the weighting function is 'best'.

238

scale raster maps (e.g. the maps shown in Figures 7.7 and 7.8) or on a colour raster screen or printer. The interpolated grid cell values can be used as an overlay in a raster database for cartographic modelling. Alternatively, another computer program can be used to thread isolines (lines of equal value) through the interpolated surface; these isolines can be drawn with a pen plotter.

7.4.3 Optimum interpolation methods using spatial autocovariance

The local weighted moving average interpolation methods discussed above provide reasonable results in many cases, but they leave several important questions open. In particular, the methods generate the following important uncertainties:
- how large should the domain or window be?
- what shape and orientation should it have for optimum interpolation?
- are there better ways to estimate the w_i weights than as a simple function of distance?
- what are the errors (uncertainties) associated with the interpolated values?
These questions led the French geomathematician Georges Matheron and the South African mining engineer D.G. Krige to develop optimum methods of interpolation for use in the mining industry generally referred to as 'kriging'. The methods have recently been used in ground-water mapping, soil mapping and related fields. The following discussion draws on material presented by Webster & Burgess (1983) and Giltrap (1983). Other useful references are Corsten (1985) and Webster (1985).

An optimum policy is a rule in dynamic programming for choosing the values of a variable to optimize a particular criterion function (Bullock & Stallybrass 1977). The interpolation methods developed by Matheron are optimum in the sense that the interpolation weights w_i are chosen to optimize the interpolation function, i.e. to provide a Best Linear Unbiased Estimate (BLUE) of the average value of a variable at a given point.

In kriging, the essential step is that the model fitted to the experimental semivariogram is used to estimate the weights w_i. The semivariogram is thus the source of the information about the shape and size of the window and the weights that must be used to estimate the value of Z at an unsampled point x_0. We have:

$$\hat{Z}(x_0) = \sum_{i=1}^{n} w_i Z(x_i) \qquad \text{Equation 7.27}$$

with $\sum_i w_i = 1$. The weights w_i are chosen so that the estimate $\hat{Z}(x_0)$ is unbiased and the estimation variance σ_e^2 is less than for any other linear combination of the observed values. The minimum variance of $\hat{Z}(x_0) - Z(x_0)$ is obtained when the w_i satisfy (for all i)

$$\sum_{j=1}^{n} w_j \gamma(x_i, x_j) + \psi = \gamma(x_i, x_0) \qquad \text{Equation 7.28}$$

where the quantity $\gamma(x_i, x_j)$ is the semivariance of Z between the sampling points x_i and x_j, and $\gamma(x_i, x_0)$ is the semivariance between the sampling point x_i and the unvisited point x_0.

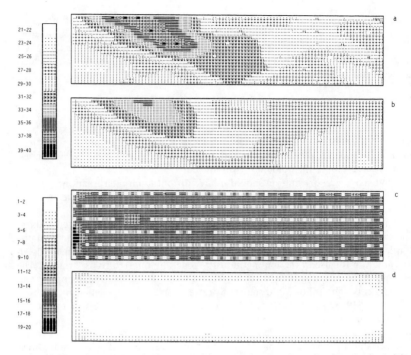

Figure 7.18 The results obtained by mapping percentage clay top-soil using a: point kriging. b: block kriging. The variances for point and block kriging are shown in c and d, respectively. Note that the point-kriging estimates are zero at the data points and how missing values affect the local estimation variances.

Both these quantities are obtained from the fitted semivariogram. Further, ψ is a Lagrange multiplier (an additional constant introduced for technical reasons). Equation 7.28 and the equation $\Sigma_{i=1}^{n}\ w_i = 1$ together form $n+1$ linear equations with $n+1$ unknowns and can thus be solved for the unknown w_i $(i = 1,...,n)$ and ψ by the standard methods of linear algebra. The minimum variance is then

$$\sigma_e^2 = \Sigma_{j=1}^{n}\ w_j\ \gamma(x_j,\ x_0) + \psi \qquad \text{Equation 7.29}$$

Corsten (1985) presents an elegant formula for predicting $Z(x_0)$ without introducing the Lagrange multiplier.

Kriging is an exact interpolator in the sense that when the equations given above are used, the interpolated values, or best local average, will coincide with the values at the data points. In mapping, values will be interpolated for points

240

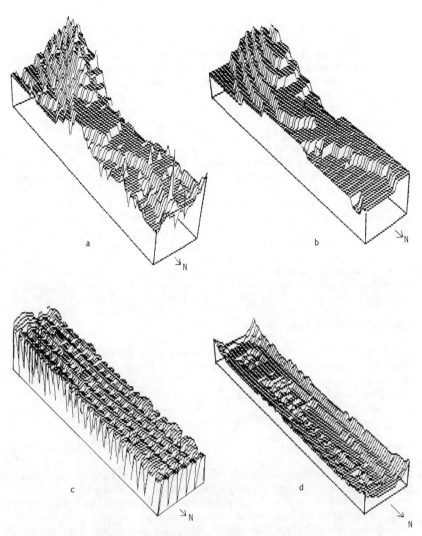

Figure 7.19 The results are the same as Figure 7.18, but they are displayed as three-dimensional surfaces.

on a regular grid that is finer than the spacing used for sampling, if that was determined by a grid. The interpolated values can then be converted to a contour map using the techniques already described. Similarly, the estimation error σ_e^2 can also be mapped to give valuable information about the reliability of the interpolated values over the area of interest.

Clearly, kriging fulfils the aims of finding better ways to estimate interpolation weights and to provide information about errors. The resulting map of interpolated values may not be exactly what is desired, however, because the point-kriging (or simple-kriging) equations 7.28 and 7.29 imply that all interpolated values relate to an area or volume that is equivalent to the area or volume of an original sample. Very often, as in soil sampling, this sample is only a few centimetres across. Given the often large, short-range nature of soil variation or nugget variance, simple kriging results in maps that have many sharp spikes or pits at the data points (Figures 7.18a and 7.19a). The error map will also return atypical values at the data points because there the error variance will equal the nugget variance – elsewhere it will be larger (Figures 7.18c and 7.19c). This can be overcome by modifying the kriging equations (see Webster 1985) to estimate an average value of Z over a block B (a block is a region of a particular size). This is useful if one wishes to estimate average values of Z for experimental plots of a given area, or to interpolate values for a relatively coarse grid matrix that could be used with other rasterized maps in overlay cartographic modelling (Burrough 1986). Figures 7.18b and 17.19b show the results of point kriging and block kriging on the same data.

The estimation variances obtained for block kriging are usually substantially lower than for point kriging, and the resulting smoothed interpolated surface is free from the pits and spikes of point kriging (Figures 7.18a, 7.18d).

7.5 Optimum sampling

A frequent problem in environmental surveys is to determine for any given area the size of the sample needed to estimate the average value of an attribute Z within given limits of reliability. To date, the most usual approach has been to assume that the map unit within which the area lies is homogeneous; the variance of the attribute is then estimated by locating a number of sample sites to obtain an unbiased estimate. The classical formula

$$n = t_\alpha^2 s^2 / (g - \mu)^2 \qquad \text{Equation 7.30}$$

is then used to estimate the sample size n needed to estimate the mean μ of the area within limits $\mu - g$ and $\mu + g$ (t_α is Student's t at probability level α).

Very often, particularly with soil properties, the value of n has been so large that investigators have been forced either to make do with too small a sample for comfort, to incur large costs for analysis, to bulk samples or to abandon the study. Webster & Burgess (1984b) have shown that providing the semivariogram of the attribute of interest is known, considerable reductions in the size of n may be possible. The secret of their method is to choose a combination of sample-point numbers and sample-point spacing in such a way as to minimize the kriging variance for the block of land for which the estimate of the mean value of the attribute is required. This is explained in the remainder of this section.

Consider the case in which it is required to sample a 1 ha square block to obtain the best estimate of the mean of an attribute Z. The maximum dimension (the diagonal) of a 1 ha square block is 141 m, so a model of the semivariogram is needed that extends to at least that distance. For practical purposes, sampling on a square grid may often prove the easiest way to gather a set of unbiased samples. There are two variable quantities:

- the number of samples to be located within the block (for samples on a square grid, $n = 1, 4, 9, 16, 25$)
- the interval or grid spacing between them.

For $n = 9$ samples, the nine points can be placed within the 1 ha square in many ways (Figure 7.20). The estimation variance can be calculated using the semivariogram for each configuration for a range of sample sizes, n, to yield a series of graphs of kriging estimation variance against grid spacing. Figure 7.21 presents an example of the results that can be obtained when a linear model is fitted to the experimental semivariogram; use of other semivariogram models affects the precise form of the relation between kriging estimation variance and the grid spacing (see Webster & Burgess 1984b) but there is always a decline in the estimation variance with increasing grid spacing, followed by a rise. For the linear model, the minimum variance is found to occur when each sample

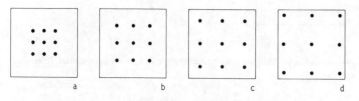

Figure 7.20 Four different ways of locating 9 points in a square.

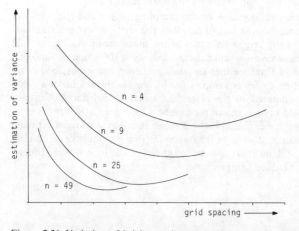
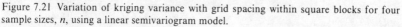

Figure 7.21 Variation of kriging variance with grid spacing within square blocks for four sample sizes, n, using a linear semivariogram model.

243

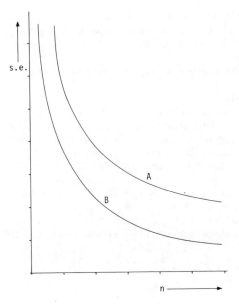

Figure 7.22 The variation of standard error versus sample size. A: classical model. B: kriging model.

point coincides with the middle of each sampling cell (Figure 7.20c). If another semivariogram model is chosen the optimum sample spacing will be slightly, but not greatly, different from that obtained with the linear model.

Knowing the optimum configuration of the sampling points and the resulting kriging estimation variances allows one to calculate the relation between standard error and sample size for the kriged estimates of the block mean. As Figure 7.22 shows, providing that the semivariogram does not only show 100% nugget variance, use of the kriging method results either in more precision being obtained from a given number of samples or fewer samples need to be taken to achieve a given level of precision, as compared to the classical estimates. Webster & Nortcliff (1984) report that 2-3 fold savings in sample numbers have been achieved in some situations. Also, Webster & Burgess (1984b) demonstrate that the kriged estimate is also more precise than that of a bulked sample from an equivalent number of observations, but this claim must also be evaluated in the light of the reduced costs of analysis that the bulked samples incur.

7.6 Discussion

Autocovariation studies have two important advantages. First, they give insight into the spatial structure of the phenomenon of interest within a given area or map unit. Second the information coming from autocorrelation studies can be used for comparing the spatial variation of related phenomena, for optimum interpolation mapping and for optimizing the size of samples for estimating average values of attributes within defined areas.

The major disadvantages of kriging, and of autocovariation studies in general, relate to the large amount of calculation necessary to obtain results, but thanks to the widespread use of computers this is becoming a non-problem. What is more important is the size and cost of the sample needed to establish the autocorrelation structure of the phenomenon of interest, and so determine the form of the semivariogram or autocorrelogram. It seems that at least 50-100 samples are needed if the semivariogram is to be estimated with any reasonable degree of confidence (see Taylor & Burrough 1986). Another major problem is that of choosing a suitable semivariogram model. Besides these, there is also the problem of the data containing various sources of non-stationarity that may seriously affect the estimates of the semivariogram.

7.7 Bibliography and historical overview of interpolation

The interpolation of isopleths (lines of equal value) has long been commonplace in topographical mapping in which the variation of relief is usually represented by contour lines. Height contours on topographical maps are drawn either by direct survey in the field or from measurements made on stereoscopic pairs of aerial photographs. In both situations, the surveyor can see directly the surface to be mapped; his job is to represent this surface as accurately as possible. In contrast, in many situations in ecology, soil survey or environmental science, the surface to be mapped cannot be seen directly. Instead it must be sampled at several discrete points at which the value of the environmental variable in question is measured. The problem is then to choose a suitable spatial model to which the data obtained from the point observations can be fitted and that can be used to estimate the value of the variable at any unvisited point. This spatial model may take many forms, but for many applications in ecology and environmental science, stochastic models of spatial variation are most suitable. Once a model has been chosen and fitted to the experimentally observed data, its surface is mapped using isopleths (contours).

Fitting mathematically defined surfaces to point data requires a considerable amount of calculation, so although the principles were known in the 1930s and 1940s, there were few serious attempts at quantitative interpolation before computers were widely available. In the 1960s, effective and easily used interpolation programs became widely available (Shephard 1968, Laboratory for Computer Graphics 1968). The development and extension of multiple regression allowed regression surfaces (trend surfaces) to be computed and mapped easily (Davis 1986). Dissatisfaction with the conceptual models underlying trend surfaces and

245

other simple interpolation techniques used in mining geology led Matheron (1971) to develop his theory of regionalized variables, which is based on a study of spatial autocorrelation functions (see also David 1977 and Journel & Huibregts 1978). His work has been taken up and extended to problems in soil science by Webster (1984; 1985) and his co-workers in England, and Nielsen and others in the United States (Nielsen & Bouma 1985). Overviews of the application of various methods of spatial analysis to a wide range of problems can be found in the literature: for geology, Agterberg (1982); natural resource survey, Burrough (1986); geography, Lam (1983); and statistical ecology, Ripley (1982).

Most of the methods described in this chapter have been programmed in PASCAL for 16-bit personal computers operating under MS-DOS or IBM PC-DOS. Details of these programs can be obtained from the Geographical Institute, Utrecht University, P.O. Box 80.115, 3508 TC Utrecht, the Netherlands.

7.8 Exercises

Exercise 7.1   Trend, autocorrelogram and semivariogram

Table 7.1 shows the value of the thickness (cm) of two sedimentary layers measured at 20 points spaced 100 m apart along a transect. For each transect:

Table 7.1 Thickness of two sedimentary layers at 20 points spaced 100 m apart along a transect.

Site No.	Thickness layer 1	Thickness layer 2
1	5.2	8.3
2	5.4	9.8
3	6.1	11.2
4	3.1	12.2
5	5.4	14.2
6	6.4	13.2
7	7.2	12.4
8	6.5	11.5
9	4.5	8.7
10	3.3	9.8
11	2.2	10.5
12	5.5	12.6
13	6.2	13.5
14	6.6	14.9
15	5.5	13.8
16	4.3	14.3
17	6.3	15.7
18	5.5	16.7
19	4.2	15.4
20	4.8	17.2

246

- plot the raw data on graph paper
- compute and plot the autocorrelogram and the semivariogram for the first 10 lags.

Examine the semivariograms and autocorrelograms carefully. Note which of the two appear to show spatial dependence, and which appear to be statistically homogeneous at this sample spacing. Note that the 5% point for r with 18 degrees of freedom is 0.444.

Examine the data for linear trends by computing the regression of the thickness of each layer against the sampling position. See whether the residuals from regression display dependence by plotting them and also by computing their semivariogram and autocorrelogram.

Exercise 7.2 Anisotropy

The data in Table 7.2 show the percentage clay found in the subsoil (at depth 3 m) as measured along two transects located at right angles to each other across an

Table 7.2 Percentage clay in subsoil along two transects N–S, E–W on an old alluvial plain.

Site No.	clay N–S	clay E–W
1	29	20
2	28	22
3	29	21
4	33	21
5	38	32
6	33	36
7	36	22
8	28	16
9	25	24
10	24	27
11	22	26
12	23	31
13	30	29
14	34	38
15	38	35
16	40	19
17	46	27
18	44	62
19	51	52
20	45	61
21	46	39
22	43	34
23	36	32
24	32	26
25	25	22

old alluvial plain. Each transect has 25 points spaced 25 m apart. For each transect:

- plot the raw data on graph paper
- compute the semivariogram.

Examine the semivariogram and attempt to reconstruct the directions along which the spatial patterns at 3 m depth are aligned.

7.9 Solutions to exercises

Exercise 7.1   Trend, autocorrelogram and semivariogram

Table 7.3 gives the mean, standard deviation, minimum and maximum of each layer. Table 7.4 gives the estimated semivariances for the first ten lags. When the data are plotted as semivariograms, the first layer shows no clear range and sill, but a (pseudo)cyclic variation with a wavelength of about 7 lags (700 m). In contrast, the semivariogram of the second series has a range of about 4 lags (400 m), with thereafter a steady increase in semivariance, which reflects the trend in the original data. Table 7.5 gives the results of estimating the autocorrelation for the two series. The presence of a weak linear trend in the second series is shown by the slope and correlation coefficient of a linear regression of the original data on sample spacing (Table 7.6). As shown in Table 7.7, linear detrending has little effect on the results obtained from Series 1, but for Series 2, linear detrending has resulted in a semivariogram with a clear range and sill. Now try using polynomial regression to remove periodicity from Series 1 and see what the resulting form of the semivariogram is.

Table 7.3 Summary statistics.

Column	Mean	s.d.	Min	Max
Series 1	5.21	1.30	2.20	7.20
Series 2	12.80	2.57	8.30	17.20

Table 7.4 Semivariance analysis.

Lag	Series 1	Series 2
1	1.171	0.993
2	2.177	2.289
3	2.781	4.022
4	2.484	5.525
5	1.762	5.811
6	1.640	5.626
7	1.390	6.024
8	1.952	6.490
9	1.959	7.670
10	1.492	8.211

Table 7.5 Autocorrelation.

Lag	Series 1	Series 2
1	0.151	0.346
2	-0.130	0.237
3	-0.287	0.090
4	-0.217	-0.020
5	0.024	-0.026
6	0.083	0.015
7	0.134	0.001
8	-0.020	0.001
9	0.021	-0.101
10	0.031	-0.148

Table 7.6 Linear regressions for detrending.

	Slope	Intercept	R
Series 1	-0.015	5.363	-0.067
Series 2	0.330	9.332	0.760

Sampling	Series 1			Series 2		
position	Z	$\hat{Z}$	$Z-\hat{Z}$	Z	$\hat{Z}$	$Z-\hat{Z}$
1	5.200	5.349	-0.149	8.300	9.661	-1.361
2	5.400	5.334	0.066	9.800	9.991	-0.191
3	6.100	5.319	0.781	11.200	10.321	0.879
4	3.100	5.305	-2.205	12.200	10.651	1.549
5	5.400	5.290	0.110	14.200	10.981	3.219
6	6.400	5.276	1.124	13.200	11.311	1.889
7	7.200	5.261	1.939	12.400	11.641	0.759
8	6.500	5.246	1.254	11.500	11.970	-0.470
9	4.500	5.232	-0.732	8.700	12.300	-3.600
10	3.300	5.317	-1.917	9.800	12.630	-2.830
11	2.200	5.203	-3.003	10.500	12.960	-2.460
12	5.500	5.188	0.312	12.600	13.290	-0.690
13	6.200	5.174	1.026	13.500	13.620	-0.120
14	6.600	5.159	1.441	14.900	13.949	0.951
15	5.500	5.144	0.356	13.800	14.279	-0.479
16	4.300	5.130	-0.830	14.300	14.609	-0.309
17	6.300	5.115	1.185	15.700	14.939	0.761
18	5.500	5.101	0.399	16.700	15.269	1.431
19	4.200	5.086	-0.886	15.400	15.599	-0.199
20	4.800	5.071	-0.271	17.200	15.929	1.271

Table 7.7 Semivariances of detrended data (i.e. residuals of Table 7.6).

Lag	Series 1	Series 2
1	1.171	0.893
2	2.174	1.975
3	2.777	3.347
4	2.490	4.457
5	1.765	4.576
6	1.638	4.164
7	1.383	3.948
8	1.941	3.661
9	1.957	3.278
10	1.474	2.666

Exercise 7.2   Anisotropy

Table 7.8 gives the estimated semivariances along the two directions sampled. The sill reached in Direction 1 is about 95% clay^2, while that in Direction 2 is 140% clay^2, with a suggestion of periodicity. The range in Direction 1 is about 6 lags, or 150 m; in Direction 2 it is about 3 lags, or 75 m. From these data it can be concluded that there is a strong anisotropy in the pattern of the buried sediments, possibly of the form shown in Figure 7.23.

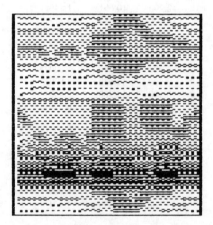

Figure 7.23 Possible pattern of anisotropy.

Table 7.8 Estimated semivariances (% clay^2).

Lag	N–S	E–W
1	10.208	59.542
2	22.696	110.826
3	44.659	142.432
4	64.452	157.095
5	79.225	162.425
6	93.974	140.684
7	95.500	147.389
8	100.176	99.000
9	95.906	122.031
10	92.767	178.300

8 Numerical methods in practice: case-studies

8.1 Spatial variation in vegetation, soil and precipitation across the Hawkesbury Sandstone Plateau in New South Wales, Australia: an analysis using ordination and trend surfaces

P.A. Burrough

8.1.1 Data collection and analysis

The Hawkesbury Sandstone Plateau, New South Wales, Australia, is an area of great scenic beauty. Through its poor, infertile soils, the plateau has been little developed for agriculture and much of it remains under natural vegetation. This case study was carried out on a small part of the Hawkesbury Sandstone Plateau, measuring roughly 25 km x 10 km from Barren Grounds Nature Reserve in the east, to Fitzroy Falls in the west (Figure 8.1). The eastern boundary of the study area is a steep cliff, where the plateau ends, some 5 km inland from the Tasman Sea. This topography is the cause of intense orographic rainfall, which varies from about 2500 mm per year in the east, falling off to about 1500 mm in the west. Although the plateau is deeply dissected by ravines, its surface is gently rolling, with an average altitude of about 600 m, which does not vary within the area studied. The vegetation, which at the time of the study was largely undisturbed, varies considerably over the area, being a low heath in the east and a closed forest in the west. There is a parallel variation in the soil, from peaty podzols in the east, through leached podzols to leached brown soils in the west.The picture is complicated by short-range relief differences; the lower parts of the plateau are occupied by swampy peat soils supporting sedges and reeds.

Early workers thought that the vegetation differences across the Hawkesbury Sandstone might be due to phosphate levels in the soil, but they had been unable to show any relation between vegetation and soil chemistry (Beadle 1954). The large trend in rainfall was only perceived in the mid-1970s, by which time data had been collected from sufficient meteorological stations for several years. The aim of this case study, reported fully by Burrough et al. (1977), was to examine the long-range spatial variations of soil, vegetation and climate to see if they covaried. Because the survey effort was limited to three people for two weeks in a not-very-accessible area, the survey had to be carefully planned and executed.

To avoid problems that could be caused by sampling the short-range variations arising from differences in relief on the plateau, the survey concentrated on 55 topographically similar sites on upper to middle slopes that were chosen by a

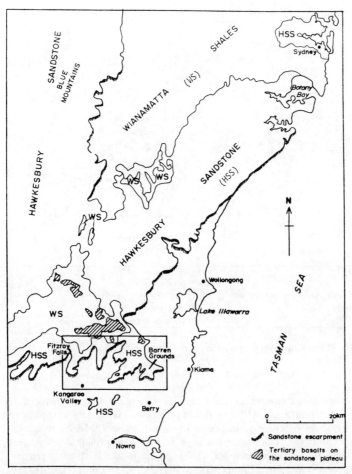

Figure 8.1 The location of the Barren Grounds study area in the state of New South Wales, Australia.

stratified random technique. The soil profile was sampled for chemical and physical properties at each site. The vegetation survey required more effort. It was decided to concentrate on the composition of the vegetation. At 18 sites (a subset of the 55), 10 m x 30 m quadrats were laid out (with the long axis down-slope) and the plant species occurring were recorded. In total, 111 different species were recorded from the 18 sites. In addition to the soil and vegetation data, climatic data were obtained for 8 meteorological stations located on or near the study area. Figure 8.2 shows the distribution of the vegetation types and the vegetation sample sites.

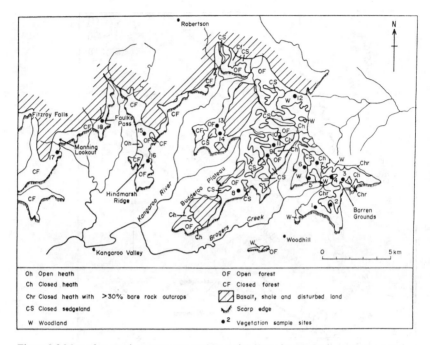

Figure 8.2 Map of vegetation types prepared by aerial-photo interpretation, and the location of vegetation sampling sites.

Because of the large number of soil variables, the soil data were reduced to their principal components, and the vegetation presence–absence data were reduced to the first two axes by correspondence analysis, otherwise known as reciprocal averaging (Section 5.2). The vector loadings for the first principal component of the soil data (which accounted for 45.7% of the total variation in the soil data) showed that it had important contributions from all soil properties strongly affected by leaching or soil moisture: colour hue, value and chroma, thicknesses of the surface organic layer, leached A2 and colour B horizons, field-moisture content and percentage organic carbon. Sites in the east had large positive scores, and sites in the west had large negative scores, for the first principal component.

Trend surfaces were computed for the height of the vegetation (observed at 55 sites), the first principal component scores for the soil data, the site scores from the first correspondence analysis axis of the vegetation data and for the mean annual rainfall. The height of vegetation and mean annual rainfall were transformed to logarithms before computing the trends. All trend surfaces were two-dimensional linear surfaces, except for rainfall, which was computed as a multiple linear regression on location and altitude. Figure 8.3 shows the results.

Figure 8.3 shows that all the linear trend surfaces slope in the same way, from

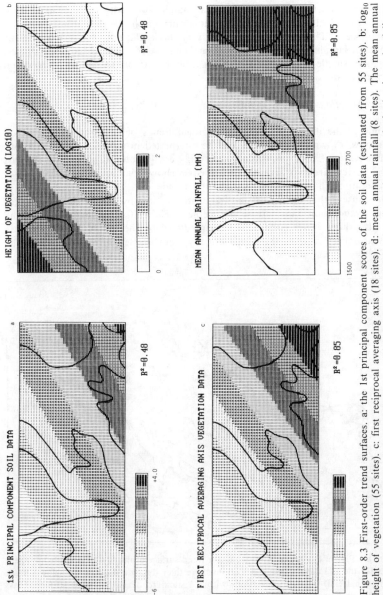

Figure 8.3 First-order trend surfaces. a: the 1st principal component scores of the soil data (estimated from 55 sites). b: log$_{10}$ height of vegetation (55 sites). c: first reciprocal averaging axis (18 sites). d: mean annual rainfall (8 sites). The mean annual rainfall was first computed using a multiple regression on log$_{10}$ data; the plot shown here was obtained using a local linear trend computed from the interpolated data at all 55 soil sampling sites.

255

south-east to north-west, a direction that matches well with the prevailing wind direction of clockwise rotating depressions that reach the area from the South Pacific Ocean. The trends appear to be very stable and not very sensitive to the number of sample points used to compute them. The R^2 values (Figure 8.3) suggest that these simple, linear trends explain a large part of the soil and vegetation variation across the plateau.

8.1.2 Deviations from the trends

Some of the deviations from the regional trends of vegetation site and soil site scores, shown in Figure 8.3, can be interpreted in terms of local variations in effective rainfall or soil drainage. As Figure 8.4 shows, most sites in the east of the Barren Grounds area near scarp edges show deviations that reflect wetter soil conditions than predicted by the simple trend.

The cause is that these parts are frequently covered by fog and low cloud that reduce evaporation and increase site wetness. The cliff edges in the west of the area also show wetter conditions than predicted by the trends. This is because the scarp there receives the winds directly from the sea through a gap in the

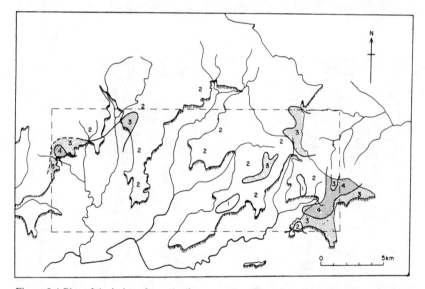

Figure 8.4 Plot of deviations from the linear trend surface of the first principal component scores for the soil data. Class 1: deviations more than 1 standard deviation below surface (drier than trend). Class 2: deviations within 1 standard deviation below surface. Class 3: deviations within 1 standard deviation above surface. Class 4: deviations more than 1 standard deviation above surface (wetter than trend). The broken-lined rectangle outlines the area for which the trend surfaces were calculated (see Figure 8.3).

hills above Brogers Creek, and also because fog and cloud tends to hang about the tops of the 200 m high waterfalls that plunge off the plateau nearby. Drier than average sites appear to be the result of better than average soil drainage; they appear to occur near places where the sandstone is more jointed and less massive than usual.

Other deviations from the trends could be less easily explained in terms of better drainage or less evaporation. Two sites on the Budderoo Plateau showed large deviations from the vegetation site score trends, but not from those of the soil. Inspection of the field data showed that the only unusual aspect of these sites was that the sand fraction was extremely fine. A resurvey confirmed these texture anomalies and also suggested that the sites lay in the vicinity of remnants of a clay soil which was possibly derived from a former covering of the Wianamatta Shales.

8.1.3 Conclusions

Trend surfaces of ordination axes indicated that the long-range changes of soil and vegetation covaried in a similar way as the mean annual precipitation, and that it is reasonable to see a causal link between them, at least for the upper to mid-slope sites on the plateau. This interpretation is supported by those deviations from the regression surface that also suggested a definite link between moisture, vegetation and soil. Other deviations from the regression that could not easily be explained by the information available caused the authors to revisit the study area, where they acquired new insights. It should be realized that these interpretations can only be made for two reasons – one is the dominance of the climatic variation across the area; the other is the deliberate way in which the field survey was set up to exclude short-range variations that would have interfered with the aim of the study.

8.2 Processing of vegetation relevés to obtain a classification of south-west European heathland vegetation

J.A.F. Oudhof and A. Barendregt

8.2.1 Introduction

In 1981 the section Vegetation Science of the Department of Plant Ecology, University of Utrecht, made a study of the heathlands of Spain and Portugal. This resulted in 200 new relevés for the north of Spain and 250 for Portugal. In addition, the department had at its disposal unpublished data collected by J.T. de Smidt in 1966: 135 relevés for Spain and 100 for Portugal.

After these data had been classified by traditional hand methods (Mueller-Dombois & Ellenberg 1974), we encountered difficulties in fitting the vegetation types into the classification system for the heathlands of the Iberian Peninsula (Rivas-Martinez 1979), and, consequently, in interpreting the results. Since it was not clear from the outset whether the data gave new syntaxonomic information about the Iberian heathlands, which might make the classification system of Rivas-Martinez inapplicable, we decided to process the 700 relevés together with all the existing Spanish and Portuguese heathland data from the literature, comprising 600 relevés. In the classification system for European heathlands, the heathlands of southern England and Ireland, Brittany, Les Landes, northern Spain and Portugal are considered to belong to the same unit (alliance), the *Ulicion minoris* P.Duvign. 1944. To obtain an overall picture of the south-west European heathlands we decided to collect from the literature as many relevés as possible for the Atlantic heathlands around the Bay of Biscay. Therefore we were also able to process other unpublished relevés collected in England, Ireland, Brittany and Les Landes (300) by J.T. de Smidt.

The purpose of the investigation was to make, with computer techniques such as cluster and ordination analysis, a syntaxonomic division of the main groups of the south-west European heathlands on the basis of 1000 new relevés and as many relevés as possible from the literature. We shall discuss the problems and the decisions we had to make to arrive at this classification.

8.2.2 Preparation of the set of data

In collecting the data, problems arose straight away about how heathland vegetation was to be described syntaxonomically. To the south, the Atlantic heathlands gradually change into a Mediterranean dwarf shrub vegetation. At many places there is a transition to woodlands and moors. The coastal heathlands show transitions to other formations. We decided to adopt a very broad definition for heathland vegetation. This enabled us to make reasoned descriptions later and reduced the likelihood that we would omit information as result of dubious decisions made at the outset.

The relevés of the same vegetation types from the literature show large differences in the number of species reported. Many of the discrepancies are due to the inclusion

258

or exclusion of mosses, liverworts and lichens. Others are the result of variation in the surface area of the relevés, from 5 m² to 200 m². Sometimes it seemed as if only a few dominant species had been recorded. Sometimes seasonal aspects, such as plants that flower in the spring, had been overlooked, because of the time of the year at which the relevé was made (Dupont 1975). These differences had to be taken into account during the processing and the interpretation of the results.

The data from the literature had to be processed into one set of data in a uniform way. As most data were presented in tables, we decided to input them as tables; these were later amalgamated into one table. Because the various tables from the literature included many synonyms, the species names had to be standardized. The angiosperms and ferns were named according to Flora Europaea (Tutin et al. 1964-1980), the mosses according to Corley et al. (1981), the liverworts according to Grolle (1976) and the lichens according to Hawksworth et al. (1980). Every species or subspecies was assigned an easily recognizable code of 8 letters, consisting of the first four letters of the generic and the species name. The names of the cryptogams were all assigned the letter X, so they could be easily collected together by computer and rearranged alphabetically or deleted if necessary. Another difficulty met when arranging a uniform set of data is the cover-abundance scale. The scale must be standardized, and we opted for a ten-part scale, 0-9 (van der Maarel 1979b; see also Subsection 2.4.2), which is easy to input as a single column. Every input table was given a number by which its geographical origin could be immediately deduced.

After we had checked the tables we had input for possible errors, the individual tables were amalgamated into one table with a program that also provided an alphabetical species list with frequencies of occurrence. This again lets us check the species code for inputting errors. In this phase species and subspecies can be united. After these procedures we prepared one table in Cornell condensed format, a table representing species and relevés by numbers. This format is suitable for processing by the computer programs TWINSPAN (Hill 1979b), DECORANA (Hill, 1979a) and FLEXCLUS (van Tongeren 1986). For the south-west European heathlands this table consisted of 3000 relevés and 1700 species.

8.2.3 Operations applied to the set of data

There are three distinct phases in the processing of the set of data (see flow diagram, Figure 8.5):
- operations applied to the total set of data resulting in a division into smaller main groups
- processing of the main groups until homogeneous clusters are formed
- a synoptical characterization of the clusters and after that a grouping of these clusters.

259

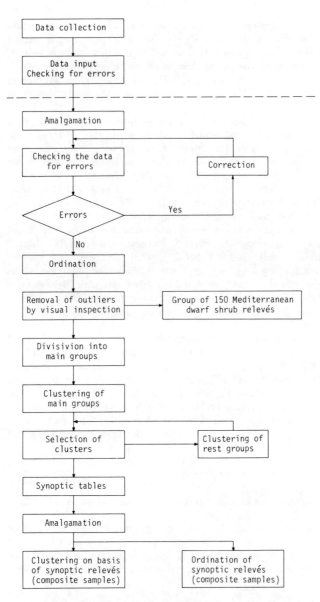

Figure 8.5 Flow diagram of operations applied to the set of data of south-west European heathlands.

Operations applied to the total set of data

As the amount of main memory used by most computer programs to handle a set of data of this volume is too large, it was necessary to split up the set into smaller parts. We tried to form main groups in such a way that relevés strongly resembling each other were gathered into one group, preferably with as little overlap as possible between the main groups. To reveal the major variation in the set of data, ordination (correspondence analysis, Section 5.2) was carried out with the program DECORANA (Hill 1979a). This program was specially adapted to the size of the total set of data. The ordination of all relevés showed a clear distinction on the first two ordination axes (Figure 8.6) between about 150 relevés, which appeared to be real Mediterranean dwarf shrub vegetation, and the other relevés of more Atlantic regions. The relevés for the Atlantic heathlands appeared to be grouped in a gradient from southern and dry to northern and wet. The 150 relevés for the Mediterranean dwarf shrub vegetation were temporarily put aside. It was known that the remaining relevés varied predominantly along a north–south geographical line.

The TWINSPAN method (Hill 1979b) was used to divide the set of data into main groups. On our computer this program could handle, depending on the

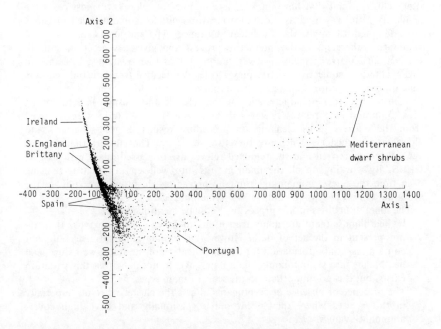

Figure 8.6 Ordination diagram of a correspondence analysis of 3000 relevés of the south-west European heathlands. Eigenvalue Axis 1: 0.758. Eigenvalue Axis 2: 0.597.

261

number of species, a maximum of 1000 relevés. Using the information about the most important variation in the set of data, we composed, from south to north, a series of sets of data for 600-700 relevés. To prevent that similar relevés would end up in other main groups by the partitioning of the total set of data, we made overlapping sets of data. The results made it possible to split up the total set of data into main groups. It turned out, for instance, that some heathland relevés for Spain and Portugal, which contained for example *Erica umbellata* L. , differed from the other relevés in the set of data. The coastal and wet heathlands were also found to form quite distinct groups. The main groups contained at most 500 relevés, which could all be handled by the clustering program FLEXCLUS (van Tongeren 1986).

Processing of the main groups to homogeneous clusters

From the main groups, homogeneous clusters were formed using the clustering program FLEXCLUS. FLEXCLUS is an interactive clustering program: a quick initial clustering is followed by an interactive process of either relocations or fusion of the clusters that resemble one another most closely, or division of the most heterogeneous clusters. This process is stopped when clusters are formed that are internally as homogeneous as possible, that differ as much as possible from other clusters, and that have an adequate number of relevés. In most cases the similarity ratio was used as a distance measure, but for some detailed divisions we attempted to apply the Euclidean Distance. The output of the program summarizes the steps in the interactive process and gives a vegetation table of the last cluster arrangement. The vegetation table is then evaluated by hand. In this evaluation, subjective criteria play a role. We decided to admit only clusters that differed in more than one differentiating species. Moreover, all relevés in the cluster had to be characteristic for that cluster and clusters had to consist of a minimum of 5 relevés. Relevés that did not fit the criteria were removed from the clusters. This cleaning up procedure results in well-separated and characterized clusters, which are therefore set aside. The clusters that are not yet well characterized and the removed relevés are processed again with FLEX-CLUS. By working up all the main groups into well-separated clusters, using each time the relevés remaining from the preceding clustering, we split up the set of data into 211 clusters, plus several remaining relevés.

The main difficulties in this process will now be mentioned:
- In heathlands, there are many transitions to other vegetation types that are not present in the set of data. However, these marginal influences do not contain essential information for the classification of the south-west European heathlands into syntaxonomic main types. We do not deny that this variation exists, but it is not important on the scale of main types.
- The presence or absence of cryptogams affects the clustering result, especially in those cases where they occur only occasionally and have high cover-abundance values.
- Furthermore it is striking that the relevés of one author of a heathland vegetation usually produce one particular set of clusters, but other relevés of the same

vegetation by other authors produce different clusters. This may be the result of the highly subjective Braun–Blanquet method or of a very limited geographical distribution of the different heathland communities.

Synoptical characterization and grouping of clusters

From the 211 clusters obtained, we computed a synoptic (cf. Section 6.6) table using the presence classes of Braun–Blanquet (1964), and a table of mean cover-abundance values in each cluster; the means are expressed on the ten-part scale mentioned above. These synoptic tables can be processed again with a clustering program to obtain a classification on a higher level. The main aim here is to classify these synoptic clusters in a small number of homogeneous groups. We attempted to characterize the results of the classification in terms of groups of differentiating species. The first division in main groups has nothing to do with the groups that result from this classification (the main groups in the first step of the analysis were only needed to get smaller groups that could be handled by the clustering programs).

8.2.4 Preliminary results

The result of a detrended correspondence analysis (Hill 1979a) on the set of data of 211 clusters (composite samples) is shown in Figure 8.7. The classification of these clusters resulted in 8 main groups. They are plotted on the first and second ordination axes. The first axis can be interpreted as a wet–dry gradient, with the samples containing *Erica tetralix* and *Sphagnum* species on the left side and the driest Portuguese heathlands on the right side. The interpretation of the second axis is more complicated, but this axis seems to be related to the availability of minerals or nutrients. The 'rich' coastal heathlands are all on the upper side of the graph, while the 'poor', wet heathlands (moorlands), the *Erica australis* heathlands (Iberian Mountains) and the *Calluna vulgaris* & *Erica cinerea* heathlands are on the lower side. The *Ulex* dominated vegetation types show an intermediate position. This might be explained by the nitrification that takes place in the root bulbs of these papilionaceous plants. The 8 main groups are briefly characterized in Table 8.1.

8.2.5 Concluding remarks

The whole process described above makes use of calculation procedures that one could call 'objective', but these so-called objective procedures are followed by several subjective decisions. The processing of such a large set of data is a cyclical process, each cycle consisting of the same actions, involving calculations, studying the output and interpreting it (Figure 8.5). Interpretation, in particular, takes a lot of time. There is also a great deal of administration involved. Since the input tables, species, relevés and clusters all comprise numeric information it is absolutely essential to adopt a systematic working procedure.

What has been achieved? In 9 months, with approximately 2 people working on the project, 3000 relevés were processed and reduced to 211 clusters, which together give a main syntaxonomic classification of the south-west European heathlands. A total review of all the important literature was made, showing the geographical, ecological and syntaxonomic relations of the vegetation types. In short, an enormous amount of clearly structured information has been assembled. This information may serve as a basis for future research on heathland ecology.

Table 8.1 Preliminary scheme showing the 8 main types of heathlands, with their geographic and floristic characteristics. E = England; I = Ireland; F = France; S = Spain; P = Portugal.

Type of heathland	Distribution and features	Some characterizing species
I Wet heathlands	E,I,F,S,P	*Erica tetralix* *Molinia caerulea* *(Erica ciliaris)*
a) very wet	I,F,S,P	*Narthecium ossifragum* *Drosera rotundifolia* *Sphagnum species*
b) moderate wet	E,I,F,S,P	*Carex binervis* *Ulex gallii* *Erica cinerea*
II Coastal heathlands	E,F,S	*Holcus lanatus* *Festuca rubra* *Lotus corniculatus* *Daucus carota*
a) England (left side)		*Scilla verna* *Plantago coronopus*
b) France (right side)		*Armeria maritima* *Leontodon taraxacoides*
c) Spain (all over the group)		*Cirsium filipendulum* *Erica vagans* *Brachypodium pinnatum*
III *Ulex galli* heathlands	E,F,S	*Ulex gallii* *Erica ciliaris*
IV *Calluna-Erica cinerea*	E,F	High cover-abundance values for *Calluna vulgaris* & *Erica cinerea*. Some clusters *Ulex minor*. Relatively more moss species with high cover-abundance values except with respect to the wet Irish clusters, which are all in Group I. *Hypnum cupressiforme* *Dicranum scoparium* *Pleurozium schreberi* *Pseudoscleropodium purum*

V	Erica umbellata-Ulex minor heathlands	S,P	Erica umbellata Halimium alysoides Ulex minor
VI	Dry Erica umbellata Chamaespartium heathlands	P	Erica umbellata Halimium alysoides Ulex micrantha Chamaespartium tridentatum Genista triacanthos Tuberaria globularifolia
VII	Very dry Erica umbellata-Halimium ocymoides heathlands	P	Erica umbellata Halimium alysoides Halimium ocymoides Erica australis Chamaespartium tridentatum
VIII	Erica australis heathlands	S,P	Erica australis Deschampsia flexuosa Vaccinium myrtillus

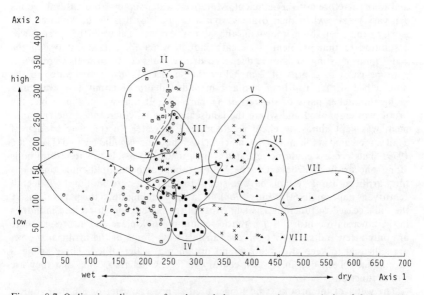

Figure 8.7 Ordination diagram of a detrended correspondence analysis of 211 relevés (composite samples) of the south-west European heathlands. Eigenvalue Axis 1: 0.618. Eigenvalue Axis 2: 0.365. For a brief description of the groups see text. Symbols: +: Ireland; □◨■England; ○◐●France; ×⋇ Spain; △ Portugal. Where groups overlap a different symbol to indicate membership of the different groups is used. Group I, II: +,○,□, ×, △ Group III: ●◨⋇ Group IV: ●■ Group V, VI, VII, VIII: ×△

8.3 Restoration and maintenance of artificial brooks in the Veluwe, the Netherlands

R.H.G. Jongman and T.J. van de Nes

8.3.1 Introduction

Water has always been used intensively in the Netherlands, not only in the polders, but in the higher parts of the country too. Since the Middle Ages the waters of the Veluwe have been used to drive water-mills. To provide water for this purpose, an artificial brook system – the 'spreng' – was developed. Despite it being artificial, it can be considered as a semi-natural brook system. Recent developments have led to a change in the historical functions of nature in society, causing it to diminish in extent and deteriorate in quality. One of the semi-natural ecosystems that have deteriorated in this process is the spreng.

The artificial brooks studied in this project are situated in the Veluwe, in the province of Gelderland. The Veluwe is a system of hills that are outwash deposits of the Saalian Ice Age. Now they rise to a height of 60-100 m. Originally, water-mills were situated along natural lowland streams. They used undershot wheels and were therefore not very efficient. Moreover, water supply from natural brooks can vary greatly when their drainage basin is small, as it is in the Veluwe. That is why people set about improving the system. An artificial well (the spreng) was a solution to that problem. To create enough water for permanent use of the water-mills, existing wells were deepened or new, fully artificial wells were made, in some cases to depths of 7 m below the surface. To bring the water to the mill, a bed with a width of about 2 m and a depth of about 1 m was built along the higher parts of the terrain (Figure 8.8). If that was not possible, the brook was embanked and where the landscape gradually descended the embankment was kept almost level, to create a head of water. In this way enough of a fall was created at the mill so that efficient overshot wheels could be used (IJzerman 1979). This process could be repeated, which made it possible to site several mills along one artificial brook. On the Veluwe about two hundred mills on about forty artificial-brook systems have existed. Most of them were built in the sixteenth century, as paper-mills (Hardonk 1968). Because of scaling-up in the paper industry the mills ceased to function as such. Most of them were converted into laundries in the nineteenth century. But here too scaling-up took place and at present there are only a few mills left that are used in this way. Some of the artificial brooks are still used to keep castle moats and ornamental waters filled. Three artificial brooks have been dug in the last century to feed water to a canal and they are still in function.

The water of the brooks is no longer used as process water or a power source; most of the brooks no longer have any industrial or agricultural significance. Even though a few of the brooks still fulfil such functions, their ecological and cultural–historical significance is greater. Since the 1950s, land-use and the use of ground and surface water has intensified in the Netherlands – also in the Veluwe. Those changes are a threat to the existence of the brooks. The major threat is

266

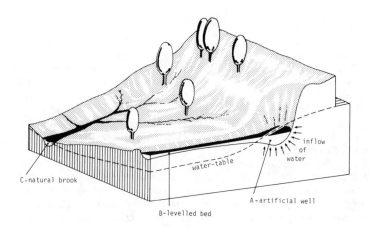

inflow
of
water

water-table

C-natural brook

A-artificial well

B-levelled bed

Figure 8.8 Landscape with an artificial brook.

a decrease in the water supply: because of a falling water-table, between 1968 and 1978 several brooks had problems with their water supply (Kant, 1982). As artificial brooks are partly situated above the water-table, the uphill parts of the brooks must be impermeable. Loss of the water may cause a dry brook down the valley. Leakage can be caused by running dry for a long period of time (irreversible desiccation) or by incorrect maintenance, for example through using machines on a vulnerable reach. Because of the weak current in the well and the levelled reaches silting up can occur easily. The welling-up and transport of water will then be hindered. Artificial brooks, being semi-natural elements, are therefore dependent on management by man.

The loss of function resulted in a loss of management, sometimes for a period of about 30 years. To restore and conserve these brook systems insight into the functioning of an artificial-brook system is essential. To study the problems involved, in 1981 the Province of Gelderland set up the working party 'Sprengen en Beken op de Veluwe' [Artificial and natural brooks in the Veluwe]; the study is split into the four parts (Jongman & van de Nes 1982):

– geohydrological research, to gain insight into the water supply and water quality of the brooks in relation to management of ground-water in general
– ecological research, to determine ecological characteristics of the brooks to lay down guiding principles for specific management
– maintenance analysis, to understand the actual state of maintenance and to be able to define future possibilities
– policy analysis, to find out how the involved interests are related and if and what changes in policy are necessary to restore and preserve the brooks.

Here we will emphasize the ecological research. The results have been the reference for the process of decision-making on priorities for restoration and a guideline for maintenance. In the following subsections, the methods used in the project and the results are discussed.

267

8.3.2 Ecological inventory

For the brooks in this project aquatic macrofauna are the most characteristic organisms. This is because we are dealing with shady streaming waters in which microflora and macroflora play a minor role. We therefore compiled data on the macrofauna of the brooks.

Part of the data come from recent publications and part were collected for this project. The available environmental data differ greatly in exactitude and completeness. They can be characterized as either binary (yes/no) or ordinal (ranking or estimation scale).

Nominal variables are environment, substrate and water type. Ordinal variables are velocity of the stream, depth, width, thickness of detritus layer, shadow, amount of iron and maintenance. Many environmental variables are lacking in the data from the literature. Except for maintenance, most of the ordinal variables are merely estimated values, not measured values. In data from the literature, fluctuation in season and internal differences between the sites are mostly not taken into account.

The sites from the data from the literature differed in size and sometimes in kind of organisms sampled. It was therefore necessary to select the samples on their usefulness for this research. Based on this, sites were selected for a supplementary sampling programme. The 181 sites from the literature were supplemented in this project by 93 sites.

8.3.3 Data analysis

Before data can be analysed they must be screened for peculiarities. Four samples were omitted because they had an extremely low quantity of organisms. In the original data the number of specimens per taxon varied from 1 to 999. However one specimen of *Polycelis felina* has another significance than one specimen of *Tubifex* species. A transformation into biomass would probably be the best approach but it was not possible in this project. Instead we used a simple ordinal transformation (Table 8.2), comparable with the transformation of van der Maarel (Table 2.1), which is used in hydrobiological research in the Netherlands.

It turned out to be necessary to fuse some taxa because of

Table 8.2 Ordinal transformation of data used in the spreng project (Cuppen & Oosterloo 1980).

Quantity of organisms	Transformation value
1 - 3	1
4 - 10	2
11 - 20	3
21 - 49	4
≥50	5

- lack of restricted knowledge about the taxonomy of some invertebrates;
- apparent lack of acquaintance of some researchers whose data we used with new taxonomic literature;
- special attention given by a few researchers to certain taxa, resulting in the determination of an organism at the level of species, genus and even family (for instance *Lumbricus variegatus*, *Lumbricus* sp. and *Lumbriculidae*, respectively), depending on the source of the data in the literature. Fusion was necessary in 24 cases. After fusion, 475 taxa were left. A great number of these had a low frequency of occurrence.

It was decided not to use taxa occurring in less than two or three percent of the sites (depending on the bulk of the data) for classification. This saves much computing time and hardly influences the final classification. The initial computing was done with all data. Based on these results the data were divided into subsets that were used in further analysis.

8.3.4   Results

The total set of data was divided in three main groups with cluster analysis. Named after their most important representative, the main groups were called *Chironomus* (C), *Asellus* (A) and *Gammarus* (G). All groups were analysed separately. The main group, *Chironomus*, was subdivided into two types, *Asellus* into four types and *Gammarus* into five types (Popma 1982). We restrict here

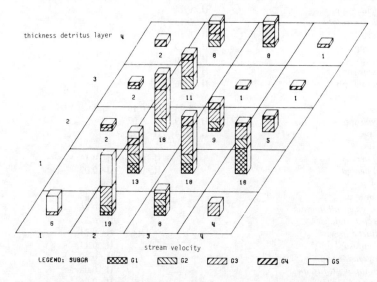

Figure 8.9 Relation of Gammarus types with environmental variables, stream velocity and thickness of detritus layer.

Table 8.3 Synoptic table of the *Gammarus* classification. Presence class (I-V) and a dominance index are indicated. The dominance index is the weight of a taxon in a cluster multiplied by its frequency in the cluster.

Type	G5	G1	G3	G2	G4
number of sites	20	19	57	26	31
mean number of taxa	21.4	20.3	21.5	26.7	19.3
Gammarus pulex	V 4.45	IV 1.66	V 4.33	V 2.36	II 0.13
Prodiamesa olivacea	IV 1.54	II 0.29	IV 0.79	V 2.01	III 0.29
Dicranota spp.	III 0.32	III 0.44	III 0.27	II 0.09	II 0.08
Paratendipes spp.	III 0.67				
Paratanytarsus spp.	II 0.28				
Herpobdella octoculata	IV 0.88	II 0.06			
Dendrocoelum lacteum	II 0.19	II 0.07			
Gammarus fossarum	III 1.50	IV 2.25			
Polycelis felina		V 2.26			
Dixa maculata		IV 1.19			
Agapetus fuscipes		III 0.33			
Eusimulium costatum		II 0.32			
Micropterna sequax		II 0.15			
Simulium ornatum		II 0.14			
Thaumaleidae spp.		II 0.14			
Crunoecia irrorata		II 0.13			
Elmis aenea		II 0.10			
Tipulidae spp.		II 0.10			
Baetis vernus		II 0.06			
Psychodidae spp.		II 0.06			
Krenopelopia spp.		II 0.06			
Helodes spp.		V 1.46	II 0.10		
Sericostoma spp.		III 0.49	II 0.07		
Halesus spp.		II 0.04	II 0.25		
Anabolia nervosa			II 0.28	II 0.14	
Lymnaea perega			II 0.12	II 0.09	
Tubificidae spp.				III 0.48	
Paratendipes gr. *albimanus*				III 0.46	
Micropsectra trivialis				II 0.37	
Stempellinella brevis				II 0.21	
Sialis fuliginosa				II 0.14	
Ceratopogonidae spp.				II 0.09	
Rheocricotopus gr. *fuscipes*				II 0.08	
Phaenopsectra spp.				II 0.06	

Species	1	2	3	4	5
Limnephilidae spp.				III 0.46	II 0.12
Corynoneura spp.				II 0.06	III 0.64
Tanytarsus spp.				III 0.60	III 0.85
Agabus spp.				II 0.15	II 0.10
Helophorus brevipalpis				II 0.07	II 0.10
Zavrelimyia spp.				II 0.24	II 0.34
Heterotanytarsus apicalis				II 0.12	II 0.33
Heterotrissicladus marcidus				II 0.18	II 0.36
Polypedilium gr. *nubuculosa*				II 0.29	II 0.25
Micropsectra gr. *praecox*	V 4.25		II 0.10		
Procladius spp.	III 0.47			III 0.63	
Hydracarina spp.	III 0.56				II 0.15
Lumbriculidae spp.		II 0.08		III 0.39	
Oligochaeta spp.	V 2.75	II 0.06	II 0.10		
Eukiefferiella gr. *discoloripes*	II 0.38	II 0.31	II 0.16		
Limnophila spp.	II 0.06	II 0.12	II 0.16		
Glossiphonia complanata	III 0.55	II 0.07		II 0.05	
Sialis lutaria	IV 1.24		III 0.27	II 0.13	
Chironomus spp.	II 0.28			II 0.09	II 0.06
Micropsectra spp.		II 0.24		IV 1.92	II 0.30
Conchapelopia spp.			III 0.22	V 2.14	II 0.21
Apsectrotanypus trifascipennis			III 0.48	IV 1.21	II 0.26
Nemoura cinerea			III 0.74	IV 0.93	V 2.73
Chaetopteryx spp.	III 0.36	III 0.58	III 0.63	III 0.46	
Pisidium spp.	III 0.50	II 0.12	II 0.26	II 0.12	
Polycelis tenuis/nigra	II 0.28	II 0.10	II 0.07	II 0.12	
Asellus aquaticus	IV 1.57		III 0.38	III 0.26	II 0.18
Limnephilus lunatus	IV 1.01		III 0.75	IV 0.78	II 0.10
Macropelopia spp.	IV 0.65		II 0.23	II 0.25	IV 0.99
Velia spp.		II 0.11	IV 0.68	III 0.31	IV 0.81
Anacaena globulus		II 0.06	II 0.09	II 0.05	II 0.08
Orthocladiinae spp.		II 0.29	II 0.11	II 0.20	II 0.33
Brillia modesta		III 0.22	III 0.21	II 0.19	III 0.28
Plectrocnemia conspersa		III 0.35	II 0.13	II 0.10	IV 1.34

our attention to the *Gammarus* group. The clustering (Ward's method, Subsection 6.2.5) of the *Gammarus* data led to a classification in five clusters (G1-G5). The results, which represent different brook ecosystems, were summarized in a synoptic table (Table 8.3). Only taxa that occurred in 20% or more of the sites were included in the table.

The results of the classification were related to the environmental variables. The five *Gammarus* types differ in stream velocity and thickness of detritus layer as shown in Figure 8.9. Type G1 is characterized by a small detritus layer and a high stream velocity. G5 is restricted to low stream velocity and a small detritus layer. G4 is mainly found in combination with a thick detritus layer. G2 and G3 are related to intermediate situations, with G3 related to higher stream velocities than G2. G3 appears to be intermediate between G1 and G2. G1 occurs in natural brooks. G5 originates from ponds and castle moats that are connected with the natural brooks. The types G2, G3 and G4 are found in the artificial brooks. Because of its position between G1 and the other brook communities, the G3 type is considered as the optimum community in an artificial brook system.

8.3.5 *Application of the results*

The results of the data analysis served as a basis for a list of priorities for restoration. Moreover, the typology of brook communities is still used as a reference for water management.

Before it was possible to indicate priorities for the restoration of the brooks, some problems had to be solved:
– which methods can be used for decision-making?
– which criteria can be used for decision-making?

The Province of Gelderland has a great deal of experience with this sort of multicriteria decision-making (van de Nes 1980). To assist policy-makers, a model has been developed based on the methods of Saaty (1977), called GELPAM (Gelders Policy Analysis Model), which compares the consequences of alternative plans, projects or objects on the basis of several criteria. It is described in detail in the literature by Ancot & van de Nes (1981).

In this project the criteria were water supply, maintenance and ecological characteristics. Hydrological research by Kant (1982) provided a number of sub-criterea for water supply: ground-water, hydraulic head (height of the water-table in relation to the floor of the brook), natural changes in ground-water levels, lowering of levels by pumping, and loss of water in the transporting reaches of the brook system. There were two sub-criteria for maintenance: the size of the areas to be maintained and costs of the restoration. An ecological characterization of the brooks could have been made for the actual situation or for a situation that could have been expected if all environmental factors were optimal. Emphasis was given to the actual situation, because the potential situation is given weight indirectly through the priorities among the sub-criteria of water supply and maintenance. Types G1 and G5 were only found in natural brooks. Therefore we concluded that the G3 type of the *Gammarus* group indicates the most characteristic artificial brook. In a similar way we inferred that the *Chironomus*

272

group indicates the least characteristic artificial brook. Based on this sequence, weights were given to community types and these led to weights for the results on geohydrology and maintenance. In the decision-making process, geohydrology was considered to be more important than maintenance. The factor ecology was only considered to be a weak criterion, again because it was already represented in the weights of the other criteria. Of the sub-criteria of geohydrology, the hydraulic head, as an indicator for regional ground-water potential, was considered to be the major one.

In the beginning, each sub-criterion was considered separately. All projects were

Table 8.4 Priorities for restoration of brooks in the eastern part of the Veluwe: Group 1 has the highest priority; Group 4 has the lowest priority. A low concordance means that the project is sensitive to different weighting; a high concordance means that it is not.

No.		Project	Weight	Concordance	Group
1	6	N-Horsth. beek (N)	0.0603	0.9127	
2	11	Tongerense beek	0.0509	0.8349	
3	4	M-Heerder beek (O)	0.0486	0.9535	Group 1
4	17	Steenbeek	0.0481	0.5499	
5	8	M-Horsth. beek	0.0479	0.9513	
6	26	Eerbeekse beek (W)	0.0458	0.9262	
7	16	Beek Orderveen (Z)	0.0455	0.7046	
8	18	Koppelsprengen	0.0450	0.3566	
9	5	Z-Heerder beek	0.0449	0.5347	
10	7	N-Horsth. beek (Z)	0.0436	0.9180	
11	21	Oosterhuizenspreng	0.0420	0.4104	Group 2
12	2	N-Heerder beek	0.0405	0.2539	
13	9	E-Horsth. beek	0.0395	0.1644	
14	19	Ugchelense beek	0.0393	0.1488	
15	12	Hartense molenbeek	0.0383	0.1380	
16	1	Molecatense beek	0.0375	0.0329	
17	15	Beek Orderveen (N)	0.0335	0.2925	
18	27	L-Soerense beek (N)	0.0289	0.5432	Group 3
19	23	Loenen (Zilven)	0.0281	0.4430	
20	22	Loenen (Stroobroek)	0.0275	0.9510	
21	13	Meibeek (N)	0.0274	0.6180	
22	10	Vlasbeek	0.0267	0.9613	
23	20	Winkewijert	0.0253	0.6660	
24	3	M-Heerder beek (W)	0.0247	0.6705	Group 4
25	24	Loenen (Steenput)	0.0233	0.6166	
26	14	Meibeek (Z)	0.0197	0.7373	
27	25	Eerbeekse beek (N)	0.0167	0.9735	

Overall Concordance: 0.6641

compared in pairs for each criterion on a nine-point scale expressing relative priority. From each matrix of these paired comparisons, weights were extracted for the projects in very much the same way as scores are extracted in ordination. The sub-criteria were also compared in pairs to obtain their relative importance. By applying the same ordination method to the matrices of comparisons, we were also able to derive weights for the sub-critera. Subsequently, the weights of each sub-criterion for a project were combined with the weights of just the sub-criteria to give overall weights for the projects; the weights express the priority of the projects judged (Table 8.4). In this method, a concordance measure is also given (Ancot & van de Nes 1981): a low concordance value indicates that the weight can change if priorities among criteria are changed.

Based on this list of priorities, several brook systems were restored: three of the first group, three of the second group and one of the third group. This was implemented in part by the Polder Board and in part by private initiative. One brook system has been restored, although it was relegated to the third group, indicating that serious problems had to be solved before restoration could be completed. Because diminished extraction of ground water by a paper-mill caused the water-table in this area to rise, the water supply improved so much that it made restoration possible.

The results of this research were also used to help in the decision-making for physical planning on a local scale and to issue permits for water extraction. It is also used as a point of reference to evaluate the management of the brooks.

References

Agterberg, F.D., 1982. Recent developments in Geomathematics. Geo-Processing 2: 1-32.

Aitchison. J. & J.A.C. Brown, 1969. The lognormal distribution. Cambridge University Press, Cambridge, 176 pp.

Alderdice, D.F., 1972. Factor combinations: responses of marine poikilotherms to environmental factors acting in concert. In: O. Kinne (Editor): Marine ecology, vol 1, part 3. Wiley, New York, p. 1659-1722:

Alvey, N.G. et al., 1977. GENSTAT: a general statistical program. Rothamsted Experimental Station, Harpenden.

Ancot, J.P. & T. J. van de Nes, 1981. Integral evaluations of alternative water management scenarios in East Gelderland. In: Water resources management on a regional scale. Committee for hydrological research TNO, Proceedings and Information no 27. The Hague, p. 129-164.

Anderberg, M.R., 1973. Cluster analysis for applications. Academic Press, London, 373 pp.

Anderson, A.J.B., 1971. Ordination methods in ecology. Journal of Ecology 59: 713-726.

Anderson, S., A. Auquier, W.W. Hunck, D. Oakes, W. Vandaele & H.I. Weisberg, 1980. Statistical methods for comparative studies. Wiley, New York, 302 pp.

Austin, M.P., 1971. Role of regression analysis in plant ecology. Proceedings of the Ecological Society of Australia 6: 63-75.

Austin, M.P., 1976. Performance of four ordination techniques assuming three different non-linear species response models. Vegetatio 33: 43-49.

Austin, M.P., 1980. Searching for a model for use in vegetation analysis. Vegetatio 42: 11-21.

Austin, M.P., R.B. Cunningham & P.M. Fleming, 1984. New approaches to direct gradient analysis using environmental scalars and statistical curve-fitting procedures. Vegetatio 55: 11-27.

Baker, R.J. & J.A. Nelder, 1978. The GLIM system, Release 3. Numerical Algorithms Group, Oxford.

Ball, G.H., 1966. A comparison of some cluster seeking techniques. Stanford Research Institute, California.

Barkman, J.J., H. Doing & S. Segal, 1964. Kritische Bemerkungen und Vorschläge zur quantitativen Vegetationsanalyse. Acta Botanica Neerlandica 13: 394-419.

Barlow, R.E., D.J. Bartholomew, J.M. Bremner & H.D. Brunk, 1972. Statistical inference under order restrictions. Wiley, New York, 388 pp.

Bartlein, P.J. & T. III Webb, 1985. Mean July temperature at 6000 yr B.T. in

Eastern North America: regression equations from fossil-pollen data. Syllogeus 55: 301-342.

Bartlein, P.J., I.C. Prentice & T. III Webb, 1986. Climatic response surfaces from pollen data for some eastern North American taxa. Journal of Biogeography 13: 35-57.

Battarbee, R.W., 1984. Diatom analysis and the acidification of lakes. Philosophical Transactions of the Royal Society of London 305: 451-477.

Batterink M. & G. Wijffels, 1983. Een vergelijkend vegetatiekundig onderzoek naar de typologie en invloeden van het beheer van 1973 tot 1982 in de Duinweilanden op Terschelling. Report Agricultural University, Department of Vegetation Science, Plant Ecology and Weed Science, Wageningen. pp 101.

Beadle, N.C.W., 1954. Soil phosphate and the delimitation of plant communities in eastern Australia. Ecology 35: 370-375.

Beals, E.W., 1985. Bray-Curtis ordination: an effective strategy for analysis of multivariate ecological data. Advances in Ecological Research 14: 1-55.

Beckett, P.H.T. & P.A. Burrough, 1971. The relation between cost and utility in soil survey. IV. Comparison of the utilities of soil maps produced by different survey procedures and to different scales. Journal of Soil Science 22: 466-480.

Becking, R.W. 1957. The Zürich-Montpellier school of Phytosociology. Botanical Review 23: 411-488.

Belsley, D.A., E. Kuh & R.E. Welsch, 1980. Regression diagnostics. Wiley, New York, 292 pp.

Benzécri, J-P. et al., 1973. L'analyse des données. II: L'analyse des correspondances. Dunod, Paris, 619 pp.

Bie, S.W. & P.H.T. Beckett, 1973. Comparison of four independent soil surveys by air photo interpretation, Paphos Area (Cyprus). Photogrammetrica 29 (6): 189-202.

Böcker, R., I. Kowarik & R. Bornkamm, 1983. Untersuchungen zur Anwendung der Zeigerwerte nach Ellenberg. Verhandlungen der Gesellschaft für Ökologie 11: 35-56.

Box, G.E.P., 1976. Science and statistics. Journal of the American Statistical Association 71: 791-799.

Box, G.E.P., W.G. Hunter & J.S. Hunter, 1978. Statistics for experimenters. Wiley, New York, 671 pp.

Braun-Blanquet, J.J., 1964. Pflanzensoziologie, Grundzüge der Vegetationskunde. 3rd Edition. Springer, Vienna, New York, 865 pp.

Bray, J.R. & J.T. Curtis, 1957. An ordination of the upland forest communities of southern Wisconsin. Ecological Monographs 27: 325-349.

Brown, G.H., 1979. An optimization criterion for linear inverse estimation. Technometrics 21: 575-579.

Brown, P.J., 1982. Multivariate calibration. Journal of the Royal Statistical Society, Series B 44: 287-321.

Buckland, S.T., 1982. Statistics in ornithology. Ibis 124: 61-66.

Bullock, A. & O. Stallybrass, 1977. Fontana dictionary of modern thought. Fontana Books, London, 694 pp.

276

Bunce, R.G.H., C.J. Barr & H. Whittaker, 1983. A stratification system for ecological sampling. In: R.M. Fuller (Editor): Ecological mapping from ground, air and space, I.T.E. symposium 10: 39-46.

Burrough, P.A., 1983a. Multiscale sources of spatial variation in soil I. The application of fractal concepts to nested levels of soil variation. Journal of Soil Science, 34: 577-597.

Burrough, P.A., 1983b. Multiscale sources of spatial variation in soil II. A non-Brownian fractal model and its application in soil survey. Journal of Soil Science 34: 599-620.

Burrough, P.A., 1986. Principles of geographical information systems for natural resource analysis. Oxford University Press, Oxford.

Burrough, P.A., L. Brown & E.C. Morris, 1977. Variations in vegetation and soil pattern across the Hawkesbury Sandstone Plateau from Barren Grounds to Fitzroy Falls, New South Wales. Australian Journal of Ecology 2: 137-159.

Campbell, N.A. & W.R. Atchley, 1981. The geometry of canonical variate analysis. Systematic Zoology 30: 268-280.

Carroll, J.D., 1972. Individual differences and multidimensional scaling. In: R.N. Shepard, A.K. Romney & S.B. Nerlove (Editors): Multidimensional scaling. Theory and application in the behavioral sciences. Vol. 1, Seminar Press, New York, p. 105-155.

Chatfield, C., 1981. The analysis of time series. Chapman and Hall, London. 268 pp.

Christiansen, F.B. & T.M. Fenchel, 1977. Theories of populations in biological communities. Springer Verlag, Berlin, 144 pp.

Chutter, F.M., 1972. An empirical biotic index of the quality of water in South African streams and rivers. Water Research 6: 19-30.

Clarke, G.M., 1980. Statistics and experimental design. 2nd Edition. Edward Arnold, London, 196 pp.

Clements, F.E., 1928. Plant succession and indicators. Wilson, New York, 453 pp.

Clymo, R.S., 1980. Preliminary survey of the peat-bog Hummell Knowe Moss using various numerical methods. Vegetatio 42: 129-148.

Cochran, W.G., 1963. Sampling techniques. 2nd Edition. Wiley, New York, 430 pp.

Cochran, W.G., 1977. Sampling techniques. 3rd Edition. Wiley, New York, 444 pp.

Cochran, W.G., 1983. Planning and analysis of observational studies. Wiley, New York, 158 pp.

Cochran, W.G. & G.M. Cox, 1957. Experimental Designs. New York, Wiley, 630 pp.

Coetzee, B.J. & M.J.A. Werger, 1975. On associatian analysis and the classification of plant communities. Vegetatio 30: 201-206

Cohen, J., 1973. Statistical power and research results. American Educational Research Journal 10: 225-229.

Cohen, J., 1977. Statistical power analysis for the behavioural sciences. Revised edition. Academic Press, London, 489 pp.

Cook, R.D. & S. Weisberg, 1982. Residuals and influence in regression. Chapman and Hall, London, 230 pp.

Corley, M.F.V., A.C. Crundwell, R. Dull, M.O. Hill & A.J.E. Smith, 1981. Mosses of Europe and the Azores; an annotated list of species, with synonyms from the recent literature. Journal of Bryology 11: 609-689.

Corsten, L.C.A., 1985. Current statistical issues in agricultural research. Statistica Neerlandica 39(2): 159-168.

Corsten, L.C.A. & K.R. Gabriel, 1976. Graphical exploration in comparing variance matrices. Biometrics 32: 851-863.

Cox, D.R., 1958. Planning of experiments. Wiley, New York, 315 pp.

Cox, D.R. & E.J. Snell, 1981. Applied statistics. Chapman and Hall, London, 197 pp.

Cramer, W, 1986. Vegetation dynamics on rising sea shores in eastern central Sweden. Acta Universitatis Uppsaliensis. Comprehensive summaries of Uppsala dissertations from the Faculty of Science 25. Almqvist & Wiksell International, Stockholm, 34 pp.

Cuppen, H.P.J.J. & W. Oosterloo, 1980. Een oriënterend onderzoek naar de macrofauna van de Nijmolense beek (Gemeente Epe). Report Regionale Milieuraad Oost Veluwe en Zuiveringsschap Veluwe, Apeldoorn, 50 pp.

Curtis, J.T. & R.P. McIntosh, 1951. An upland forest continuum in the prairie-forest border region of Wisconsin. Ecology 32: 476-496.

Dale, M.B., 1975. On objective methods of ordination. Vegetatio 30: 15-32.

Dargie, T.C.D., 1984. On the integrated interpretation of indirect site ordinations: a case study using semi-arid vegetation in south-eastern Spain. Vegetatio 55: 37-55.

David, M., 1977. Geostatistical ore reserve estimation. Developments in Geomathematics 2. Elsevier, Amsterdam, 364 pp.

Davies, P.T. & M.K-S. Tso, 1982. Procedures for reduced-rank regression. Applied Statistics 31: 244-255.

Davis, J., 1986. Statistics and Data Analysis in Geology. 2nd Edition. Wiley, New York, 550 pp.

Davis, R.B. & D.S. Anderson, 1985. Methods of pH calibration of sedimentary diatom remains for reconstructing history of pH in lakes. Hydrobiologia 120: 69-87.

Davison, M.L., 1983. Multidimensional scaling. Wiley, New York, 242 pp.

Dawkins, H.C., 1983. Multiple comparisons misused: why so frequently in response-curve studies? Biometrics 39: 789-790.

Dent, B.D., 1985. Principles of thematic map design. Addison-Wesley, Reading, Massachusetts.

De Sarbo, W.S. & V.R. Rao, 1984. GENFOLD2: a set of models and algorithms for the general unfolding analysis of preference/dominance data. Journal of Classification 1: 147-186.

de Wit, A.E., J. Spuij & P.J.C. van Boheemen, 1984. Slootplanten als indicatoren: een onderzoek naar indicatoren en hun bruikbaarheid in een biologisch meetnet. Landschap 1: 219-227.

Dixon, W.J., 1981. BMDP statistical software. University of California Press, Berkeley, 726 pp.

Dobson, A.J., 1983. Introduction to statistical modelling. Chapman and Hall, London, 125 pp.

Draper, N. & H. Smith, 1981. Applied regression analysis. 2nd Edition. Wiley, New York, 709 pp.

Dunn, G. & B.S. Everitt, 1982. An introduction to mathemathical taxonomy. Cambridge University Press, Cambridge.

Dupont, P., 1975. Les limites altitudinales des landes atlantiques dans les montagnes cantabriques (Nord de l'Espagne). In: J.M. Géhu (Editor): La vegetation des landes d'Europe occidentale Lille 1975, Colloques phytosociologiques 2: 47-58. J. Cramer Vaduz.

Efron, B., 1982. The jackknife, the bootstrap and other resampling plans. SIAM, Philadelphia, 92 pp.

Egerton, F.N., 1968. Leeuwenhoek as a founder of animal demography. Journal of Historical Biology 1: 1-22.

Ellenberg, H., 1948. Unkrautgesellschaften als Mass für den Säuregrad, die Verdichtung und andere Eigenschaften des Ackerbodens. Berichte über Landtechnik, Kuratorium für Technik und Bauwesen in der Landwirtschaft 4: 130-146.

Ellenberg, H., 1979. Zeigerwerte der Gefässpflanzen Mitteleuropas. Scripta Geobotanica 9, Göttingen, 121 pp.

Ellenberg, H., 1982. Vegetation Mitteleuropas mit den Alpen in Ökologischer Sicht. 3rd Edition. Ulmer Verlag, Stuttgart, 989 pp.

Everitt, B.S., 1980. Cluster analysis. 2nd Edition. Heinemann, London, 122 pp.

Everitt, B.S., 1984. An introduction to latent variable models. Chapman and Hall, London, 107 pp.

Faith, D.P., 1983. Asymmetric binary similarity measures. Oecologia 57: 287-290.

Fasham, M.J.R., 1977. A comparison of nonmetric multidimensional scaling, principal components and reciprocal averaging for the ordination of simulated coenoclines and coenoplanes. Ecology 58: 551-561.

Feoli, E. & L. Orloci, 1979. Analysis of concentration and detection of underlying factors in structured tables. Vegetatio 40: 49-54.

Finney, D.J., 1964. Statistical methods in biological assay. Griffin, London, 668 pp.

Fisher, R.A., 1954. Statistical Methods for Research Workers. 12th Edition. Oliver & Boyd, London, 356 pp.

Fresco, L.F.M., 1982. An analysis of species response curves and of competition from field data sets: some results from heath vegetation. Vegetatio 48: 175-185.

Gabriel, K.R., 1971. The biplot graphic display of matrices with application to principal component analysis. Biometrika 58: 453-467.

Gabriel, K.R., 1978. Least squares approximation of matrices by additive and multiplicative models. Journal of the Royal Statistical Society, Series B 40: 186-196.

Gauch, H.G., 1979. COMPCLUS – A FORTRAN program for rapid initial clustering of large data sets. Cornell University, Ithaca, New York.

Gauch, H.G., 1982. Multivariate analysis in community ecology. Cambridge

279

University Press, Cambridge, 298 pp.

Gauch, H.G. & G.B. Chase, 1974. Fitting the Gaussian curve to ecological data. Ecology 55: 1377-1381.

Gauch, H.G., G.B. Chase & R.H. Whittaker, 1974. Ordinations of vegetation samples by Gaussian species distributions. Ecology 55: 1382-1390.

Gauch, H.G., R.H. Whittaker & S.B. Singer, 1981. A comparative study of nonmetric ordinations. Journal of Ecology 69: 135-152.

Gause, G.F., 1930. Studies on the ecology of the Orthoptera. Ecology 11: 307-325.

Gifi, A., 1981. Nonlinear multivariate analysis. DSWO-press, Leiden, 451 pp.

Giloi, W.K., 1978. Interactive Computer Graphics – Data Structures, algorithms, languages. Prentice Hall, Englewood Cliffs, N.Y., 354 pp.

Giltrap, D., 1983. Computer Production of soil maps I. Production of grid maps by interpolation. Geoderma 29: 295-311.

Gittins, R, 1985. Canonical analysis. A review with applications in ecology. Springer-Verlag, Berlin, 351 pp.

Glacken, C.J., 1967. Traces on the Rhodian shore. Nature and culture in western thought from ancient times to the end of the eighteenth century. University of California Press, Berkely, 762 pp.

Gleason, H.A., 1926. The individualistic concept of the plant association. Bulletin of the Torrey Botanical Club 53: 7-26.

Goodall, D.W., 1954. Objective methods for the classification of vegetation. III. An essay in the use of factor analysis. Australian Journal of Botany 1: 39-63.

Goodall, D.W., 1970. Statistical Plant ecology. Annual review of Ecology and Systematics I: 99-124.

Goodall, D.W., 1973. Numerical methods of classification. In: R.H. Whittaker (Editor): Handbook of vegetation science, Volume 5: Ordination and Classification of plant communities. Junk, the Hague: 575-615.

Goodall, D.W. & R.W. Johnson, 1982. Non-linear ordination in several dimensions. A maximum likelihood approach. Vegetatio 48: 197-208.

Gordon, A.D., 1981. Classification. Methods for the exploratory analysis of multivariate data. Chapman and Hall, London, 193 pp.

Gounot, M., 1969. Méthodes d'étude quantitative de la végétation. Masson, Paris, 314 pp.

Gourlay, A.R. & G.A. Watson, 1973. Computational methods for matrix eigen problems. Wiley, New York, 130 pp.

Gower, J.C., 1966. Some distance properties of latent root and vector methods used in multivariate analysis. Biometrika 53: 325-338.

Gower, J.C. 1974. Maximal predictive classification. Biometrics 30: 643-654.

Green, R.H., 1979. Sampling design and statistical methods for environmental biologists. Wiley, New York, 257 pp.

Greenacre, M.J., 1984. Theory and applications of correspondence analysis. Academic Press, London, 364 pp.

Greig-Smith, P., 1971. Application of numerical methods to tropical forests. In: G.P. Patil, E.C. Pielou & W.E. Waters (Editors): Statistical ecology. Pennsylvania State University. 3: 195-204.

Grolle, R., 1976. Verzeichnis der Lebermoose Europas und benachbarter Gebiete. Feddes Repertorium 87: 171-279.

Guillerm, J.L., 1971. Calcul de l'information fournie par un profil écologique et valeur indicatrice des espèces. Oecologia Plantarum 6: 209-225.

Hajdu, L.J., 1982. Graphical comparison of resemblance measures in phytosociology. Vegetatio 48: 47-59.

Hardonk, R., 1968. Koornmullenaers, Pampiermaeckers en Coperslaghers. Korte historie van de watermolens van Apeldoorn, Beekbergen en Loenen. Historisch Museum Moerman, Apeldoorn, 256 pp.

Harrell, F., 1980. The LOGIST procedure. In: SAS supplemental library user's guide, 1980 Edition. SAS Institute Inc., Cary, N.C., p. 83-89.

Hartigan, J.A., 1975. Clustering algorithms. Wiley, New York, 364 pp.

Harvey, T., 1981. Environmental intervention: The monitoring paradigm. I. The monitoring concept and practice of descriptive monitoring. The environmentalist 1: 283-291.

Hawksworth, D.L., P.W. James & B.J. Coppins, 1980. Check-list of Britisch lichenforming, lichenocolous and allied fungi. Lichenologist 12: 1-115.

Heiser, W.J., 1981. Unfolding analysis of proximity data. Thesis. University of Leiden, Leiden, 273 pp.

Heiser, W.J., 1986. Undesired nonlinearities in nonlinear multivariate analysis. In: E. Diday et al. (Editors): Data analysis and Informatics 4. North Holland, Amsterdam, p. 455-469.

Heiser, W.J., 1987. Joint ordination of species and sites: the unfolding technique. In: P. Legendre & L. Legendre (Editors): Developments in numerical ecology. Springer Verlag, Berlin, p. 189-221.

Hill, M.O., 1973. Reciprocal averaging: an eigenvector method of ordination. Journal of Ecology 61: 237-249.

Hill, M.O., 1974. Correspondence analysis: a neglected multivariate method. Applied Statistics 23: 340-354.

Hill, M.O., 1977. Use of simple discriminant functions to classify quantitative phytosociological data. In: E. Diday, L. Lebart, J.P. Pagès & R. Tomassone (Editors): First International Symposium on Data Analysis and Informatics. Vol 1: 181-199. Institut de Recherche d'Informatique et d'Automatique, Le Chesnay, 181-199.

Hill, M.O., 1979a. DECORANA – A FORTRAN program for detrended correspondence analysis and reciprocal averaging. Cornell University Ithaca, N.Y., 52 pp.

Hill, M.O., 1979b. TWINSPAN – A FORTRAN program for arranging multivariate data in an ordered two-way table by classification of individuals and attributes. Cornell University Ithaca, N.Y., 90 pp.

Hill, M.O., R.G.H. Bunce & M.W. Shaw, 1975. Indicator species analysis, a divisive polythetic method of classification and its application to a survey of native pinewoods in Scotland. Journal of Ecology 63: 597-613.

Hill, M.O. & H.G. Gauch, 1980. Detrended correspondence analysis, an improved ordination technique. Vegetatio 42: 47-58.

281

Hocking, R.R. & O.J. Pendleton, 1983. The regression dilemma. Communications in Statistics – Theory and Methods 12: 497-527.

Hogeweg, P., 1976. Iterative character weighing in numerical taxonomy. Comput. Biol. Med. 6: 199-211.

Howe, S.E. & T.III Webb, 1983. Calibrating pollen data in climatic terms: improving the methods. Quaternary Science Reviews 2: 17-51.

Hurlbert, S.H., 1984. Pseudoreplication and the design of ecological field experiments. Ecological Monographs 54: 187-211.

Ihm, P. & H. van Groenewoud, 1984. Correspondence analysis and Gaussian ordination. COMPSTAT lectures 3: 5-60.

IJzerman, A.J., 1979. Sprengen en sprengenbeken op de Veluwe. Een onderzoek naar beheer, onderhoud en watervoorziening in historisch perspectief. Rapport ten behoeve van de commissie bestudering waterhuishouding Gelderland, Arnhem, 110 pp.

Innis, G.S., 1979. Letters to the editor. Bulletin of the Ecological Society of America 60: 142.

Israëls, A.Z., 1984. Redundancy analysis for qualitative variables. Psychometrika 49: 331-346.

Jaccard, P., 1912. The distribution of the flora of the alpine zone. New Phytologist 11: 37-50.

Janssen, J.G.M., 1975. A simple clustering procedure for preliminary classification of very large datasets of phytosociological relevées. Vegetatio 30: 67-71.

Johnson, S.B. & R.D. Berger, 1982. On the status of statistics in phytopathology. Phytopathology 72: 1014-1015.

Jongman, R.H.G. & Th.J. van de Nes (Editors), 1982. Beken op de Veluwe. Een onderzoek naar de mogelijkheden voor herstel en behoud. Begeleidingscommissie Proefgebied Nationaal Landschap Veluwe, Arnhem, 112 pp.

Journel, A.J. & C.J. Huijbregts, 1978. Mining geostatistics. Academic Press, New York, 600 pp.

Jowett, G.H. & G. Scurfield, 1949. A statistical investigation into the distribution of Holcus mollis L. and Deschampsia flexuosa L. Trin. Journal of Ecology 37: 68-81.

Kalkhoven, J. & P. Opdam, 1984. Classification and ordination of breeding bird data and landscape attributes. In: J. Brandt & P. Agger (Editors): Methodology in Landscape ecological research and planning. Universitetsforlag, Roskilde Volume III: 15-26.

Kanal, L., 1974. Patterns in pattern recognition 1968-74. IEEE Transactions on Information Theory 20: 697-722.

Kant, G.R., 1982. Beken op de Veluwe. Hydrologische aspecten in relatie tot de watervoering van de beken op de Oost- en Zuid Veluwe. Basisrapport 1 ten behoeve van de werkgroep Sprengen en beken. Provincie Gelderland, Arnhem, 84 pp.

Kendall, D.G., 1971. Abundance matrices and seriation in archaeology. Zeitschrift für Warscheinlichkeitstheorie 17: 104-112.

Kilic, M., 1979. Soil variability near Follonica, Italy. MSc. Thesis. MSc Course in Soil Science and Water Management, Agricultural University, Wageningen.

Kirkpatrick, J.B., P.R. Minchin & J.B. Davies, 1985. Floristic composition and macroenvironmental relationships of Tasmanian vegetation containing bolster plants. Vegetatio 63: 89-96.

Kooijman, S.A.L.M., 1977. Species abundance with optimum relations to environmental factors. Annals of System Research 6: 123-138.

Krebs, C.J., 1978. Ecology, the experimental analysis of distribution and abundance. Harper and Row, New York, 648 pp.

Kruijne, A.A., D.M. de Vries & H. Mooi, 1967. Bijdrage tot de oecologie van de Nederlandse graslandplant (with english summary). Verslagen Landbouwkundige Onderzoekingen 696. Pudoc, Wageningen, 65 pp.

Kruskal, J.B., 1964. Multidimensional scaling by optimizing goodness of fit to a nonmetric hypothesis. Psychometrika 29: 1-27.

Kruskal, J.B. & J.D. Carroll, 1969. Geometric models and badness-of-fit functions. In: P.R. Krishnaiah (Editor): Multivariate analysis II. Academic Press, New York, p. 639-671.

Laboratory for Computer Graphics, 1968. SYMAP Manual, Graduate School of Design, Harvard.

Lachenbruch, P.A., 1975. Discriminant analysis. Hafner Press, New York, 128 pp.

Lam, N.S., 1983. Spatial interpolation methods: a review. The American Cartographer 10: 129-149.

Lambert, J.M. & M.B.Dale, 1964. The use of statistics in phytosociology. Advances in Ecological Research 2: 59-99.

Lambert, J.M., S.E. Meacock, J. Barrs & P.F.M. Smartt, 1973. AXOR and MONIT: two new polythetic-divisive strategies for hierarchial classification. Taxon 22: 1733-176.

Lance, G.N. & W.T. Williams, 1967. A general theory of classificatory sorting strategies. I Hierarchical systems. Computer Journal 9: 373-380.

Loucks, O.L., 1962. Ordinating forest communities by means of environmental scalars and phytosociological indices. Ecological Monographs 32: 137-166.

Louppen, J.W.M. & E. van der Maarel, 1979. CLUSLA: a computer program for the clustering of large phytosociological data sets. Vegetatio 40: 107-114.

Lozano, J.A. & J.D. Hays, 1976. Relationship of Radiolarian assemblages to sediment types and physical oceanography in the Atlantic and Western Indian Ocean. In: R.M. Cline & J.D. Hays (Editors): Investigation of late Quaternary paleoceanography and paleoclimatology. Geological Society of America Memoir 145: 303-336.

Mandelbrot, B.B., 1982. The fractal geometry of nature. Freeman, San Francisco, 460 pp.

Mardia, K.V., J.T. Kent & J.M. Bibby, 1979. Multivariate analysis. Academic Press, London, 521 pp.

Matheron, G., 1971. The theory of regionalized variables and its applications. Les Cahiers du Centre de Morphologie Mathematique de Fontainebleu. Ecole Nationale Superieure des Mines de Paris.

McBratney, A.B., R. Webster, R.G. McLaren & R.B. Spiers, 1982. Regional variation of extractable copper and cobalt in the topsoil of south-east Scotland.

Agronomie 2 (10): 969-982.

McCullagh, P. & J.A. Nelder, 1983. Generalized linear models. Chapman and Hall, London, 261 pp.

McIntosh, R.P., 1981. Succession and ecological theory. In: D.C. West, H.H. Shugart & D.B. Botkin (Editors): Forest succession. Concepts and application. Springer Verlag, New York, p. 10-23.

Mead, R. & D.J.P. Pike, 1975. A review of response surface methodology from a biometric viewpoint. Biometrics 31: 803-851.

Meijers, E., 1986. Defining confusions-confusing definitions. Environmental monitoring and assessment 7: 157-159.

Meijers, E., W.J. ter Keurs & E. Meelis, 1982. Biologische meetnetten voor het beleid. Mededelingen van de Werkgemeenschap Landschapsecologisch Onderzoek 9: 51-58.

Mertz, D.B. & D.E. McCauley, 1980. The domain of laboratory ecology. Synthese 43: 95-110.

Meulman, J. & W.J. Heiser, 1984. Constrained multidimensional scaling: more directions than dimensions. COMPSTAT 1984, Physica-Verlag, Vienna, p. 137-142.

Minchin, P.R., 1987. An evaluation of the relative robustness of techniques for ecological ordination. Vegetatio 67: 1167-1179.

Montgomery, D.C. & E.A. Peck, 1982. Introduction to linear regression analysis. Wiley, New York, 504 pp.

Morrison, D.F., 1976. Multivariate statistical methods. 2nd Edition. McGraw-Hill, Tokyo, 415 pp.

Mosteller, F. & J.W. Tukey, 1977. Data analysis and regression. Addison-Wesley, Reading, 588 pp.

Mueller-Dombois, D. & H. Ellenberg, 1974. Aims and methods of vegetation ecology. Wiley, New York, 547 pp.

Naes, T. & H. Martens, 1984. Multivariate calibration. II Chemometric methods. Trends in analytical chemistry 3: 266-271.

Neef, E., 1982. Stages in development of landscape ecology. In: S. Tjallingi & A. de Veer (Editors): Perspectives in Landscape Ecology. Proceedings of the International symposium of Landscape ecology at Veldhoven Netherlands. Pudoc, Wageningen, p. 19-27.

Nelder, J.A. & R.W.M. Wedderburn, 1972. Generalized linear models. Journal of the Royal Statistical Society, Series A 135: 370-384.

Nielsen, D.R. & J. Bouma (Editors), 1985. Soil spatial variability. Proceedings of a workshop of the ISSS and the SSSA, Las Vegas, USA, 30 Nov. – 1 Dec. 1984. Pudoc, Wageningen, 243 pp.

Nishisato, S., 1980. Analysis of categorical data: dual scaling and its applications. Toronto University Press, Toronto, 276 pp.

Noy-Meir, I., 1973. Data transformation in ecological ordination. I. Some advantages of non-centering. Journal of Ecology 61: 329-341.

Noy-Meir, I., 1974. Catenation: quantitative methods for the definition of coenoclines. Vegetatio 29: 89-99.

Noy-Meir, I., D. Walker & W.T. Williams, 1975. Data transformations in ecological ordination. II. On the meaning of data standardization. Journal of Ecology 63: 779-800.

Odum, E.P., 1971. Fundamentals of ecology. 3rd Edition. Saunders, Philadelphia.

Orlóci, L., 1966. Geometric models in ecology. I. The theory and application of some ordination methods. Journal of Ecology 54: 193- 215.

Orlóci, L., 1967. An agglomerative method for the classification of plant communities. Journal of Ecology 55: 193-206.

Parker, R.E., 1979. Introductory statistics for biology. Studies in biology, No 43. Arnold, London, 122 pp.

Pavlidis, T., 1982. Algorithms for graphics and image processing. Springer Verlag, Berlin, 416 pp.

Pielou, E.C., 1969. An introduction to mathematical ecology. Wiley, New York, 286 pp.

Pielou, E.C., 1976. Population and community ecology. Gordon and Breach, New York, 432 pp.

Pielou, E.C., 1977. Mathematical ecology. 2nd Edition. Wiley, New York, 385 pp.

Pielou, E.C., 1984. The interpretation of ecological data. A primer on classification and ordination. Wiley, New York, 263 pp.

Popma, J., 1982. Beken op de Veluwe. Numerieke classificatie van macrofauna gemeenschappen van sprengen en beken op de Veluwe. Basisrapport 2 ten behoeve van de werkgroep sprengen en beken. Provincie Gelderland, Arnhem, 132 pp.

Popma, J., L. Mucina, O. van Tongeren & E. van der Maarel, 1983. On the determination of optimal levels in phytosociological classification. Vegetatio 52: 65-76.

Prentice, I.C., 1977. Non-metric ordination methods in ecology. Journal of Ecology 65: 85-94.

Prentice, I.C., 1980. Vegetation analysis and order invariant gradient models. Vegetatio 42: 27-34.

Quinn, J.F. & A.E. Dunham, 1983. On hypothesis testing in ecology and evolution. The American Naturalist 122: 602-617.

Ramensky, L.G., 1930. Zur Methodik der vergleichenden Bearbeitung und Ordnung von Pflanzenlisten und anderen Objekten, die durch mehrere, verschiedenartig wirkende Faktoren bestimt werden. Beiträge zur Biologie der Pflanzen 18: 269-304.

Rao, C.R., 1964. The use and interpretation of principal component analysis in applied research. Sankhya A 26: 329-358.

Rao, C.R., 1973. Linear statistical inference and its applications. 2nd Edition. Wiley, New York, 625 pp.

Ripley, B., 1981. Spatial statistics. Wiley, New York, 252 pp.

Risser, P.G., J.R. Karr & R.T.T. Forman, 1984. Landscape ecology directions and approaches. A workshop held at Allerton Park, Piatt. County Illinois 1983. Illinois Natural History Survey Special Publication Number 2, 16 pp.

Rivas-Martínez, S., 1979. Brezales y jarales de Europa occidental. Lazaroa 1: 5-127.

Rogister, J.E., 1978. De ecologische mR- en mN-waarden van de kruidlaag en de humuskwaliteit van bosplantengezelschappen. Werken Proefstation Waters en Bossen (Groenendaal- Hoeilaart) reeks A 20: 1-29.

Roux, G. & M. Roux, 1967. A propos de quelques méthodes de classification en phytosociologie. Revue de statistique appliquée 15: 59-72.

Saaty, T.L., 1977. A scaling method for priorities in hierarchical structures. Journal of mathematical psychology 14: 234-281.

Salton, G. & A. Wong, 1978. Generation and search of cluster files. Association for computing machinery transactions on database systems 3: 321-346.

SAS Institute Inc., 1982. SAS user's guide: statistics, 1982 Edition. SAS Institute Inc., Cary, N.C., 584 pp.

Schaafsma, W. & G.N. van Vark, 1979. Classification and discrimination problems with application, part IIa. Statistica Neerlandica 33: 91-126.

Scheffé, H., 1973. A statistical theory of calibration. Annals of Statistics 1: 1-37.

Schiffman, S.S., M.L. Reynolds & F.W. Young, 1981. Introduction to multi-dimensional scaling. Theory, methods and applications. Academic Press, London, 413 pp.

Schwartz, L.M., 1977. Nonlinear calibration. Analytical chemistry 49: 2062-2068.

Seber, G.A.F., 1977. Linear regression analysis. Wiley, New York, 465 pp.

Sheenan, P.J., 1984. Effects on community and ecosystem structure and dynamics. In: P.J. Sheenan, D.R. Miller, G.C. Butler & Ph. Bourdeau (Editors): Effects of pollutants at the ecosystem level. Wiley, New York, p. 51-99.

Shepard, R.N., 1962. The analysis of proximities: multidimensional scaling with an unknown distance function. II. Psychometrika 27: 219-246.

Shepard, D., 1968. A two-dimensional interpolation function for irregularly spaced data. In: Proc. Ass. Comput. Mach. 517-523.

Sibson, R., 1972. Order invariant methods for data analysis. Journal of the Royal Statistical Society, Series B 34: 311-349.

Simberloff, D., 1983. Competition theory, hypothesis-testing, and other community ecological buzzwords. The American Naturalist 122: 626-635.

Slob, W., 1987. Strategies in applying statistics in ecological research. Free University Press, Amsterdam, 112 pp.

Sneath, P.H.A. & Sokal, R.R., 1973. Numerical taxonomy. W.H. Freeman, San Fransisco.

Snedecor, G.W. & W.G. Cochran, 1980. Statistical methods. 7th Edition. The Iowa State University Press, Ames, 507 pp.

Sokal, R.R. & C.D. Michener, 1958. A statistical method for evaluating systematic relationships. University of Kansas Science bulletin 38: 1409-1438.

Sokal, R.R. & F.J. Rohlf, 1981. Biometry. 2nd Edition. Freeman, San Fransisco, 877 pp.

Sørensen, T., 1948. A method of establishing groups of equal amplitude in plant sociology based on similarity of species content. Det. Kong. Danske Vidensk. Selsk. Biol. Skr. (Copenhagen) 5 (4): 1-34.

Southwood, T.R.E., 1978. Ecological methods, with particular reference to the study of insect populations. 2nd Revised edition. Methuen, London, 391 pp.

Strong jr., D.R., 1980. Null hypotheses in ecology. Synthese 43: 271- 285.

Swain, P.H., 1978. Fundamentals of pattern recognition in remote sensing. In: P.H. Swain & S.M. Davies (Editors): Remote sensing: the quantitative approach. McGraw-Hill, New York.

Swaine, M.D. & P. Greig-Smith, 1980. An application of principal components analysis to vegetation change in permanent plots. Journal of Ecology 68: 33-41.

Tacha, T.C., W.D. Warde & K.P. Burnham, 1982. Use and interpretation of statistics in wildlife journals. Wildlife Society Bulletin 10: 355-362.

Taylor, C.C. & P.A. Burrough, 1986. Multiscale sources of spatial variation in soil III. Improved methods for fitting the nested model to one-dimensional semivariograms. Mathematical Geology 18: 811-821.

ten Berge, H.F.M., L. Stroosnijder, P.A. Burrough, A.K. Bregt & M.J. de Heus, 1983. Spatial variability of physical soil properties influencing the temperature of the soil surface. Agricultural Water Management 6: 213-216.

ter Braak, C.J.F., 1982. DISCRIM – A modification of TWINSPAN (Hill, 1979) to construct simple discriminant functions and to classify attributes, given a hierarchical classification of samples. Report C 82 ST 10756. TNO Institute of Mathematics, Information processing and Statistics, Wageningen.

ter Braak, C.J.F., 1983. Principal components biplots and alpha and beta diversity. Ecology 64: 454-462.

ter Braak, C.J.F., 1985. Correspondence analysis of incidence and abundance data: properties in terms of a unimodal reponse model. Biometrics 41: 859-873.

ter Braak, C.J.F., 1986a. Canonical correspondence analysis: a new eigenvector technique for multivariate direct gradient analysis. Ecology 67: 1167-1179.

ter Braak, C.J.F., 1986b. Interpreting a hierarchical classification with simple discriminant functions: an ecological example. In: E. Diday et al. (Editors): Data analysis and informatics 4. North Holland, Amsterdam, p. 11-21.

ter Braak, C.J.F., 1987a. The analysis of vegetation-environment relationships by canonical correspondence analysis. Vegetatio 69: 69-77.

ter Braak, C.J.F., 1987b. CANOCO – a FORTRAN program for canonical community ordination by [partial] [detrended] [canonical] correspondence analysis, principal components analysis and redundancy analysis (version 2.1) Agricultural Mathematics Group, Wageningen, 95 pp.

ter Braak, C.J.F., 1988. Partial canonical correspondence analysis. In: H.H. Bock (Editor): Classification and related methods of data analysis. North Holland, Amsterdam, p. 551-558.

ter Braak, C.J.F. & L.G. Barendregt, 1986. Weighted averaging of species indicator values: its efficiency in environmental calibration. Mathematical Biosciences 78: 57-72.

ter Braak, C.J.F. & C.W.N. Looman, 1986. Weighted averaging, logistic regression and the Gaussian response model. Vegetatio 65: 3-11.

Titterington, D.M., G.D. Murray, L.S. Murray, D.J. Spiegelhalter, A.M. Skene, J.D.F. Habbema & G.J. Gelpke, 1981. Comparison of discrimination techniques applied to a complex data set of head injured patients (with discussion). Journal of the Royal Statistical Society, Series A 144: 145-175.

Toft, C.A. & P.J. Shea, 1983. Detecting community-wide patterns: estimating power strengthens statistical inference. The American Naturalist 122: 618-625.

Torgerson, W.S., 1958. Theory and methods of scaling. Wiley, New York, 460 pp.

Tso, M.K-S., 1981. Reduced-rank regression and canonical analysis. Journal of the Royal Statistical Society Series B 43: 89-107.

Tutin, T.G. et al. (Editor), 1964-1980. Flora Europaea. Cambridge University Press, Cambridge.

van Dam, H., G. Suurmond & C.J.F. ter Braak, 1981. Impact of acidification on diatoms and chemistry of Dutch moorland pools. Hydrobiologia 83: 425-459.

van de Nes, T.J., 1980. Introduction to a system approach for water management in Gelderland, The Netherlands. In: Water resources management on a regional scale. Committee for hydrological research TNO, Proceedings and Information no 27, p. 9-22.

van der Maarel, E., 1979a. Multivariate methods in phytosociology, with reference to the Netherlands. In: M.J.A. Werger (Editor):The study of vegetation. Junk, The Hague, pp. 161-226.

van der Maarel, E., 1979b. Transformation of cover-abundance values in phytosociology and its effects on community similarity. Vegetatio 39: 97-114.

van der Maarel, E., L. Orlóci & S. Pignatti, 1976. Data processing in phytosociology, retrospect and anticipation. Vegetatio 32: 65-72.

van der Maarel, E., J.G.M. Janssen & J.M.W. Louppen, 1978. TABORD, a program for structuring phytosociological tables. Vegetatio 38: 143-156.

van der Meijden, R., E.J. Weeda, F.A.C.B. Adema & G.J. de Joncheere, 1983. Flora van Nederland. 20th Edition. Wolters-Noordhoff, Groningen, 583 pp.

van der Steen, W.J., 1982. Algemene methodologie voor biologen. Bohn, Scheltema and Holkema, Utrecht, 115 pp.

van Tongeren, O., 1986. FLEXCLUS, an interactive program for classification and tabulation of ecological data. Acta Botanica Neerlandica 35: 137-142.

Varmuza, K., 1980. Pattern recognition in chemistry. Springer Verlag, Berlin, 217 pp.

Vevle, O. & K. Aase, 1980. Om bruk ov oekologiske faktortall i norske plantesamfunn. In: K. Baadsvik, T. Klokk & O.I. Rönning (Editors): Fagmoete i vegetasjonsoekologi pa Kongvoll 16- 18.3.1980. Kgl. Norske Videnskabers Selskab Museet, Botanisk serie 1980-5, p. 178-201.

Vink, A.P.A., 1963. Planning of Soil Surveys in Land Development. Publication of the International Institute for Land Reclamation and Improvement (ILRI) 10, 53 pp.

von Humboldt, A., 1806. Ideen zu einer Physiognomik der Gewächse. Cotta, Stuttgart.

Webster. R., 1984. Elucidation and characterization of spatial variation in soil using regionalized variable theory. In: G. Verly et al. (Editors): Geostatistics for Natural Resources Characterization, Part 2, Reidel, Dordrecht, p. 903-913.

Webster, R., 1985. Quantitative spatial analysis of soil in the field. Advances in

Soil Science, Volume 3. Springer-Verlag, New York.

Webster, R. & T.M. Burgess, 1983. Spatial variation in soil and the role of kriging. Agricultural Water Management 6: 111-122.

Webster, R. & T.M. Burgess, 1984. Sampling and bulking strategies for estimating soil properties in small regions. Journal of Soil Science 35: 127-140.

Webster, R. & S. Nortcliff, 1984. Improved estimation of micronutrients in hectare plots of the Sonning Series. Journal of Soil Science 35: 667-672.

Westhoff V. & E. van der Maarel, 1978. The Braun Blanquet approach, In: R.H. Whittaker (Editor): Classification of plant communities. Junk, The Hague, p. 287-399.

Whittaker, J., 1984. Model interpretation from the additive elements of the likelihood function. Applied Statistics 33: 52-65.

Whittaker, R.H., 1956. Vegetation of the Great Smoky Mountains. Ecological Monographs 26: 1-80.

Whittaker, R.H., 1967. Gradient analysis of vegetation. Biological Reviews 49: 207-264.

Whittaker R.H., 1973. Approaches of classifying vegetation. In: Whittaker R.H. (Editor): Ordination and classifiction of communities. Junk, The Hague, p. 325-354.

Whittaker, R.H., S.A. Levin & R.B. Root, 1973. Niche, habitat and ecotope. American Naturalist 107: 321-338.

Wiens, J.A. & J.T. Rotenberry, 1981. Habitat associations and community structure of birds in shrub steppe environment. Ecological Monographs 51: 21-41.

Williams, B.K., 1983. Some observations on the use of discriminant analysis in ecology. Ecology 64: 1283-1291.

Williams, E.J., 1959. Regression analysis. Wiley, New York, 211 pp.

Williams, W.T. (Editor), 1976a. Pattern analysis in agricultural science. CSIRO Melbourne, Australia.

Williams, W.T., 1976b. Hierarchical divisive strategies. In: W.T. Williams (Editor): Pattern analysis in agricultural science. CSIRO, Melbourne, Australia.

Williams, W.T. & J.M. Lambert, 1959. Multivariate methods in plant ecology. I. Association analysis in plant communities. Journal of Ecolology 47: 83-101.

Williams, W.T. & J.M. Lambert, 1960. Multivariate methods in plant ecology. II The use of an electronic digital computer for association-analysis. Journal of Ecology 48: 689-710.

Williams, W.T. & J.M. Lambert, 1961. Multivariate methods in plant ecology. III Inverse association-analysis. Journal of Ecology 49: 717-729.

Williams, W.T., J.M. Lambert & G.N. Lance, 1966. Multivariate methods in plant ecology. V. Similarity-analysis and information-analysis. Journal of Ecology 54: 427-445.

Williamson, M., 1972. The analysis of biological populations. Arnold, London, 180 pp.

Wishart, D., 1978. CLUSTAN user manual. Program library unit. Edinburgh University Press, Edinburgh, 175 pp.

Wold, H., 1982. Soft modelling: the basic design and some extensions. In: K.G.

Jöreskog & H. Wold (Editors): Systems under indirect observation. II. North Holland, Amsterdam, p. 1-54.

Yarranton, G.A., 1969. Plant ecology: a unifying model. Journal of Ecology 57: 245-250.

Yarranton, G.A., 1970. Towards a mathematical model of limestone pavement vegetation. III. Estimation of the determinants of species frequencies. Canadian Journal of Botany 48: 1387-1404.

Zelinka, M. & P. Marvan, 1961. Zur Präzisierung der biologischen Klassifikation der Reinheit fliessender Gewässer. Archiv für Hydrobiologie 57: 389-407.

Zhang, L., 1983. Vegetation ecology and population biology of Fritillaria meleagris L. at the Kungsängen nature reserve, eastern Sweden. Acta Phytogeographica Suecica 73: 1-92.

Index

Numbers in bold type indicate the pages on which the term is explained.

canonical eigenvalue 141 (see also eigenvalue)

canonical ordination 6, 93, **136-151**, 154, 155

canonical variate analysis (CVA) **148-149**, 151, 157

causality 22, 26, 29, 155

CCA, see canonical correspondence analysis

centring **20**, 130

centroid **100**, 142, 169 (see also weighted average)

centroid clustering **189**

checking of data 259

chi-square distance **152**

chi-square test **43-44**, 60, 69, 202

choice of methods 5-7, 153-156, 180-183, 219-220, 245

chord distance **178**, 181-183, 185

choropleth map **213**

classical scaling 152

classification 8, 10, 23, **174**, 258 (see also cluster analysis)

classification rule **80**

classification system 258

cluster analysis 4, 6, 8, 23, 155, **174-212**, 258, 269

cluster optimality 198

coefficient of community **177**, 181-183

coefficient of determination **37**

coefficient of variation **19**

community composition 64, 78, 174

community ecology 1, 2

community types 174, 273

competition between species 60

complementary variables **53**

complete-linkage clustering **188**, 192, 193, 204

component analysis 91 (see also principal component analysis)

composite clustering **197**

composite gradient 154, 155

composite sample 263

computer programs 47, 158, 193, 197, 201, 245-246

concordance measure 274

confidence interval **35**, 40, 65-66, 67, 72-74, 173

constained ordination, see canonical ordination

continuous variable **18**, 19, 202 (see also quantitative variable)

cophenetic correlation **189**

COR, see canonical correlation analysis

cord distance **178**, 181-183, 185

correlation 24, 26, 37, 40, 81, 129, 132, 147 (see also regression)

correlation matrix 131, 163

correlation ratio 157

correspondence analysis (CA) 93, **95-116**, 154, 157, 160-161, 165-166, 194, 254, 261

cosine coefficient **178**, 181-183

counts 8, 20, 50, 184

covariables **156**

covariance 146, 159, **225**, 229

covariance biplot 124, **129**, 159

covariance matrix 131

critical value **35**, 40, 72, 75

cross-product matrix 159, 160

cut-level, see pseudo-species

CVA, see canonical variate analysis

data approximation 93

data transformation **20-21**, 49, 61, 103, 130, 151, 184-186, 216, 229

DCA, see detrended correspondence analysis

DCCA, see detrended canonical correspondence analysis

decision-making 12, 272-274

DECORANA 103, 110, 112, **158**

degrees of freedom **35-36**, 44, 49, 53

dendrogram **186-192**, 199

detection 23, 24

detrended correspondence analysis 93, **105-109**, 115-117, 154, 158, 167, 263, 265

detrending **106**, 108, 115, 139

detrending-by-polynomials **108**

deviance, see residual deviance

deviance test **49**, 56, 59, 60, 70, 72, 77

293

deviation 257 (see also residual)
d.f., see degrees of freedom
diagonal structure **63**, 97, 103, 196
dichotomized ordination analysis **195**
differencing **229**
differential species 174, 194
dimensionality reduction, see matrix approximation
direct gradient analysis 29, **64**, 91 (see also regression and canonical ordination)
discrete variable 18
DISCRIM 201
discriminant analysis **79**, 87, 148 (see also canonical variate analysis)
discriminant functions 201
dispersion of site scores **100**, 170
dispersion of species scores 95, **101**, 110, 137-138, 157, 160, 169-171
dissimilarity 94, 151-152, **176**, 178-183
distance-weighted least-squares 238
distribution, see frequency distribution and also response curve
divisive methods **191-196**
domain **238**, 239
dominant species 103, 131, 183
down-weighting of rare species 110, 186, 194
dual scaling 157
dummy variable **58**, 140, 150
Dune Meadow Data 5, 62, 97-102, 107, 113, 115-117, 120-124, 127-130, 133-136, 139, 145-146, 149-150, 166, 187-193, 196, 200-201

E, see expected response
ecology 1, 2, 8, 9
ecological amplitude 42, 78, 84, 115, 203 (see also tolerance)
edge effects 108, 221
eigenvector **101**, 122, 161, 169 (see also singular vector)
eigenvalue 100, **101**, 122, 139, 141, 147, 159, 160-164, **169**
EM algorithm 158
environmental scalar 151

error **30**, 148, 216, 223, 239 (see also residual)
error map 242
error part of response model **30**, 33
error sum of squares, see residual sum of squares
error sum of squares clustering **191**
estimation of parameters 14, 30, **34**, 39, 45, 51
Euclidean Distance 152, **178**, 181-183, 185, 205
Euclidean Distance biplot **129**, 147, 160
evaluation of methods 60-61, 153-156, 219-220, 245
exact interpolator 240
expected abundance 109 (see also expected response)
expected response 29, 32, 35, 37
experiments 12-15, 22, 24
explanatory power 53, 77
explanatory variable 11, 12, 26, **29-30**, 51-59, 154
exploratory data analysis 23
exponential curve 33, 40, **44**
exponential model for semivariogram **231**
exponential transformation **184**

factor analysis 91 (see also principal component analysis)
F distribution 35, 53
F test 35, **53**, 219
field survey 2, 3, 4, 15, 24
first-order moment **224**
fitted value **34**, 36, 40, 67, 122-124
FLEXCLUS **197**, 200, 259
fraction of variance accounted for 34, **37**, 41, 124, 130, 141
frequency distribution 16, **18-20**
frequency of occurrence 43
furthest-neighbour clustering **188**
fusion of clusters 262
fusion of taxa 268

Gaussian curve, see Gaussian response curve

passive species and sites 166
PCA, see principal components analysis
PCA of instrumental variables 157
PCA of y with respect to x 157
PCO, see principal coordinate analysis
percentage distance **179**
percentage variance accounted for, see fraction of variance accounted for
percentage similarity **177**, 181-183
periodic variation 233, 234
permutation test 156
Petrie matrix **103-104**, 169
plane **51-52**, 125, 127
point kriging **242**
Poisson distribution **19-20**, 50, 111, 114, 157, 184
polar ordination 156
policy analysis 267
polynomial regression **41**, 65, 87, 217
polythetic **193**
posterior distribution **80**
power, see statistical power analysis
power algorithm **157**, 158
prediction 29 (see also calibration)
prediction interval 34, **36**, 67, 73
preference score 194, 205
presence–absence data 8, 18, 31, 43-49, 53-62, 65, 79-85, 97, 104, 112, 142, 176-177, 184
principal component **118**, 122, 125, 254
principal component analysis (PCA) 93-94, **116-132**, 144, 152, 154-160, 166, 193
principal coordinate analysis (PCO) **152**, 158
prior distribution **80**, 86
priorities for restoration 272-274
probability of occurrence **43-44**, 79, 82, 84, 88-89, 109
proof 14, 23-24
pseudo-cyclic variation 237
pseudo-species **50**, 194

quadrat 8
quantitative variable **17**, 37, 44, 65, 81, 213

Q-mode algorithm **131**, 152, 159

R^2 156, 256 (see also coefficient of determination)
range in semivariogram **230-232**, 234, 250
rank sum test **202**
rare species 82, 109-110, 130, 183
ratio scale **17**, 179
RDA, see redundancy analysis
reallocation 197
reciprocal averaging 97, 157, 254 (see also correspondence analysis)
reduced-rank regression 157
redundancy 175
redundancy analysis (RDA) **144-148**, 151, 154, 156-158, 163
refined ordination 195
regionalized variables **224**, 246
regression 5, **29-77**, 79, 86, 91, 93, 111, 115, 119, 125, 132, 135, 140, 145, 154-155
regression coefficient 37
regression diagnostics 60
regression equation **30** (see also response function)
regression model 20, 215 (see also response model)
regression sum of squares **36**, 149
relevé **8**, 258
relocation 262 (see also reallocation)
rescaling in DCA **106**, 108, 115 (see also scaling of ordination axes)
residual **34**, 36-37, 52, 122-123, 156, 218, 224
residual deviance **46**, 49, 53, 111, 115
residual standard deviation 36
residual sum of squares **34**, 36-37, 41, 53, 117, 149
residual variance **37**
response curve **32**, 60, 94 (see also unimodal response curve)
response function **29**, 41, 79, 82
response model **30**, 93, 153-155, 158
response surface **51-59**
response variable 11, 12, 26, **29-30**, 83,

135, 154
retrospective study **15**
R-mode algorithm **131**, 159
row conditional scaling 157
rule of assignment 80

sample 4, **7-8**, 10, 15
sampling 3, 4, 7, 10, **15-16**, 242-244
SAS 47, 158
scaling of ordination axes 102-103, 106, 127-129, 141, 147, 166
scatter diagram 31, 93
score, see site score and species score
seasonal aspects 259
s.d., see standard deviation
s.e., see standard error
second-order moments **225**
selection of variables 13, 56, 139, 201
semivariogram **226**, 230-237, 239, 242, 243, 246
semivariogram model **231-232**, 245
semivariance **225**
sensitivity to ..., see rare species and dominant species
Shepard diagram **152-153**
short-range variation 215, 222, 229,252
sigmoid curve 31-32, **44-47**, 65 (see also monotonic response curve)
significance tests **14**, 22, 29, 34-35, 40, 44, 49, 53, 156, 198, 202
significant optimum **49**
sill **230-232**
similarity 91, **175-183**, 196, 198, 204
similarity ratio **177**, 181-183, 187, 200, 262
single-linkage clustering **186-187**, 204, 205
singular value 157, **159**
singular-value decomposition 157, **158**
singular vector 157, **158**
site **7-8**, 24, 29, 78, 91, 175, 268
site score 96, **97**, 112, 118, 125, 127, 129, 132, 139, 145, 159
skewed distribution **18-19**, 31, 33, 216
slope parameter 37, 83, 119
smoothing 229

Sørensen coefficient **177**, 181-183
spatial (auto)covariance **214**, 215, 222, 224, 225, 239 (see also semivariogram)
spatial (auto)correlation 219, **224-229**, 245, 248
spatial independence 220
spatial interpolation **218**, 221, 237-242
spatial scale 214, 228
spatial structure 215, 223-224, 245
species-by-sites table 199 (see also table arrangement)
species-centred PCA **130**, 131, 152, 158
species composition 150, 168 (see also community composition)
species–environment correlation **139**, 140, 145, 147, 156, 167
species–environment relations 22, 29, 64-65, 91-93, 132, 136-137, 155-156, 174, 199-202
species-indicator value **61** (see also indicator value)
species packing model **85**
species score 95, 97, 112, **118**, 122, 137, 155, 159, 166
species space 125, **178**
species turnover 168 (see also standard deviation, response curve)
spherical model **231-232**
square-root transformation **184**
standard deviation, normal distribution **18-19**
standard deviation of species turnover, see standard deviation, response curve
standard deviation, response curve 103, **106-107**, 154, 168, 173 (see also tolerance and length of ordination axis)
standard deviation, sample 21, **36**
standard error 35, 65
standardization of variables 21, 140, 145, 150, 185
standardization procedure 100, 123, 165
standardized canonical coefficient **168**
standardized PCA **130**
stationarity **224**, 228-230
statistical moments **224-225**
statistical power analysis 13, **14**, 25, 60